STP 1201

Life Prediction Methodologies and Data for Ceramic Materials

C. R. Brinkman and S. F. Duffy, editors

ASTM Publication Code Number (PCN)
04-012010-09

ASTM
1916 Race Street
Philadelphia, PA 19103

Library of Congress Cataloging-in-Publication Data

Life prediction methodolgies and data for ceramic materials / C.R.
  Brinkman and S.F. Duffy, editors.
      (ASTM special technical publication ; 1201)
    Symposium sponsored by ASTM Committee C-28 on Advanced Ceramics.
    ISBN 0-8031-1864-3
    1. Ceramic materials--Testing. 2. Acceleated life testing.
  3. Non-destructive testing. I. Binkman, C. R. II. Duffy, S. F.,
  1965-   . III. ASTM Committee C-28 on Advanced Ceramics.
  IV. Series.
  TA455.C43L54   1994   93-44605
  620.1'4--dc20   CIP

Photocopy Rights

Peer Review Policy

Each paper published in this volume was evaluated by three peer reviewers. The authors addressed
all of the reviewers' comments to the satisfaction of both the technical editor(s) and the ASTM
Committee on Publications.

To make technical information available as quickly as possible, the peer-reviewed papers in this
publication were printed "camera-ready" as submitted by the authors.

The quality of the papers in this publication reflects not only the obvious efforts of the authors and the
technical editor(s), but also the work of these peer reviewers. The ASTM Committee on Publications
acknowledges with appreciation their dedication and contribution to time and effort on behalf of ASTM.

Printed in Ann Arbor, MI
January 1994

Foreword

This publication, *Life Prediction Methodologies and Data for Ceramic Materials,* contains papers presented at the symposium of the same name, held in Cocoa Beach, FL on 11–13 Jan. 1993. The symposium was sponsored by ASTM Commitee C-28 on Advanced Ceramics and the American Ceramics Society. C. R. Brinkman of Oak Ridge National Laboratories in Oak Ridge, TN and S. F. Duffy of Cleveland State University in Cleveland, OH presided as symposium chairmen and are editors of the resulting publication.

Contents

PREDICTION OF THE BEHAVIOR OF STRUCTURAL COMPONENTS

Overview

ASTM Committee C-28 on Advanced Ceramics was organized in 1986 when it became apparent that ceramics were under consideration for many new high technology applications. Proposed applications in the aerospace, biomedical, military, power generation, processing, and automotive industries were viewed as being particularly demanding in terms of property requirements and subsequently required an abundance of experimental data to guide emerging design and fabrication technologies. Hence, it became apparent that industry-oriented standards were needed for production, inspection, testing, data analysis, and probabilistic design of components in order to use the attractive features of these emerging materials, as well as minimize any shortcomings. Accordingly, Committee C-28 was organized into various subcommittees (including C-28.02 on Design and Evaluation) whose goals and objectives are reflected in the needs just mentioned. Specific responsibilities include: development of appropriate standards that address the topics of nondestructive evaluation (NDE), statistical analysis, and design of components fabricated from advanced ceramics. Early in 1990, members of this committee determined that it was appropriate to organize an international symposium aimed at presenting a state-of-the-art review. The review would focus on requisite design data and methods of generating this data, failure modeling, statistical techniques for the analysis and interpretation of this data, and probabilistic design methodologies that are a necessity in the analysis of components used in high technology applications, such as advanced heat engines. The anticipation was that the information presented at this symposium would serve as a basis in developing future standards. Time will tell whether this anticipation is fulfilled.

Twenty-seven papers were presented at the symposium, and 24 were subsequently published in this volume. The papers contained herein were grouped into three general subject areas. This selection was somewhat arbitrary, and several papers could easily be placed in more than one category. The categories include data and model development, life prediction methodologies, and prediction of the behavior of structural components. We expect that these subjects will not only be of interest to authors of future ASTM standards, but also to those interested in data generation requirements and model development. We also expect that the information contained in this publication will be pertinent for brittle monolithic ceramics as well as ceramic matrix composite materials. We note that ceramic-based material systems will be used in many advanced technologies where performance at elevated temperatures and in environments where strength degradation due to slow crack growth is of concern, or both. Thus, the articles that address this issue may be of particular interest to individuals who are involved in the development of lifing methods for advanced ceramics.

Data and Model Development

The papers in this section broadly describe generation of mechanical properties data. These data are used to identify optimized fabrication processes that minimize defects in advanced ceramics. Advanced NDE techniques such as acoustic microscopy and microfocus X-ray for pretest examination of specimens are discussed and results presented. The articles presented in this section also describe specific types of test data including: experimental procedure and equipment for determining tensile fast fracture, tensile creep, tensile cyclic fatigue, and flexure behavior used in life prediction methodology development. The influence of loading wave form, time, and temperature on the behavior of silicon nitride is described. Examples of representation and interpretation of data are given in various ways. These include: the use of fracture maps that can be used to generate stress allowables for a given application, competing Weibull analyses that delineate the probability of failure by surface or volume flaws, and constitutive equations for predicting creep and creep-rupture behavior under uniaxial and multiaxial loading conditions. Note that the presented creep loading regimes were both constant with time and also varied in a step-wise fashion. The validity and problems associated with use of flexural data for determining creep parameters are discussed. Finally, results and analyses of monotonic tensile fast fracture are presented and compared with several kinds of flexural test data. Results presented in papers found in this section should be of particular interest to the experimentalist whose focus is characterizing ceramic materials using tensile and flexural test techniques. As an example, the paper by Foley et al. contains results of over 100 tensile tests conducted at room temperature on a single material. Tensile specimens and the equipment required to successfully conduct this type of test are often prohibitively expensive. Testing in sufficient numbers to fully characterize a material with multiple flaw (or strength) distributions further increases the cost. Hence, reducing the expense in obtaining an optimum, high-quality data base was identified as being a major challenge to future experimentalists.

Life Prediction Methodologies

Metal alloys, such as those currently used for pressure vessels and gas turbines, have mechanical and physical properties that are readily available and easily implemented in an analysis of a component's response to applied boundary conditions. With metal alloys, the engineer often associates a high degree of confidence in the resulting component analysis. Factors of safety are applied to define exact stress allowables as a part of a deterministic design methodology. Furthermore, materials with high ductilities are selected for an extra margin of safety. In contrast, components fabricated from ceramic materials require new design methodologies for predicting component life that account for uncertainty in safe life expectancy. These materials are brittle by nature with an inherent scatter in strength. Life expectancies are not only controlled by the distribution and evolution of defects present after a component has been fabricated, but also by defects that may nucleate under load. Use of these materials fundamentally requires probabilistic design techniques that account for this behavior. Several articles in this section review life prediction methodologies for monolithic materials and ceramic matrix composites (CMC) and applications are provided in most instances. Various failure models are employed that are embedded in the framework of weakest-link theory and Weibull statistics. These models are exercised with a number of simple component/specimen geometries such as uniaxial three- and four-point unnotched bars. In addition, component/test specimens with more complex biaxial states of stress, such as notched beams and bars, ball-on-ring, ring-on-ring, and uniform pressure-on-disk specimens are highlighted.

In a paper by Scholten et al. a question is posed as to whether or not it is necessary to track the defect from which a brittle fracture originates when predicting multiaxial strength of ceramics. In order to examine this, a data set was developed from uniaxial and biaxial test specimens fabricated from several materials. Mixed-mode fracture criteria were compared with experimental results. Deviations from weakest-link theory were found in some instances when different fracture criteria were applied. These deviations were greatest in the more dense materials where the defect density was small. Microcracks were found nucleating during the test that the authors maintain violates the weakest-link principles. It was concluded that strength could be predicted with the introduction of a "size-independent" strength parameter.

It was further concluded that if consideration is given to experimental errors, then differences in multiaxial strength predictions for several specific loading conditions can readily be attributed to lack of precision. This underscores the importance of experimental accuracy in conducting multiaxial tests.

A methodology that predicts creep life using continuum damage mechanics is outlined by Chuang and Duffy for continuous fiber reinforced ceramic matrix composites (CFCMC). A number of potential creep-damage mechanisms in advanced ceramics are examined and the corresponding constitutive laws are outlined. The authors demonstrated that this methodology has potential for establishing estimates of creep life when stresses, temperatures, volume fractions of the constituents, and material properties are known. The work also points to the need for additional theories that allow extrapolation of short-term laboratory data to long-term service conditions.

Fracture data representing uniaxial and biaxial bend specimens fabricated from sintered alumina were compared by Chao and Shetty. Both environment (inert dry N_2 and deionized water) and strain rate were varied to determine if strength degradation due to slow crack growth in biaxial flexure can be predicted from simple uniaxial tests conducted in an inert environment. The authors concluded that this predictive approach was feasible, so long as the statistical uncertainties in both Weibull parameters (modulus and characteristic strength) and slow crack growth parameters (crack growth exponent and crack growth velocity) are properly taken into account.

Johnson and Tucker pointed to the variations observed in estimates of the Weibull parameters when two different estimation procedures are used. A data base composed of specimens with different applied boundary conditions and multiple specimen sizes was employed. A pooled sample of 137 test specimens fabricated from sintered silicon carbide were tested in six different combinations of specimen size and bending configuration. Comparisons are made with results using maximum likelihood and linear regression estimators after the estimators were applied to the uniaxial specimens in the pooled sample. A general consensus has emerged, which the authors support by their analysis, that maximum likelihood estimators are preferred since this approach offers the ability to unbias parameter estimates and establish confidence bounds on the estimated Weibull values. Furthermore, the authors emphasized the importance that high-quality fractography has on parameter estimates in the presence of multiple flaw populations.

Tucker and Johnson demonstrated that two multiaxial stochastic models recently reported in the literature (that is, the Batdorf-Heinisch and the Lamon-Evans models) yield equivalent probability of failure predictions. This allowed the authors to define a generalized size factor that accounts for geometry, loading conditions, and multiaxials stress states. The factor facilitates parameter estimation when the data base contains multiple specimen geometries and applied boundary conditions.

A method is presented by Margetson for analyzing component strength in the presence of both surface and volume flaws for a number of probabilistic models based on the principles of fracture mechanics. The methodology presented is applicable to multiaxial test configurations.

Prediction of the Behavior of Structural Components

Progress as well as difficulties encountered in predicting behavior of specific ceramic components and subcomponents are outlined in the articles found in this section. Component geometries include: notched bars, C- and O-ring specimens, ring-on-ring square plates and internally pressurized tubes. Initially, modeling a component typically requires the design engineer to resort to finite element methods to obtain accurate stress distributions and identify regions with high-stress gradients. Once gradients have been minimized and the stress state has been ascertained, component life is determined using a number of models that address different failure modes. Design concepts using principles from continuum damage mechanics, fracture mechanics, and Weibull statistics are often incorporated in life prediction codes such as the NASA computer program CARES/LIFE. Many authors emphasized the need for an adequate materials data base to properly implement this type of design approach. The data base must be constructed using carefully selected test-specimen geometries that establish requisite design parameters. Once these data are established, the results from structural component tests can be used to challenge the predictive capabilities of the models incorporated in various design codes. In order for this information to be of use to the design engineer, component level tests must represent the various service conditions encountered in real-life applications. Moreover, component tests must promote and isolate failure modes such as: fast fracture, slow crack growth, and creep.

Jadaan examined internally pressurized SiC tubes that were tested at temperatures and pressures in order to promote failure by slow crack growth and creep rupture, or both. Methodologies are presented that allow prediction of failure by either of these two mechanisms. Supporting data from standard creep-rupture, O-ring, and compressed C-ring tests are used to develop the methodology. Complications in use of these types of specimens for characterization of highly porous tubes are subsequently discussed in a paper by Krankendonk and Sinnema.

Estimating the stochastic parameters that characterize the inherent strength of a ceramic material is fundamentally important to any type of probabilistic design approach. Cuccio et al. provided an extensive treatment on estimating Weibull parameters and component reliability, providing methods to establish confidence intervals on both. The authors provide methodology for the following:

- censored analysis of competing strength distributions,
- analysis of data from specimens with multiple sizes,
- analysis of data from specimens with multiple-loading conditions,
- analysis of data from multiple temperature tests,
- calculation of confidence intervals on parameter estimates, and
- calculation of confidence intervals on reliability estimates.

The methodology is exercised using a data base from uniaxial fast-fracture tests conducted on test specimens fabricated from silicon carbide and silicon nitride.

Fabricators of ceramic components can dramatically improve their product reliability by removing prior to service components with gross or unusual defects. This is accomplished through NDE programs or proof-testing components. Highly sophisticated methods have been recently developed to perform NDE inspection. However, Brückner-Foit et al. point out that the NDE techniques suffer from several drawbacks, including cost and resolution. The authors discuss the advantages, disadvantages, and outline an approach for multiaxial proof testing. In addition, two examples are presented that illustrate the approach and typical problems associated with proof testing.

Finally, an overview of the integrated design code CARES/LIFE is presented in a paper by Nemeth et al. This public domain computer algorithm allows the design engineer to predict the time-dependent reliability of a component if the dominant failure mode is slow crack growth. The authors outline the supporting theoretical development, and two examples provide the reader with insight regarding the capabilities of the code.

Based on comments and feedback following the symposium, the chairmen felt the symposium was most successful in meeting the goals and objectives originally set forth during initial organization. Much of the information given in these papers is currently being used in formulating new ASTM standards and by the designers who are implementing advanced ceramics in many demanding applications. Furthermore, the chairs wish to express their gratitude to the authors for their efforts in preparing their manuscripts, putting on well-orchestrated and professional presentations at the symposium, and responding to reviewers comments in a thoughtful manner. In addition, the chairs are deeply indebted to the reviewers for their timely efforts and scholarly assessment of the manuscripts.

C. R. Brinkman

Martin Marietta Energy Systems; Oak Ridge, TN 37831-6154; symposium chairman and editor.

S. F. Duffy

Cleveland State University, Cleveland, OH 44115; symposium chairman and editor.

Data and Model Development

Michael R. Foley[1], Vimal K. Pujari[1], Lenny C. Sales[1], and
Dennis M. Tracey[1]

**SILICON NITRIDE TENSILE STRENGTH DATABASE FROM CERAMIC TECHNOLOGY
PROGRAM PROCESSING FOR RELIABILITY PROJECT**

REFERENCE: Foley, M. R., Pujari, V. K., Sales, L. C., and Tracey, D. M.,
"Silicon Nitride Tensile Strength Database from Ceramic Technology
Program Processing for Reliability Project," Life Prediction
Methodologies and Data for Ceramic Materials, ASTM STP 1201, C. R.
Brinkman and S. F. Duffy, Eds., American Society for Testing and
Materials, Philadelphia, 1994.

ABSTRACT: Tensile strength data generated in Norton's Ceramic
Technology Program (CTP) Processing for Reliability Project is presented
for a hot isostatically pressed (HIP'ed) 4 wt% yttria-silicon nitride
(designation NCX-5102). This database represents the result of an
extensive multi-variable experimental matrix designed to identify an
optimized process directed at eliminating or minimizing critical flaws.
The strength data follow from room temperature fast fracture tests of
net-shaped-formed, pressure cast cylindrical buttonhead tensile bars.
Results of over one hundred tensile tests coupled with detailed
fractography are summarized using Weibull statistics including competing
risk analyses. Specimen fabrication and mechanical testing issues (e.g.
machining, strain gaging) which were addressed to ensure the integrity
of the strength database will also be discussed.

KEYWORDS: silicon nitride, reliability, tensile strength, Weibull
analysis, competing risk, machining, strain gaging

A variety of strength-degrading flaws introduced during the
initial stages of traditional processing methods can produce
unacceptable mechanical reliability of structural ceramics. Impurities
in the starting material components (powder, sintering aids,
surfactants, binders) and agglomerates formed during powder processing
are but two examples of strength-degrading flaws. Forming related
cracks, voids and metallic impurities introduced during forming and
grain growth during densification are further examples of failure
originating flaws. Even if all of these intrinsic flaw types can be
minimized or eliminated, the final step of fabrication, machining, can
leave various extrinsic flaws on the surface of the component.
    Silicon nitride-based ceramics are continually being evaluated for

---

[1]Senior Research Engineer, Senior Research Associate, Senior Research
Engineer, Research Group Leader, respectively, Saint Gobain/Norton
Industrial Ceramics Corporation, Northboro Research and Development
Center, Goddard Road, Northboro, MA 01532-1545.

room and elevated temperature structural applications [1-8]. Flaws such as the ones described above not only limit the ultimate strength of these materials but increase the scatter in the strength data giving rise to low reliabilities (low Weibull modulus). Mechanical testing coupled with detailed fractography can identify specific flaw types and provide valuable feedback [9] to the powder processing, forming, densification as well as the machining unit operations. The data discussed in this paper are from the second phase of a three stage Department Of Energy (DOE)/Oak Ridge National Laboratory (ORNL) CTP program on processing for reliability [10]. This report addresses the issues of machining, bending during tensile testing and reliability statistics. Room-temperature tensile strength data are used in conjunction with detailed fractography data to perform competing risk Weibull analyses.

EXPERIMENTAL PROCEDURE

Materials

A 4 wt% yttria-doped Si_3N_4 (NCX-5102) was selected to be tested at room temperature. The net shaped formed (NSF) buttonhead tensile rod is the model component being used to evaluate all process improvements. The material is processed and pressure cast in a closed loop, aqueous powder processing operation in a clean room environment. After casting, the specimens were dried, presintered and HIP'ed to >99.5% theoretical density (ρ_{THEO} = 3.23 g/cm^3). After densification, the tensile specimens underwent a pre-machining inspection including density and runout, see Fig. 1. The individual HIP runs were qualified using flexure strength in accordance with ASTM Test Method for Flexural Strength of Advanced Ceramics at Ambient Temperature (C 1161 type B) and K_{Ic} [11] as controls. The entire flow diagram for the tensile specimen history from densification to post-mortem analysis is also shown on Fig. 1.

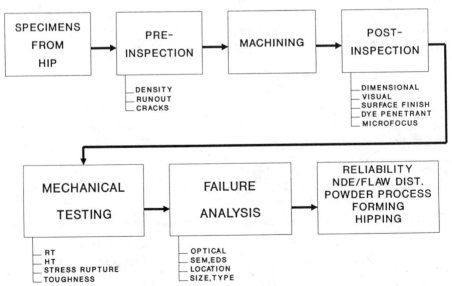

FIG. 1--Specimen flow diagram for testing and failure analysis.

Experimental Design

Several variables were examined by way of an L8 x L4 experiment. The experimental plan [10] involved 16 processed batches of silicon nitride and each of the experimental blocks was evaluated through room temperature tensile tests on approximately thirty tensile rods amounting to over 500 rods tested. The unit operation control variables evaluated in the experimental plan were:

Slurry conditions
* Binder. (2)
* Surfactant. (2)
* Casting rate. (2)
* Solids loading. (2)
* Pre-HIP treatment. (2)

HIP conditions
* HIP cycle. (2)
* Fixturing. (2)

The results of the experimental plan defined a set of optimum slurry and HIP conditions. The following additional variables were subsequently evaluated separately:

Post HIP conditions
* Machining conditions.(4)
* Post machining conditions.(2)

The four machining conditions are described in detail below and are part of a separate experiment. The two post-machining conditions involved either performing a thermal treatment aimed at annealing surface and subsurface damage created by machining or leaving the surface as-machined. Previous work [12] showed a 10% increase in room-temperature mean tensile strength by application of the surface oxidation heat treatment.

Finally, the fully optimized process was repeated for a subset of tensile specimens. The process included a 2-step HIP process, a specified procedure for machining, and thermal surface treatment. The statistical analysis, described below, was performed on this subset of tensile specimen data.

Machining

The final machined[2] cylindrical buttonhead tensile specimen is the ORNL design [8] except that the gage diameter is 6.0 $\pm$ 0.1mm and 35 mm gage length. All specimens for the main body of strength data were machined using a specifically designed standard operating procedure (SOP) (procedure #1) as shown in Table 1. It should be noted that the gage section is longitudinally ground during both the roughing and final steps.

After machining, all specimens underwent an extensive inspection including: dimensional tolerance, surface finish, microfocus X-ray and liquid dye penetrant. Customary practice expresses surface finish specification in terms of the average roughness R_a. Earlier work [9] resulted in a reduction of the R_a specification from 0.4 μm (16 μin) to 0.25 μm (10 μin) at the buttonhead radius and to 0.2 μm (8 μin) along the gage length.

Four distinct machining procedures (including the original SOP) were evaluated for their influence on tensile strength as an additional experiment. The four machining procedures differed according to

---

[2]Chand Associates, Inc., 2 Coppage Dr., Worcester, MA 01603-1252

intermediate grinding steps, diamond wheel grit size and depths of cut as outlined in Table 1. This experiment maintained the above R_a specification while altering the rough and intermediate grinding steps with the intent being to reduce subsurface damage due to the prior step. On the basis of the grinding parameters employed, the machining procedures were ranked according to operational precision from #1 to #4, with #4 being the most precise procedure. Consistent with the focus on machining damage effects, the specimens were not given post-machining heat treatments for the purpose of this study. A total of 72 specimens from 3 HIP runs were machined and tested for this study. The procedures were evaluated by tensile strength and statistical analyses.

TABLE 1--Procedure steps for machining tensile rods.

Procedure	Roughing Step	Intermediate Step(s)	Finishing Step
#1 Original SOP	180 grit	-	320 grit (0.51mm)*
#2 Experimental	320 grit	320 grit	320 grit
#3 Experimental	180 grit	320 grit	800 grit (0.15mm)
#4 Experimental	180 grit	320, 400, 600 grit - (0.05mm) (0.05mm)	800 grit (0.05mm)

*The number in parenthesis refers to amount of stock removed by that step.

Testing

The cylindrical buttonhead tensile specimens were tested at room-temperature on a commercial electro-mechanical test machine[3] utilizing commercial ,self-aligning self-contained hydraulic load train couplers[4] and straight tri-split copper collets. A double ramp loading procedure was used to test all specimens. The specimen is initially loaded to 6668 N at 39 MPa/min. This allowed time for the fully annealed copper collets to deform to match the radius of the buttonhead. After the initial ramp to 6668 N, the specimen was then loaded to failure at a stressing rate of 600 MPa/min.

The load train was checked before testing with an alignment tool for actuator/load cell alignment and a strain gaged tensile specimen for coupler alignment.

Strain Gaging of Tensile Specimens

A total of 68 specimens were strain gaged prior to testing, to study bending during testing and to determine the effect of bending on tensile strength and reliability data. Each specimen had four gages equispaced and circumferentially attached at the longitudinal center of the gage section. The percent bending (PB) was calculated in accordance with ASTM Test Method for Sharp-Notch Testing with Cylindrical Specimens (E 602) such that

---

[3] Instron Model 8562, Canton, MA

[4] Instron "Super-grip", Canton, MA

$$PB = \frac{[(\Delta g_{1,3})^2 + (\Delta g_{2,4})^2)]^{1/2}}{g_0} \; x\, 100 \qquad (1)$$

where

$$\Delta g_{1,3} = \frac{(g_1 - g_0) - (g_3 - g_0)}{2} = \frac{(g_1 - g_3)}{2} \qquad (2)$$

$$\Delta g_{2,4} = \frac{(g_2 - g_0) - (g_4 - g_0)}{2} = \frac{(g_2 - g_4)}{2} \qquad (3)$$

$$g_0 = \frac{(g_1 + g_2 + g_3 + g_4)}{4} \qquad (4)$$

and g_1, g_2, g_3, and g_4 are the strain gage readings in units of strain.

A digital bridge completion and data acquisition system[5] was used to monitor and record the strain and load data. Initially, 25 tensile specimens were strain gaged and the load-PB data were recorded without any realignment. The measured PB at failure was < 5.5% for 24 of 25 specimens with one at 12.5%. This assured proper alignment of the test fixtures and the validity of the data.

In the second part of the experiment, 10 of 20 specimens (predetermined randomly) were to be realigned if the measured PB at a preload of 6668 N was >10%. The realignment involved unloading the specimen, twisting the grip/collet load train assembly and reloading the specimen. If the measured PB was still >10%, the realignment was performed a second time and the specimen was tested to failure regardless of the measured preload PB. Load-displacement curves have a "knee" in the curve at about 5556 N (1250 lbs.) where the copper collets deform into the buttonhead radius. Therefore the preload was chosen to be greater than 5556 N to monitor the "preload" PB.

Fig. 2 shows representative load-PB curves for strain-gaged specimens. A typical curve shows an increase in PB during initial loading but as the collets seat themselves the PB begins to decrease. Fig. 2 shows that the range of PB at the preload level (6668 N) is 4 to 16, and in all cases the PB at failure is ≤6%. For two of the specimens for which realignment took place, the least PB achieved at the preload was 12%, but still resulted in low (2%) PB at failure. However, as shown in Fig. 3, and consistent with the decreasing PB level after preload, there appears to be a trend that the higher strength specimens experience a lower PB at failure. Also, the specimens that were realigned had a lower PB at failure than those without a realignment. A t-test showed no significant difference between the aligned and unaligned samples.

In an attempt to determine the effect of PB on the Weibull modulus, an additional 23 specimens were strain gaged and tested to

---

[5]Daytronic System 10 DataPAC, Miamisburg, OH

failure.   The average (g_o) and maximum strains (maximum of g_1, g_2, g_3 and g_4) were recorded for each specimen.   Two Weibull moduli were calculated.   The first modulus, equal to 10.04, was determined from the standard fracture stress (load/area) of each specimen (mean = 882 MPa). A second or predicted fracture stress (mean = 880 MPa)was determined as the product of the maximum strain times Young's Modulus (310 GPa).   A Weibull analysis was performed using the predicted strengths giving a value of 10.27.   For practical purposes the difference (2.2%) is negligible.   Therefore, 100% strain gaging was not performed.   However, a periodic sampling using a strain gaged specimen to check load train alignment was conducted.

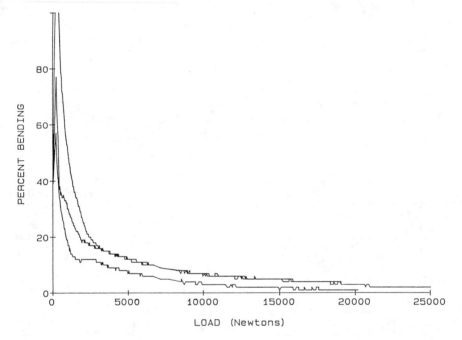

FIG. 2-- Typical percent bending vs. load curves for strain-gaged tensile specimens.

Fractography

Optical fractography was performed on all fractured specimens. Each specimen was marked on the surface (outside the gage section) relative to its position in the tensile grips.   This allowed specific location of any surface failures and helped identify possible misaligned grips.   The diameter was measured after failure near the fracture surface.   This eliminated the possibility of surface damage from the calipers prior to testing.   The macro-location of the failure origin was determined (gage vs. transition) by similar measurement.

Following optical observations, all specimens were submitted for scanning electron microscopy, SEM, and energy dispersive spectroscopy, EDS.   Low and high magnification micrographs were taken of each fracture surface and flaw origin.   The EDS was performed at each origin to determine if any impurities were present.   The location, type and size of each flaw was noted, measured and tabulated for database evaluation.

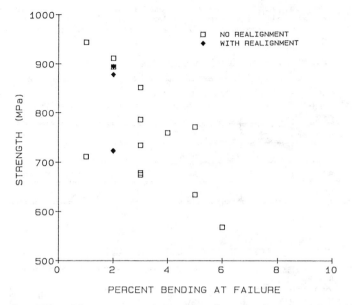

FIG. 3-- Strength vs. percent bending for strain-gaged tensile specimens.

Data Analysis

The tensile strength data were used to perform Weibull analyses with the assumptions of unimodal strength distribution, two-parameter formulation and the ranked regression technique. In addition the strength data were separated based on flaw type and a competing risk analysis using Weibull theory was also performed. Fractography did not always reveal the failure-originating flaw. Therefore, for purposes of Weibull analysis, the unknowns were assumed to be from a flaw origin identical to its nearest neighbor in strength.

RESULTS AND DISCUSSION

Tensile Strength Data Summary

The overall experimental plan involved 16 powder batches from the L8 x L4 experiment. Tensile strength data have been obtained for specimens from all batches and have been reported earlier [13]. These data represent over 500 tensile tests with strengths ranging from 548 to 1091 MPa. Mean strengths for specific batches range from 600 to 923 MPa.

The details of the concluding four batches of the optimization study are summarized in Table 2. As each subsequent batch was "cleaned up" by elimination of iron contaminating sources the mean strength increased by over 170 MPa. Optical and SEM fractography were performed on all specimens and the results are summarized in Table 3. There is a significant shift from volume failures to surface failures with the elimination of iron contamination from Batch XXX to Batch Z. The flaw type "Am", amorphous region, was found to be related to the post-machining heat treatment. The predominant flaw origin was now related to machining. The other significant flaw type "Mp", zones of microporosity, occurred in the volume and may be related to the

densification process. It is apparent from Batch Z data that surface
flaws related to machining are the major strength and reliability
limiting defects. This data prompted the machining experiment discussed
in the following.

TABLE 2--RT tensile strength data for batches XXX - Z

Batch	Mean Strength (MPa)	Std. Dev. (MPa)	Characteristic Strength (MPa)	Weibull Modulus m	No. of Specimens
XXX	707	88	746	9.0	25
X	771	105	816	8.5	18
Y	847	103	892	9.4	45
Z	878	108	924	9.4	47

TABLE 3--Fractography summary for batches XXX - Z

Batch	Location		Flaw Type (%)					Size Range (μm)
	% V	% S	I	Mp	M	Am	U	
XXX	68	32	72	0	20	0	8	>100
X	29	71	41	6	24	12	17	15-50
Y	2	98	0	2	28	68	2	10-80
Z	29	71	0	40	45	9	6	5-65

V = volume   I = iron inclusions
S = surface   Mp = microporosity
                    M = machining damage
                    Am = amorphous region at surface
                    U = unknown

Machining Experiment

     A total of 72 tensile specimens were machined and tested for this
study. To properly evaluate the differences between the four
procedures, only those specimens that failed from surface originating
flaws were considered (64 of 72). An Analysis of Variance (ANOVA) test
was performed to compare differences in the machining procedures. Of
the four machining procedures, the procedure #1 (original SOP) shows the
least variability, Table 4. In spite of the major differences in
operational precision, the resultant tensile strength data indicated
that there is no statistically significant effect of procedure on
strength as shown in Table 4. R_y (maximum peak to valley) surface
finish values were found to differ for #1 & #2 vs. #3 & #4 and are also
shown in Table 4. Since the primary consideration is assuring a high
Weibull modulus, the procedure with the least variability (#1), will be
utilized for future machining and can be used as a benchmark for further
machining experiments.

TABLE 4--Machining procedure strength data.

Procedure	Mean Strength (MPa)	Standard Error (MPa)	Surface Finish R_y (microns)	Number of Specimens
1 (SOP)	812	18	1.26	31
2 (Experimental)	767	35	1.33	10
3 (Experimental)	847	30	0.77	14
4 (Experimental)	845	42	0.68	9

Strength-Flaw Size Correlation

A limited amount of fractography data from an early batch was quantitatively analyzed to establish relationships between strength and flaw size. Strength data for ten specimens which failed from normal machining related surface damage are plotted as a function of $1/\sqrt{a}$ in Fig. 4. The maximum depth of the surface flaw (not the surface length) is plotted as the dimension a. These ten tests cover a 586 - 899 MPa tensile strength range and a 20 - 60 μm crack depth (a) range. The linear regression fit of the data is drawn along with two fracture mechanics predictions which follow

$$\sigma_f = \alpha * K_{Ic}/\sqrt{a} \qquad (5)$$

based upon the batch average K_{Ic} value of 5.35 MPa$\sqrt{m}$. The semi-circular crack and long crack models have values of 0.71 [14] and 0.50 [15], respectively, for the shape factor α in eqn (5). There is excellent agreement between the regression line and the semi-circular surface crack fracture mechanics prediction. The prediction following from the long surface crack model can be seen to fall significantly below the actual data. This suggests that the normal strength limiting machining damage has the form of semi-circular cracks, as is shown in the next section.

Competing Risk Weibull Statistics

The 53 tensile specimens fabricated with the "optimized process" have a mean strength of 888 ± 111 MPa with strengths ranging from 548 to 1091 MPa. The corresponding tensile strength distribution is plotted in Figure 5. Competing risk Weibull distributions were generated for surface and volume flaws and are shown in Figure 6. The Weibull moduli are 9.3 and 9.9 for surface and volume flaws, respectively. It is interesting to note that at any given applied stress level the probability of failure is greater for a surface flaw than for a volume flaw.

Detailed fractography observed that within the surface flaw population there were 3 flaw origin types: surface amorphous regions and two types of machining damage. The volume flaws were of one type and were characterized as agglomerate-porous zones. Figures 7-10 show micrographs of these flaw types. Included are normal machining damage, atypical machining damage, Si-rich amorphous regions and zones of microporosity.

Based on the fractographic data, another competing risk Weibull analysis was performed using the 3 surface flaw and the 1 volume flaw strength distribution data. For clarity, only the three surface distributions are shown in Fig. 11. Since the strength distribution data (mean = 889 MPa and m = 9.9) from the volume failures are from only one flaw type, the modulus does not change. However, the three distributions from the surface flaw data show very different results.

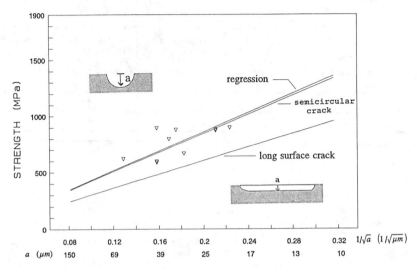

FIG. 4-- Strength-flaw correlation for machining damage failure origins.

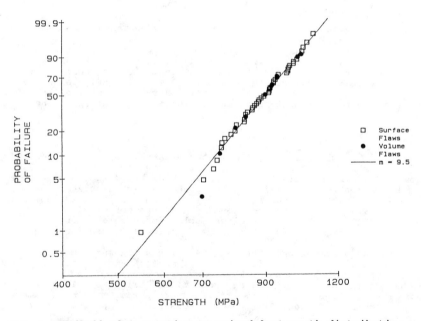

FIG. 5-- Weibull plot assuming an unimodal strength distribution.

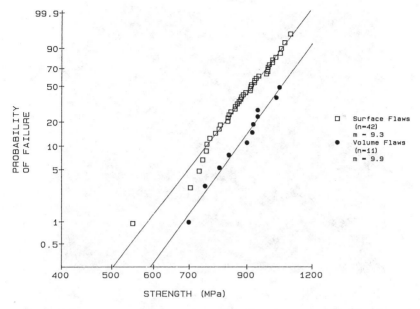

FIG. 6-- Weibull plot using competing risk for surface and volume failure.

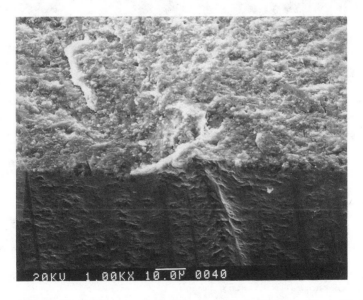

FIG. 7-- SEM of the failure origin of a specimen fractured at RT (σ_f = 685 MPa) showing atypical machining damage.

FIG. 8-- SEM of the failure origin of a specimen fractured at RT (σ_f = 988 MPa) showing an amorphous region.

FIG. 9-- SEM of the failure origin of a specimen fractured at RT (σ_f = 909 MPa) showing normal machining damage.

FIG. 10-- SEM of specimen fractured at RT (σ_f = 1040 MPa) showing a volume defect containing microporosity.

As expected, the lowest strengths and Weibull Modulus (mean = 696 MPa and 6.4) were from specimens having atypical machining damage as the critical flaw origin. The middle strength values (mean = 867 MPa and m = 11.6) were from specimens with Si-rich amorphous origins. These defects developed during the annealing process and their precursors may have been from local inhomogeneities in the microstructure or from subsurface cracks formed during machining. These strengths overlap both the atypical machining damage and normal machining damage strengths. The third and highest set of strengths and Weibull Modulus (954 MPa and 15.2) were from specimens that had surface flaws from normal machining (procedure #1).

SUMMARY

1. Tensile test data on NCX-5102 silicon nitride fabricated using a closed-loop aqueous system were analyzed for strength and reliability.

2. Bending due to load-train misalignment was measured on strain-gaged tensile specimens and was not found to be significant at failure.

3. A machining procedure was chosen based on minimizing variability in strength data rather than surface finish alone.

4. Five flaw types were identified using SEM fractography. With processing improvements metallic inclusions were eliminated and surface flaws dominated the strength distributions.

5. Competing risk Weibull analyses showed that the probability of failure from surface flaws was greater than the probability of failure from volume flaws at any given applied stress.

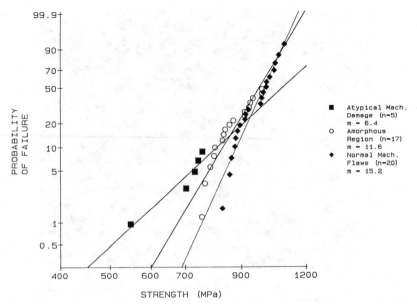

FIG. 11-- Weibull plot using competing risk for all three surface flaw types.

6. Plans for the remainder of the program are to verify the optimized process by testing over 300 tensile specimens at room temperature with goals of over 900 MPa mean strength and a Weibull Modulus of 20.

ACKNOWLEDGEMENTS

This research was sponsored by the U.S. Department of Energy, Assistant Secretary for Conservation and Renewable Energy, Office of Transportation Technologies, as part of the Ceramic Technology Project of the Materials Development Program, under contract DE-AC05-84OR21400 with Martin Marietta Energy Systems, Inc.

REFERENCES

[1]   Fang, H.T., Cuccio, J.S., Wade, J.C., Seybold, K.G., and Jenkins, M., "Progress in Life Prediction Methodology for Ceramic Components of Advanced Heat Engines", Proceedings of the Annual Automotive Technology Development Contractors' Coordination Meeting, 1991, Society of Automotive Engineers, Warrendale, PA, June 1992, pp 261-272.

[2]   Wiederhorn, S.M., Krause, R., and Cranmer, D.C., "Tensile Creep Testing of Structural Ceramics", Proceedings of the Annual Automotive Technology Development Contractors' Coordination Meeting, 1991, Society of Automotive Engineers, Warrendale, PA, June 1992, pp 273-280.

[3]   Sankar, J., Kelkar, A.D., Vaidyanathan, R., and Gao, J, "Creep Testing of SNW-1000 Sintered Silicon Nitride", Proceedings of the

Annual Automotive Technology Development Contractors' Coordination Meeting, 1991, Society of Automotive Engineers, Warrendale, PA, June 1992, pp 293-308.

[4]   Khandelwal, P.K., and Vaccari, D.L., "Life Prediction Methodology of Ceramic Engine Components", Proceedings of the Annual Automotive Technology Development Contractors' Coordination Meeting, 1991, Society of Automotive Engineers, Warrendale, PA, June 1992, pp 253-260.

[5]   Hecht, N.L., Goodrich, S.M., Chuck, L., and McCullum, D.E., "Mechanical Testing of Candidate Si_3N_4 Ceramics", Proceedings of the Annual Automotive Technology Development Contractors' Coordination Meeting, 1991, Society of Automotive Engineers, Warrendale, PA, June 1992, pp 397-402.

[6]   Liu, K.C., Pih, H., Stevens, C.D., and Brinkman, C.R., "Tensile Creep Behavior and Cycle Fatigue/Creep Interaction of HIP'ed Si_3N_4," Proceedings of the Annual Automotive Technology Development Contractors' Coordination Meeting, 1990, Society of Automotive Engineers, Warrendale, PA, June 1991, pp 213-220.

[7]   Pasto, A.E., Natansohn, S., Avella, F., Cotter, D., Dodds, G., et.al., "Development of Improved Processing Methods for High Reliability Structural Ceramics for Advanced Heat Engines", Oak Ridge National Laboratory Report ORNL/Sub/89-SD548/1, July 1992.

[8]   Ferber, M.K., and Jenkins, M.G., "Rotor Data Base Generation", Ceramic Technology for Advanced Heat Engines, Oak Ridge National Laboratory, Oak Ridge, TN, Semiannual Progress Report for October 1990 through March 1991, pp 340-361, July 1991.

[9]   Foley, M.R., and Pujari, V.K., "Tensile Testing in the Development of Processing Methods for High Strength/High Reliability Silicon Nitride," Ceramic Engineering and Science Proceedings, Sept - Oct 1992, 16th Annual Conference on Composites and Advanced Ceramic Materials.

[10]   Pujari, V.K., Tracey, D.M., Foley, M.R., Corbin, N.D., Sales, L.C., et.al., "Reliability Improvements of High Strength Silicon Nitride Through Process Optimization and Control," Proceedings of the Annual Automotive Technology Development Contractors' Coordination Meeting, 1991, Society of Automotive Engineers, Warrendale, PA, June 1992, pp. 137-144.

[11]   Chantikul, P., Anstis, G.R., Lawn, B.R., Marshall, D.B., "A Critical Evaluation of Indentation Techniques for Measuring Fracture Toughness: II, Strength Method", Journal of American Ceramic Society, 64, 9, 539-543, 1981.

[12]   Pujari, V.K., Tracey, D.M., Foley, M.R., Sales, L.C., Paille, N.I., et.al, "Improved Processing" in Ceramic Technology Project Semiannual Progress Report for October 1991 through March 1992, Oak Ridge National Laboratory, TN, ORNL/TM-12133, September 1992, pp 75-113.

[13]   Pujari, V.K., Tracey, D.M., Foley, M.R., Sales, L.C., Pelletier, P.J., et. al., "Processing Methodology for the Production of Reliable High Strength Silicon Nitride," presented at the Annual Automotive Technology Development Contractors' Coordination Meeting, Dearborn, MI, Nov 1992.

[14] Tracey, D.M., "3D Elastic Singularity Element for Evaluation of K Along an Arbitrary Crack Front," International Journal of Fracture, 9, pp. 340-343, 1973.

[15] Ewalds, H.L. and Wanhill, R.J.H., Fracture Mechanics, Edward Arnold Publishers, Ltd., Baltimore, MD, pp 42-43, 1984.

Jagannathan Sankar,[1] Srikanth Krishnaraj,[2] Ranji Vaidyanathan,[2] Ajit D. Kelkar,[1]

ELEVATED TEMPERATURE BEHAVIOR OF SINTERED SILICON NITRIDE UNDER PURE TENSION, CREEP, AND FATIGUE

REFERENCE: Sankar, J., Krishnaraj, S., Vaidyanathan, R., and Kelkar, A. D., "Elevated Temperature Behavior of Sintered Silicon Nitride Under Pure Tension, Creep, and Fatigue," Life Prediction Methodologies and Data for Ceramic Materials, ASTM STP 1201, C. R. Brinkman and S. F. Duffy, Eds., American Society for Testing and Materials, Philadelphia, 1994.

ABSTRACT: Pure tensile fast fracture, tensile creep, and tensile cyclic fatigue/creep interaction data are reported for GTE SNW-1000 sintered silicon nitride (Si_3N_4) which is being investigated as a candidate material for advanced heat engine applications.

Pure uniaxial tensile tests conducted at room temperature and at elevated temperatures indicate that the tensile strength of this material was retained to 1100°C, above which there was a sharp decrease in strength.

Tensile creep tests performed at 1100°C and 1200°C showed that the steady state creep rate was dependent on both the temperature and the applied stress, the effect of temperature being more dominant than the applied stress. Further, creep induced deformation by linking of small pores, and changes in the chemical contents were observed.

The effect of cyclic loading on creep, and residual tensile strength were also studied at 1200°C. Test results showed that precycling can dramatically increase creep resistance. Cyclic loading at 1200°C also increases the strength of the material.

KEYWORDS: ceramic, silicon nitride, tension, creep, fatigue, fracture

[1]Associate Professor, Department of Mechanical Engineering, North Carolina A & T State University, Greensboro, North Carolina 27411.

[2]Graduate Student and Research Associate respectively, Department of Mechanical Engineering, North Carolina A & T State University, Greensboro, North Carolina 27411.

19

INTRODUCTION

Significant progress has been made in the recent years in developing sintered silicon nitride structural ceramics for heat engine applications. This material exhibits excellent strength and oxidation resistance at elevated temperatures, resistance to thermal shock, low thermal expansion, and low thermal conductivity that allow newer and innovative solutions to component design problems [1-2]. Further, sintered silicon nitride is less expensive than other forms of silicon nitride, and the technology exists to produce components in a commercial quantity.

Although considerable progress has been made in increasing both the strength and working temperature of advanced engineering ceramics, problems of unpredictable failures persist due to random flaws inherent to these materials. Further, use of structural ceramics in heat engines involves simultaneous exposure to more than one "hostile" environment [3]. In addition, ceramics are inherently weaker in tension due to latent defects. This necessitates that appropriate design methodologies and design data must be established through suitable test methods especially under pure tension to assure the material's reliability and durability. Hence, uniaxial tensile behavior at various temperatures, tensile creep behavior, and tension-tension cyclic fatigue properties at elevated temperatures have to be determined and understood.

This paper summarizes the results of study on the mechanical behavior of sintered silicon nitride at elevated temperatures. Strength, creep, and fatigue tests were conducted. Results are correlated with the microstructure of the material, and failure mechanisms are hypothesized.

TEST MATERIAL AND SET UP

Material

The material, GTE SNW-1000, a commercial grade of sintered silicon nitride was procured from GTE Wesgo Division, Belmont, CA as buttonhead tensile specimens (Figure 1). Each specimen contained a nominal chemical composition of 6 wt.% Y_2O_3, 2 wt.% Al_2O_3, and the rest silicon nitride. The specimens were approximately 152.4 mm (6.0 inches) long and had a nominal gage dimension of 25.0 mm (0.984 inches) in length by 6.0 mm (0.2362 inches) in diameter . All specimens had a ground surface finish of 2 to 3 μm (80 to 120 μ inches).

Mechanical Investigations

A servo-hydraulic mechanical test system[a] was used for conducting the uniaxial tensile, creep, and fatigue tests. All tests were performed in load control mode using in-house developed programs. This automated system consists of a test processor interfaced to the control electronics, a computer and its associated peripheral devices,

---

[a]MTS 880 System, MTS, Minneapolis

and softwares to operate the system.

Specimens were gripped using a self aligning hydraulic grip system with eight free floating pistons originally designed by the Oak Ridge National Laboratory (ORNL). A detailed description of the gripping system is given elsewhere [4-5].

A box shaped, low-profile, resistance-heating furnace with silicon carbide heating elements[b] was used to heat the specimens. The advantage of this furnace was that only the gage section of the specimen was exposed directly to the heat. The ends of the specimen were connected to metal extension rods(René 41 super alloy), which were water-cooled. The specimens were coated with boron nitride lubricoat[c] to avoid bonding between the pullrod sleeves and the specimen.

The longitudinal deformation of the specimen was measured using a non-contact laser telemetric system[d]. A laser beam was directed through a window in the furnace toward the gage section of the specimen. The specimens were instrumented with a set of flags made of platinum that were attached to the gage section of the specimen using a high temperature alumina based adhesive[e]. Any change in the distance between the flags was sensed by the laser extensometer and was converted into an analog signal. The signal was then converted into digital elongation using an A/D channel.

All elevated temperature tests were conducted in air, after 4 hours of heating to ensure thermal equilibrium within the furnace prior to application of the load. This procedure was followed to minimize the errors introduced in strain measurements due to air density fluctuations in the furnace [6].

Microstructural Investigations

All optical microscopy was performed using a Polaroid MP4 microscope. Fractographical studies were conducted using a scanning electron microscope (SEM)[f], while microchemical analyses were performed using an x-ray energy dispersive spectrum (EDS)[g] analyzer attached to the SEM.

The sections used for analyses in the gage section were, a) fracture surface, and b) a polished section 2 mm below the fracture surface along the transverse (perpendicular to stress) direction.

---

[b]Model 3450, ATS Corporation, Butler, PA

[c]ZYP Coating Inc., Oak Ridge, TN 37831

[d]Model 121, Zygo Corporation, Middleton, CT

[e]Ceramabond 569, Aremco Products, Ossining, NY 10562

[f]Model ISI SS-40, Topcon Technologies, Pleasonton, CA

[g]Model TN-5400, Noran Instruments, Middleton, WI

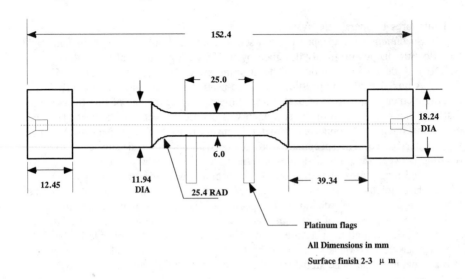

FIG. 1--GTE SNW-1000 tensile specimen configuration

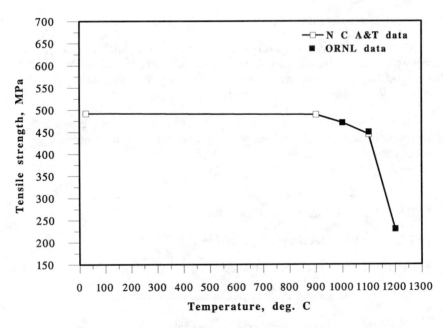

FIG. 2--Tensile strength vs. temperature for GTE SNW-1000

TEST CONDITIONS AND RESULTS

Room Temperature Tensile Tests

Tensile tests were performed at five different stressing rates: 3.75, 7.5, 450, 900, and 2100 MPa/min. A total of twenty-nine specimens were tested. Test results indicated no effect of stressing rates on the strength of the material. Material showed an average tensile strength of 492 MPa at room temperature.

Elevated Temperature Tensile Tests

A total of fourteen specimens were tested at 900°C and 1100°C at stressing rates of 7.5 and 2100 MPa/min. Average tensile strengths of 489 MPa, and 448 MPa were obtained at 900°C, and 1100°C respectively. The 1200°C tests performed by Liu et al., at Oak Ridge National Laboratory (ORNL) reported [7] an average tensile strength of 230 MPa for this material. The tensile strength data of GTE SNW-1000 versus temperature are shown in Figure 2.

Creep Tests

A total of eight silicon nitride specimens were tested in creep at 1100°C and 1200°C under various loads. Three specimens were tested at 1100°C at applied stresses of 224 MPa, 314 MPa, and 381 MPa that are 50%, 70%, and 85% of the tensile strength of the material at this temperature respectively. The strain versus time for all the specimens tested at 1100° C is shown in Figure 3.

Five specimens were tested at 1200° C, at applied stresses of 57 MPa, 92 MPa, 115 MPa, 138 MPa, and 161 MPa, that are 25%, 40%, 50%, 60%, and 70% of the tensile strength of the material at this temperature respectively. The strain versus time for all the five specimens tested at 1200° C are shown in Figure 4. The steady state creep rate versus applied stress for the specimens tested at both 1100°C, and 1200°C are shown in Figure 5, while plots of natural log of steady state creep rate versus natural log of rupture time (Monkman - Grant relationship) for the broken samples and for all tested samples are shown in Figure 6.

Creep and Tensile tests after Cycling

All tests were performed at 1200°C in uniaxial tension. Three types of investigations were performed on eleven(11) specimens to:
1. understand the effects of precycling on the creep behavior.
2. investigate the effects of precycling and creep loading on the residual tensile strength.
3. study the effects of precycling on the residual tensile strength of the material.
The test parameters for all tests were as follows.

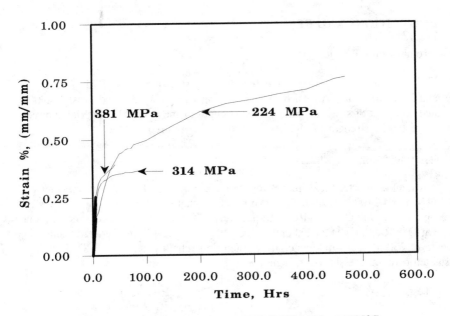

FIG. 3--Creep strain vs. time for GTE SNW-1000 at 1100⁰C

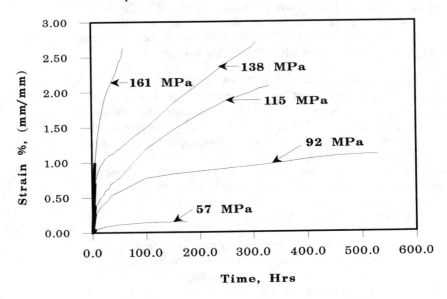

FIG. 4--Creep strain vs. time for GTE SNW-1000 at 1200⁰C

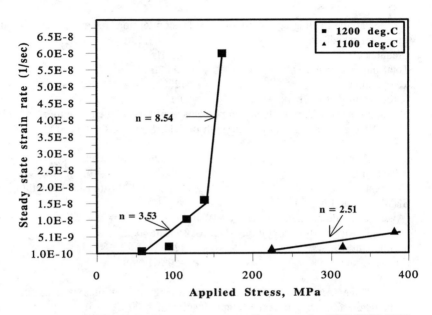

FIG. 5--Steady state creep rate versus applied stress for GTE SNW-1000

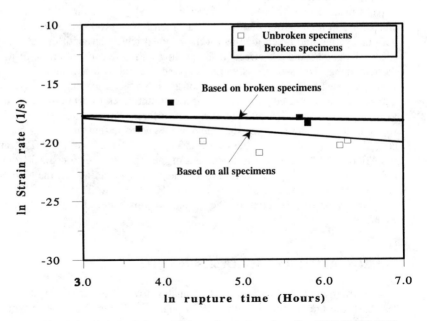

FIG. 6--Ln creep strain rate versus ln rupture time for GTE SNW-1000

Fatigue loading conditions--Cyclic tension-tension under load control with a triangular waveform.

Frequency: 0.5 Hz
Total number of cycles: 50 000
Temperature: 1200°C
R-Ratio (min. stress/max. stress): 0.1

Maximum stress: Various levels in percentage of the fatigue strength(σ_f) of GTE SNW-1000 at 1200°C. (Fatigue strength of GTE SNW-1000 at 1200°C = 140 MPa [7]).

Creep loading conditions--Uniaxial constant tensile load.

Temperature: 1200°C
Constant load: 161 MPa (70% tensile strength(σ_t) of GTE SNW-1000 at 1200°C)

(Average tensile strength of GTE SNW-1000 at 1200°C = 230 MPa [7]).

Tensile loading conditions--Uniaxial tensile load

Temperature: 1200°C
Stressing rate: 450 MPa/min

Three specimens were tested in creep (Figure 7) in uniaxial tension, immediately following uniaxial fatigue loading at 1200°C. Two (2) specimens were cycled each using a constant amplitude triangular waveform (Figure 8). Load was applied in tension with maximum stresses of 50%σ_f (70 MPa) and 70%σ_f (98 MPa). The third specimen was cycled from low (50%σ_f) to high (70%σ_t) stress levels in steps, a method known as "coaxing" (Figure 8) [8]. The specimen was then tested in creep for 290 hours, followed by pulling in tension at 1200°C to study the effects of both precycling and creep loading on the residual tensile strength of the material.

To understand the effects of precycling on the residual strength of the material, the remaining eight specimens were precycled and pulled in tension. The specimens were precycled in four different levels of tension-tension loading stress patterns at 1200°C. It was expected that precycling a constant block of 50 000 cycles at different loading patterns would reveal the effects of different fatigue stresses on the residual strength of the material. Two specimens were cycled in tension in each pattern, the conditions being maximum stresses of 25%σ_f (35 MPa), 50%σ_f (70 MPa), and 70%σ_f (98 MPa) at 1200°C. Two specimens out of the eight were also coaxed from low (50% σ_f) to high (70% σ_f) stress levels in steps. After cycling, all the eight specimens were tested in tension at a stressing rate of 450 MPa/min. This was performed to study the effect of precycling on the strength of this material and results are shown in Figure 9.

MICROSTRUCTURAL ANALYSES RESULTS

Microscopy and fracture surface analyses were conducted for all specimens tested. These are shown in Figures 10 through 13. Figure 10 shows a typical fracture surface morphology of a specimen tested in pure tension and at room temperature, while the fracture surface of a specimen tested in creep at 1200°C is seen in Figure

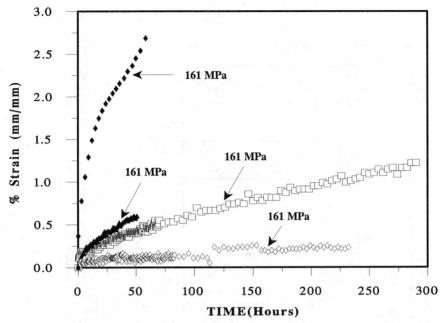

50000 cycles at 50% fatigue strength (70 MPa), R=0.1, and creep at 161 MPa
50000 cycles at 70% fatigue strength (98 MPa), R=0.1, and creep at 161 MPa
50000 cycles coaxed from 50% to 70% fatigue strength, and creep at 161 MPa
Pure creep at 1200 deg. C and 161 MPa

FIG. 7--Effect of fatigue on creep behavior of GTE SNW-1000

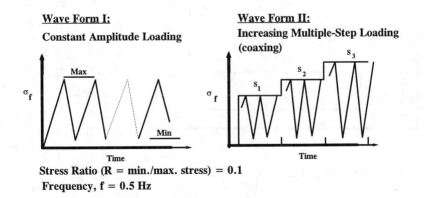

FIG. 8--Waveforms used in cyclic loading

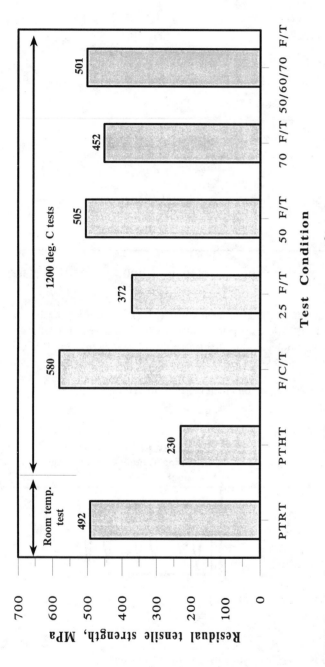

PTRT: Pure tension at room temperature. All other tests were conducted at 1200⁰C.

PTHT: Pure tension at high temperature (1200⁰C).

F/C/T: 50000 cycles coaxed from 50% to 70% fatigue strength, R = 0.1, creep at 70% σ (161 MPa) - 290 hrs, and pure tension.

25 F/T: 50000 cycles at 25% fatigue strength (35 MPa), R = 0.1, and pure tension

50 F/T: 50000 cycles at 50% fatigue strength (70 MPa), R = 0.1, and pure tension

70 F/T: 50000 cycles at 70% fatigue strength (98 MPa), R = 0.1, and pure tension

50/60/70 F/T: 50000 cycles coaxed from 50% to 70% fatigue strength, R = 0.1, and pure tension

FIG. 9--Tensile strength comparison of GTE SNW-1000 under different test conditions

FIG. 10--Fracture surface under tension
(room temperature, SNW-1000)

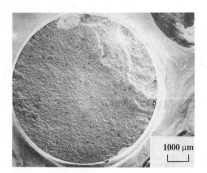

FIG. 11--Fracture surface after
creep (1200⁰C, SNW-1000)

FIG. 12--Fracture surface after cyclic fatigue
and tension (1200⁰C, SNW-1000)

FIG. 13--Creep tested sample
showing linking of pores (arrowed)

11. The fracture surface of a specimen that was precycled at 1200°C and pulled in tension at a stressing rate of 450 MPa/min at the same temperature is shown in Figure 12. Figure 13 is a micrograph of an area observed in the polished transverse section of a specimen tested in creep at 1200°C.

EDS work done on the polished section to determine the variation in silicon content across the diameter of a typical pore in both the creep tested and untested specimens is given in Figure 14. Observed variation in the yttrium content across a typical pore is shown in Figure 15.

DISCUSSION OF TEST RESULTS

<u>Room and Elevated Temperature Tensile Tests</u>

From Figure 2, it can be observed that the average tensile strength of this material at room temperature is 492 MPa, and it retains this value till 900°C. The decrease in tensile strength was moderate by 10%, as temperature increased to 1100°C. However, the material exhibited a sharp decrease in tensile strength by 50%, as temperature further increased to 1200° C. Tensile strength data obtained by Oak Ridge National Laboratory (ORNL) are also included. Only the average tensile strength values of the data at each test temperature are shown.

<u>Creep Tests</u>

Since the tensile strength changes significantly at 1100°C and 1200°C, studies on creep behavior of this material were concentrated at these temperatures. Figures 3 and 4 show that SNW-1000 exhibited a high creep rate with a relatively short primary stage, followed by a low strain rate with an extended secondary stage of creep at both temperatures. However, no tertiary creep stage prior to failure was observed. The data plotted (Figure 5) exhibited a clear trend of increasing steady state creep rate with increasing stress at both 1100°C, and 1200°C. For creep tests conducted at 1200°C, the slope of the relationship between steady state creep and applied stress changed drastically above a certain applied stress ($0.6\sigma_f$) from a slope value of 3.53 to 8.54. Similar results have also been reported by Wiederhorn and others [9-10] for materials such as NT-154 , GN-10 (a HIP'ed silicon nitride using Y_2O_3 and BaO as additives), and whisker-reinforced silicon nitride [11]. They indicated the possibility of creep being controlled by cavitation processes at high stresses and temperatures.

Figure 6 shows that all creep rupture data (both curves) for this material fall approximately on a single curve when the creep rate is plotted as a function of time to failure, irrespective of temperature or applied stress. The behavior was similar when the relationship was plotted for (a) only the specimens tested to failure and, (b) all the specimens. This relationship between the rupture time and steady state creep rate, known as the Monkman-Grant equation, in combination with the creep curves can be used to establish allowable stress for this material for high temperature applications.

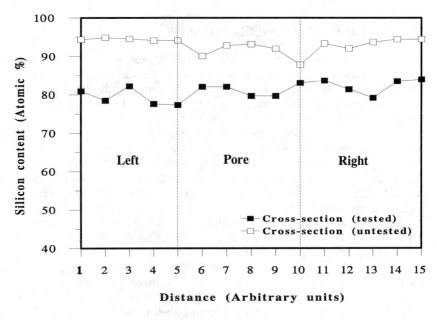

FIG. 14--Variation of silicon content across a typical pore

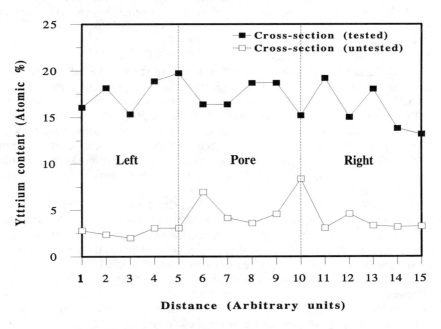

FIG. 15--Variation of yttrium content across a typical pore

Creep and Tensile tests on Precycled specimens

Figure 7 shows that cyclic loading at 1200^0C prior to creep loading lowers the primary creep strain and reduces steady state creep rates as compared to those of non-fatigued creep curves (Figure 4). For comparison purposes, the data for a pure creep specimen tested at 161 MPa (from Figure 4) is also included in Figure 7. A comparison of the three creep curves (Figure 7) of the specimens that were precycled at different stress levels and patterns, showed that the coaxed specimen yielded the least primary creep strain among the precycled specimens. From the results, it can be seen that coaxing enhances the creep resistance of the material. A specimen that was precycled and tested in creep without fracture for 290 hours showed the maximum residual tensile strength of 580 MPa when tested in tension at 1200^0C. This is a significant enhancement above the average tensile strength of the as-sintered material at 1200^0C that is 230 MPa, and the strength of the precycled/crept specimen is even higher than its average room temperature tensile strength of 492 MPa as reported by Vaidyanathan et al., [5].

Residual tensile strengths of SNW-1000 at 1200^0C with different fatigue backgrounds are compared in Figure 9 with average monotonic tensile strengths obtained at room temperature and 1200^0C. The results indicate that the tension-tension cyclic loading at 1200^0C raised the strength of the material considerably. Further, the strength improved when fatigued at stresses $> 25\%$ σ_f. It is quite possible that there is a threshold stress above which the effect would be more pronounced and this material can be made strong by a simple precycling operation at elevated temperature.

Microstructural Analyses

Examination [12] showed that fractures were initiated normally from pores as indicated by arrows in Figure 10, and occasionally from pre-existing inclusions. The fracture morphology of specimens tested in creep was characterized by a large area of flat fracture region as shown in Figure 11, while the precycled and tensile tested sample showed (Figure 12) a small flat fracture mirror region surrounded by a slow crack growth (SCG) zone. Figure 13, taken from the polished creep tested specimen, showed creep induced deformation processes linking pores (arrowed). This is different from that of the untested and tensile tested samples where pores were much smaller and separate [5].

EDS analyses results across pores (Figures 14 and 15) showed clearly that there are differences in silicon and yttrium contents between creep tested and untested specimens. The yttrium content is higher and the silicon content is lower in and around the pores in the creep tested specimen as compared to the untested specimens. Wiederhorn et al., [9], through Transmission Electron Microscopy (TEM) on NT-154 material reported that secondary crystalline phases relatively rich in yttrium were formed during creep, while the interphase remained rich in silicon.

From the precycling/residual tensile strength test results, it can be said that the improvement in the strength of the material could probably be due to the transformation of the intergranular glassy phase at elevated temperatures to a stronger crystallized phase by cyclic loading [8]. Mechanical behavior of silicon nitride is known to be influenced by the crystalline structure of the grain boundary material [13]. Research [14] has also shown that amorphous intergranular phase of silicon nitride can be devitrified by thermal annealing in the temperature range above 1350°C or by cyclic loading at an appropriate temperature. The creep process (equivalent to thermal annealing) besides the precycling process could have aided in devitrification. This could be why the specimen with the history of precycling and creep exhibited a good tensile strength. More work has to be done to clarify these effects in terms of both mechanical testing and detailed high resolution microscopy studies.

CONCLUSIONS

Based on pure uniaxial tensile, tensile creep, and fatigue/creep interactions tests at room and elevated temperatures on SNW-1000 sintered silicon nitride, the following conclusions can be made.

1. The tensile strength of the material is not affected by stressing rates from room to 1100°C.

2. The strength of the material is retained upto 1100°C. However, as temperature increased to 1200°C, the strength decreased by 50%.

3. The steady state creep rate increased with increasing stress at both 1100°C and 1200°C. For creep tests conducted at 1200°C, the slope of relationship between steady state creep versus applied stress changed drastically from a slope value of 3.53 to 8.54 above a certain applied stress $> 0.6 \, \sigma_t$.

4. All creep rupture data for this material could be represented by a single curve when creep rate is plotted as a function of time to failure, irrespective of temperature or applied stress. This relationship known as the Monkman-Grant equation, in combination with the creep curves can be used to establish allowable stress for this material for high temperature applications.

5. Optical microscopy and SEM done on the creep tested specimens showed clear evidence of linking of pores during creep.

6. EDS analyses indicated high yttrium and low silicon content in and around the pores in the creep tested specimens as compared to the untested specimens.

7. Cyclic loading at 1200°C prior to creep loading resulted in lowering both primary creep strain as well as steady state creep rates as compared to those of creep curves for the as-sintered specimens.

8. Precycling at 1200°C enhanced residual tensile strength of the material by a factor of about 2 compared to that of the as-sintered material tested at 1200°C.

ACKNOWLEDGEMENTS

This research was sponsored by the U.S. Department of Energy, Office of Transportation Technologies, as part of the Ceramic Technology for Advanced Heat Engines Project of the Advanced Materials Development Program, under contract DE-AC05-840R21400 with Martin Marietta Energy Systems, Inc.
The authors thank Dr. Ray Johnson and Dr. Ken Liu of Oak Ridge National

Laboratory for their continuing support and many helpful technical suggestions throughout this program.

REFERENCES

[1]   Neil, J. T., "The Big Three in Structural Ceramics," Materials Engineering 99(3), March 1984, pp 37-41.

[2]   Govila, R. K., "Strength Characterization of Yttria-doped Sintered Silicon Nitride," Journal of Materials Science, 20(12), 1985, pp 4345-53.

[3]   Courtney, T. H., Mechanical Behavior of Materials, McGraw Hill, New York, 1990.

[4]   Liu, K. C., and Brinkman, C. R., "Tensile Cyclic Fatigue of Structural Ceramics," Proceedings of the Twenty Third Automotive Technology Development Contractors' Coordination Meeting, P-165, Society of Automotive Engineers, Warrendale, PA, Oct. 1985, pp 279-284.

[5]   Vaidyanathan, R., Sankar, J., and Avva, V. S., "Testing and Evaluation of Si_3N_4 in Uniaxial Tension at Room Temperature," Proceedings of the Twenty Fifth Automotive Technology Development Contractors' Coordination Meeting, P-209, Society of Automotive Engineers, Warrendale, PA, Oct. 1987, pp 175-186.

[6]   Sankar, J., Kelkar, A. D., and Vaidyanathan, R., "Mechanical Properties and Testing of Ceramic Fiber - Ceramic Composites," Proceedings of the Fourth Annual Fossil Energy Materials Conference, U.S. Department of Energy and ASM International, 1990, pp 51-60.

[7]   Liu, K. C., and Brinkman, C. R., "Dynamic Tensile Cyclic Fatigue of Si_3N_4," Proceedings of the Twenty Fifth Automotive Technology Development Contractors' Coordination Meeting, P-209, Society of Automotive Engineers, Warrendale, PA, Oct. 1987, pp 189-197.

[8]   Liu, K. C., Pih, H., Stevens, C. O., and Brinkman, C. R., "Tensile Creep Behavior and Cyclic Fatigue/Creep Interaction of Hot-Isostatically Pressed Si_3N_4," Proceedings of the Annual Automotive Technology Development Contractors' Coordination Meeting, P-243, Society of Automotive Engineers, Warrendale, PA, Oct. 1990, pp 213-220.

[9]   Wiederhorn, S. M., Hockey, B. J., Cranmer, D. C., and Yeckley, R., "Tensile Creep Behavior of Hot Isostatically Pressed Silicon Nitride, "Journal of Materials Science, Vol. 28, 1985, pp 445-453.

[10]   Wiederhorn, S. M., Krause, R., and Cranmer, D. C., "Tensile Creep Testing of Structural Ceramics," Proceedings of the Annual Automotive Technology Development Contractors' Coordination Meeting, P-256, Society of Automotive Engineers, Warrendale, PA, Oct. 1991, pp 273-280.

[11]   Hockey, B. J., Wiederhorn, S. M., Liu, W., Baldoni, J. G., and Buljan, S. -T., "Tensile Creep of Whisker-reinforced Silicon Nitride," Journal of Materials Science, Vol. 26, 1991, pp 3931-3939.

[12]   Quinn, G. D., and Braue, W. R., "Secondary Phase Devitrification Effects Upon Static Fatigue Resistance of Sintered Silicon Nitride," Ceramic Engineering and Science Proceedings, Vol. 11, Nos. 7-8, 1990, pp 616-632.

[13]   Tsuge, A., Nishida, K., and Komatsu, M., "Effect of Crystallizing the Grain-Boundary Glass Phase on the High Temperature Strength of Hot-Pressed Silicon Nitride Containing Y_2O_3," Journal of American Ceramic Society, Vol. 58, Nos. 7-8, 1975, pp 323-26.

[14]   Cinibulk, M. K., Thomas, G., and Johnson, S. M., "Grain-Boundary-Phase Crystallization and Strength of Silicon Nitride with a YSiAlON Glass," Journal of American Ceramic Society, Vol. 73, No. 6, June 1990, pp 1606-1612.

Sheldon M. Wiederhorn[1], George D. Quinn[2] and Ralph Krause[3]

FRACTURE MECHANISM MAPS: THEIR APPLICABILITY TO SILICON NITRIDE

REFERENCE: Wiederhorn, S. M., Quinn, G. C., and Krause, R., "Fracture Mechanism Maps: Their Applicability to Silicon Nitride," Life Prediction Methodologies and Data for Ceramic Materials, ASTM STP 1201, C. R. Brinkman and S. F. Duffy, Eds., American Society for Testing and Materials, Philadelphia, 1994.

ABSTRACT: Fracture mechanism maps provide a means of assessing the structural reliability of ceramics at elevated temperatures. They can be used to summarize large quantities of data dealing with effects of load, temperature and environment on component lifetime. They also can be used to generate a design envelope that defines stress allowables for a given application. In this paper, we review the history and philosophy behind fracture mechanism maps and then discuss methods of obtaining such maps in an efficient manner. Based on data obtained in simple tensile tests, these methods are illustrated for one of the newer grades of silicon nitride. The map is then used to compare this material with a high temperature structural alloy, and another, older grade of silicon nitride. Finally, we discuss the use of fracture mechanism maps for design.

KEYWORDS: silicon nitride, fracture mechanism maps, creep, creep rupture, ceramics, lifetime prediction, reliability.

Improvements in the mechanical properties of silicon nitride over the past 20 years have finally resulted in a family of materials that can be used for complicated components in gas turbines at elevated temperatures. Methods of processing have been developed so that complex shapes, such as turbine rotors, can be made with

---

[1] Senior NIST Fellow, Materials Science and Engineering Laboratory, The National Institute of Standards and Technology, Gaithersburg, MD 20899

[2] Ceramic Engineer, Ceramics Division, The National Institute of Standards and Technology, Gaithersburg, MD 20899

[3] Research Chemist, Ceramics Division, The National Institute of Standards and Technology, Gaithersburg, MD 20899

confidence. With these new materials, experimental gas turbines have been built and operated at full power at 1300°C [1]. These results look promising for the eventual development of commercial turbines for vehicular transportation.

Structural reliability is one problem that remains to be treated for the safe use of silicon nitride at elevated temperatures. For promising grades of this material, a database of mechanical properties as a function temperature, stress and environment is needed. The importance of fatigue and corrosion to mechanical reliability has to be assessed. Finally, a methodology for predicting lifetime has to be developed. This paper deals with the last of these questions. A specific, relatively simple method is recommended for establishing stress and temperature allowables for different grades of silicon nitride. The history of the method is discussed; its use on a modern grade of silicon nitride is illustrated, and an extension of the method to account for measurement uncertainty is suggested.

FRACTURE MECHANISM MAPS - THEIR HISTORY

The Origin of Fracture Mechanism Maps

The use of fracture mechanism maps for materials selection was first suggested by Ashby, Gandhi and Taplin for f.c.c. metals and alloys [2] and was expanded to other materials by Gandhi and Ashby [3]. The idea behind such maps is relatively simple. On a plot of tensile stress versus temperature, regions of failure are defined within which one failure mechanism dominates the failure process. The stress and temperature axes are usually represented as a normalized tensile stress (stress, σ, divided by Youngs modulus, E) and a homologous temperature (temperature, T, divided by the melting temperature, T_m). These axes tend to reduce fracture mechanism maps into classes of materials. Lines of constant failure time are plotted on the map, so that for a given stress and temperature, failure time can be estimated for a given application. An example of a fracture mechanism map for monel, a high temperature alloy, is given in Fig. 1.

Ashby et al. [2] considered six potential failure mechanisms. At low temperatures (T < $0.3 \cdot T_m$), these include: cleavage and intergranular brittle fracture; plastic growth of voids (either transgranular or intergranular) and rupture by necking or shearing off. At creep temperatures, (T > $0.3 \cdot T_m$) mechanisms include: intergranular creep fracture by either void or wedge crack growth; growth of voids by power-law creep (either transgranular or intergranular) and rupture due to dynamic recovery or recrystallization. Mechanisms are identified by the fractographic analysis of tensile test specimens after fracture. Hence, fractographic analysis is an essential element of materials evaluation and of the development of mechanism maps. Some of these mechanisms are indicated in Fig. 1.

Of the mechanisms considered, the most important ones for structural ceramics are intergranular creep fracture at elevated temperature, and intergranular brittle fracture at lower temperatures. For ceramics and other brittle

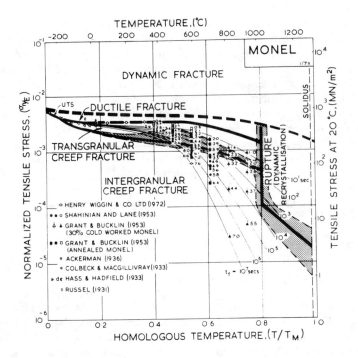

FIG. 1--Fracture mechanism map for monel, a high temperature alloy. This map illustrates the original concept by Ashby et al. [2].

materials, intergranular brittle fracture can be divided into three sub-mechanisms depending on the role of plastic deformation during fracture [3]. Intergranular brittle fracture I requires pre-existing cracks and occurs in the absence of general plasticity. Intergranular brittle fracture II requires plastic deformation to nucleate cracks, as for MgO at low temperatures. Intergranular brittle fracture III occurs in the presence of plasticity and slip at grain boundaries.

Ghandi and Ashby were also the first to present a fracture map for silicon nitride, Fig. 2. The map was developed from flexural strength data obtained at a number of laboratories. Fracture mechanisms given on the map were determined by microstructural analysis of broken specimens. The data were not sufficient to plot lines of constant failure time. Also, materials studied contained considerable glass as a sintering aid and so did not have very good high temperature creep resistance. Thus, intergranular creep fracture in Fig. 2 is shown to start at approximately 800°C. Two regions of cleavage are indicated: cleavage I is the same as intergranular brittle fracture I. No plasticity accompanies fracture in this region. Cleavage 3 is the same as intergranular brittle fracture III; plastic deformation accompanies crack growth in this region. Dynamic fracture delineates the theoretical strength of the material. The practical limits of strength for these

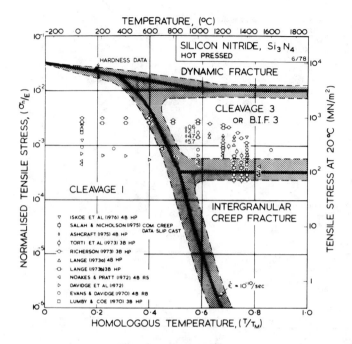

FIG. 2--Earliest fracture mechanism map for silicon nitride. Most of the data were from flexure tests. Mechanisms were identified by microscopic examination of fractured specimens. From ref. 3.

early materials are indicated by the data points. Newer materials have a much higher strength at low temperatures, approximately 1000 MPa.

Fracture Mechanism Maps for Silicon Nitride

The first complete fracture map for silicon nitride was developed by Quinn [4, 5] for NC132[4], a grade of MgO-doped silicon nitride made by Norton, Fig. 3a. The data for this map were collected by several hundred flexural stress rupture and strength tests, for exposure times as long as 20,000 h. A comparative set of flexure and tensile test data was used to convert the flexural data into tensile data, and a fracture mechanism map for tension was obtained. Four regimes of behavior are

---

[4] The use of commercial designations does not imply endorsement by the National Institute of Standards and Technology.

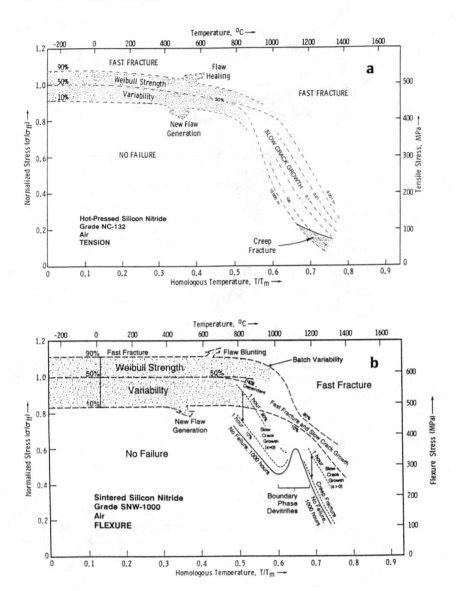

FIG. 3--(a) Fracture mechanism map for magnesia-doped hot-pressed silicon nitride in air in direct tension. The left ordinate axis is the tensile stress normalized by the mean room-temperature fast-fracture strength (σ_{ff}). From ref. 5. (b) Fracture mechanism map for a yttria-alumina doped sintered silicon nitride. Since the scatter in failure times was very high, confidence bands of 10 and 90% are shown for the fast fracture and slow crack growth (1 h failure time) regions. From ref. 10.

indicated on the map. At low temperatures, the fast-fracture regime is separated from a no fracture regime by the fracture strength. Uncertainty in the position of the strength line is indicated by plotting 10%, 50% and 90% failure probabilities. At higher temperatures the strength line broadens into lines representing failure due to slow crack growth. This region is equivalent to the region of intergranular brittle fracture 3 on the diagram by Ghandi and Ashby. Finally, at the lowest stresses and highest temperatures, a region of creep fracture is indicated. The material studied by Quinn was of better quality than that studied by earlier investigators. Consequently, the data in the slow crack growth and creep fracture regimes are shifted to higher temperatures on Quinn's map.

In addition to plotting the map, Quinn identified three sets of equations that could be used to characterize lifetime, and determine boundaries between failure regions. The fracture strength was assumed to be controlled by brittle fracture and to be influenced by the presence of cracks or some other defects in the material. The relation between fracture stress, S_i, and cumulative failure probability, P, is given by the Weibull [6] equation,

$$P = 1 - \exp\left[-\int^v \left(\frac{S_i - \sigma_u}{\sigma_o}\right)^\alpha dV\right] \tag{1}$$

This is a three parameter Weibull equation in which α is the Weibull modulus, σ_u is the threshold stress, σ_0 is a normalization parameter called the Weibull material scale parameter. V is the specimen volume.

In the slow crack growth regime, lifetime is determined by the growth of cracks from pre-existing flaws. The crack velocity, v, is determined by the stress intensity factor, K_I, at the tip of the growing crack [7],

$$v = A \cdot K_I^N \cdot \exp(-Q_{scg}/RT) \tag{2}$$

N is the stress exponent for crack growth and A is an empirical constant. Q_{scg} is the activation energy for slow crack growth. Equation 2 can be integrated to give the total time to failure, t_f, under conditions of constant tensile stress, σ_a [8],

$$t_f = B \cdot \sigma_a^{-N} \cdot S_i^{N-2} \cdot \exp(Q_{scg}/RT) \tag{3}$$

where $B = 2/[AY^N(N-2)K_{IC}^{N-2}]$. The initial strength is $S_i = K_{IC}/Y\sqrt{c_i}$, where c_i is the initial crack size and Y is a constant that depends on specimen and crack geometry. Equation 3 is used to plot the lines of constant failure time on the fracture mechanism map. As S_i depends on the failure probability, P, t_f also depends on P and, through equation 1, on the constants from the Weibull equation. Therefore, t_f also depends on σ_u, σ_0, α and V.

At the lowest stresses and highest temperatures, failure in silicon nitride is dominated by the generation and accumulation of damage due to cavity formation at grain boundaries. Within this region, component lifetime is related to creep behavior by the Monkman-Grant relation [9]. As noted originally by these authors, the time to failure can be expressed as a power function of the minimum creep rate:

$$t_f = C \cdot \dot{\epsilon}_{min}^{-m} \tag{4}$$

$\dot{\epsilon}_{min}$ is the minimum creep rate and C and m are constants. If $\dot{\epsilon}_{min}$ can be expressed in the form of an Arrhenius modified Norton equation:

$$\dot{\epsilon}_{min} = D \cdot \sigma_a^n \cdot \exp(-Q_c/RT) \tag{5}$$

then the time to failure is given by:

$$t_f = C \cdot D^{-m} \cdot \sigma_a^{-mn} \exp(m \cdot Q_c/RT) \tag{6}$$

C and D are empirical constants and Q_c is the apparent activation energy for creep. Equation 6 can be used to define lines of constant failure time in the creep rupture region of the failure mechanism map. Also, by equating equation 3 to equation 6, the boundary between the slow crack growth region and the cavitation region can be determined.

Quinn and Braue [10] subsequently prepared a fracture mechanism map for a commercial sintered silicon nitride, GTE Grade SNW 1000, containing yttria and alumina sintering aids. The fracture map, based primarily on flexure testing, is shown in Fig. 3b and has two different slow crack growth regions (one with and one without creep strain) and a creep fracture region. The map is more complicated than that for NC 132 for two reasons. First, the trends of static fatigue behavior are radically altered at 1000-1100°C by the devitrification of the boundary phase. Second, the scatter in failure times was much higher (up to four orders of magnitude for identically loaded specimens) and therefore the use of median times-to-failure may be misleading. Instead, empirically estimated 10 and 90% confidence bounds are shown for the fast fracture and slow crack growth (1 hour) regions. The no failure line is the boundary for which no failures were observed in 1000 hours.

Fracture Mechanism Maps for Design

The use of fracture mechanism maps for design was first suggested by Matsui et al. [11] who recognized that such maps have the potential of providing a design envelope which defines stress allowables for a given application. Fracture mechanism maps were first applied to a turbocharger rotor. Stress and temperature distributions were calculated for the rotor under maximum operating conditions, Fig. 4. For each point on the rotor, the temperature and stress define a point on the fracture mechanism map, Fig. 5. If these points lie within an envelope defining allowable times to failure, the design is considered acceptable. In the map shown,

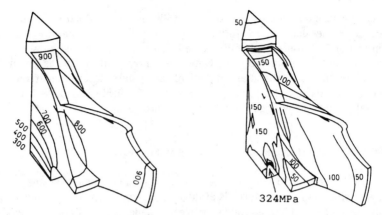

FIG. 4--Finite element analysis of a rotor, giving temperature and stress distributions in the rotor during operation. From ref. 10.

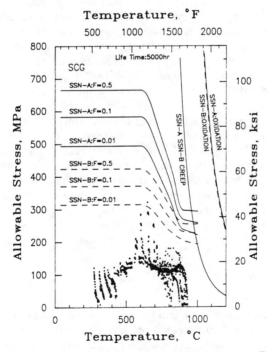

FIG. 5--Fracture mechanism map for turbocharger rotor. Two sintered silicon nitride (SSN) materials are plotted: material A and material B. F is the failure probability. Points represent temperatures and stresses at various locations on the rotor. Material A appears to be the better of the two.

the inner contour indicates an expected failure time of 5000 h. Two regions are indicated on the map: one for failure by slow crack growth; the other for failure by creep rupture. Failure probability levels are indicated on the slow crack growth curves ranging from 0.01 to 0.5. Only the central failure line is given for creep failure. As noted in the diagram, all of the points determined from the finite element analysis of stress and temperature fall within the 5000 h envelope for material A, indicating acceptability of the rotor design.

FRACTURE MECHANISM MAPS FOR NEW MATERIALS

In this section, a fracture mechanism map is developed for one of the newer grades of silicon nitride. Methods of developing such maps from experimental data are discussed first. PY6, a grade of material made by GTE Laboratories (Waltham, MA) is used to illustrate these methods. PY6 is a silicon nitride alloy with 6 wt% yttrium oxide added as a sintering aid. Final consolidation is by hot isostatically pressing (HIPing) generally resulting in a bimodal ß-silicon nitride microstructure with an amorphous to partially-crystalline yttrium disilicate intergranular phase [22]. PY6 was selected because nominally similar versions of this alloy have been tested at three different laboratories [Oak Ridge National Laboratory (ORNL), Southern Research Institute (SoRI), and the National Institute of Standards and Technology (NIST)]. Although the material tested by ORNL had been cold isostatically pressed and HIPed and while the material tested by NIST and SoRI had been injection molded and HIPed, a comparison of test results between these laboratories can be made based on the compositional similarities of the PY6 versions tested. Furthermore, the same lot of powder was used to make the material tested at NIST and SoRI, so a comparison of test results between these laboratories should be meaningful. Most of the data for the fracture mechanism map were obtained at high temperatures and low stresses, where intergranular creep fracture occurs. Hence, most of the comparison of data between laboratories is made in this region. As the collection of data in the crack growth region has not been extensive, this region of the fracture mechanism map is least reliable. Similarly, strength measurements at lower temperatures are sparse, and hence are not considered in this paper.

Tensile Test Techniques

Since the publications of Matsui et al. [11] and Quinn [4, 5, 10], significant advances have been made in test technique. Prior to 1986, most testing was done in flexure. Since that time testing techniques have advanced to the state that much of the new data are being generated in tension. Three generic grip designs are currently being used for high temperature tensile testing. One design uses hot grips in which the entire specimen is in the hot zone of the test furnace. The second uses warm grips in which the loading points on the specimen are cooler than the test furnace but warmer than ambient temperature. The third uses cold grips in which

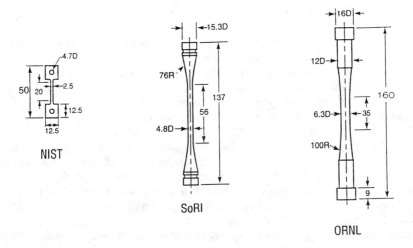

FIG. 6--Examples of specimens used in the study of PY6.

the loading points on the specimen are cooled to ambient temperature typically by flowing, chilled water.

Hot grips are being used at a number of laboratories. At NIST, flat dogbone shaped specimens were developed [12], Fig. 6. Pin loading improves alignment and reduces flexural strains during testing to less than 2% of the total creep strain. Specimens are relatively small, 50 mm overall length, are made by simple machining techniques, and are relatively inexpensive, approximately $65 per specimen. Because of the simple geometry, the size of the specimen is easily modified for testing of smaller specimens, sometimes needed for experimental batches of material. The specimen gauge length is measured in these specimens by attaching flags to the specimens, and using a laser transducer to detect the position of the flags. With this technique, gauge length measurements at temperatures as high as 1500°C were made to ±2 μm. More recently, Foley et al. [13] improved the design of the dogbone specimen by reducing the stress concentration at the end of the gauge section. They also showed that specimens with cylindrical gauge sections could be made following their design.

A different geometry for measuring creep was developed by Ohji [14]. His specimens were also completely immersed in the hot zone of the furnace. Gripping however, was along the outer edge of the specimen, by use of a clevis. This type of gripping has the advantage that higher stresses can be applied to the specimen without fear that the ends of the specimen will be torn loose. Flexure free alignment is, however, more difficult with this type of loading. Ohji had the gauge section of his specimen cylindrically machined so that there were no edges in the gauge section. A further feature of the specimen was the built in flags that eliminated potential flag slippage during testing. Because of the exacting machining tolerances on the bearing surfaces, and the round gauge section, specimen costs for this specimen were 2 to 4 times those of the NIST dogbone specimens. Furthermore, because of the more complex geometry, scaling to smaller sizes is not as easy for these specimens.

Cold and warm grip specimens must be long enough to span the hot zone of the test furnace. Consequently they are longer and more expensive than hot grip specimens. Specimen costs can be as much as 5 times those of hot grip specimens. However, because the grips are made of high temperature alloys, and are at relatively low temperatures during testing, larger forces can be applied to the specimens. Hence, the specimen cross section can be approximately 6 times that of the hot grip specimens. Cold grip specimens most commonly used today were designed by Liu et al. [15] of ORNL. They are made from rods of material approximately 160 mm long and 20 mm in diameter. They are cylindrical and are gripped by a button head in the cooler parts of the furnace, Fig. 6. Alignment accomplished via a room temperature fixture is excellent with less than 1% strain in bending. Machining costs for the ORNL buttonhead specimen typically run around $200 per specimen. A slightly different cylindrical specimen was developed earlier by Pears and Digesu [16] at SoRI, Fig. 6. This geometry is being used by SoRI, in combination with warm grips, for high temperature testing of silicon nitride. Finally, edge loaded flat specimens have also been developed for testing ceramics at high temperatures [17, 18].

Tensile Creep Data

An example of creep behavior for PY6 (made by GTE), a grade of material containing 6 w/o as a bonding phase, is shown in Fig. 7. For the run lasting 2511 h, Fig. 7a, transient creep was observed for almost 1000 h. Transient creep was observed during the entire creep test for the run lasting 23.5 h, Fig. 7b. These data are representative of much of the data presented in the literature. Tensile creep tests on silicon nitride suggest that most grades of this material exhibit a long primary creep stage, which may take up more than half the total creep life. In some grades of silicon nitride, the entire creep life may consist of primary creep [17, 19-21]; in others, steady state may be apparent after about one-half the specimen lifetime [17, 22, 23]. Devitrification of the grain boundary phase in these materials, caused by long term annealing, tends to reduce, but not eliminate the primary creep [17]. In some cases, silicon nitride exhibits tertiary creep [22, 26], usually as a gentle upward curvature on the strain-time creep curve. Strains to failure range from about 0.5 to 2 percent depending on the test conditions and the grade of silicon nitride.

Data of the type shown in Fig. 7 can be represented graphically by plotting the minimum creep rate (usually the creep rate just prior to failure) as a function of stress and temperature. When plotted in this manner, experimental data are usually found to fit equation 5. Thus, in Fig. 8, the solid straight lines represent a fit of the experimental data over a temperature range 1250°C to 1400°C. The apparent activation energy, Q_c, for the fit was 1349 kJ/mol; the stress exponent, n, was 8.7; and the logarithm of the coefficient, $\log_{10}D$, was 17.67 (for stress given in MPa and minimum creep rate given in s^{-1}).

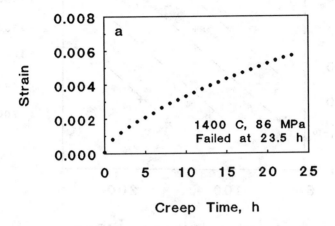

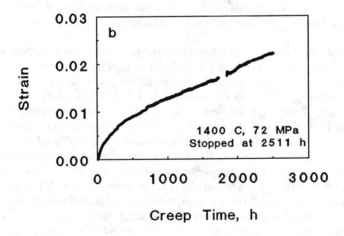

FIG. 7--Typical creep data obtained on NIST specimens. Very long primary transient creep and the absence of tertiary creep before failure is characteristic of PY6.

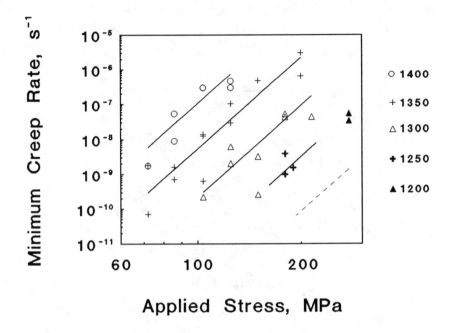

FIG. 8--Minimum creep rate of PY6 as a function of applied stress. Data from 1250°C to 1400°C fit a Norton type of equation.

The high values of the apparent activation energy and the stress exponent are typical of what is observed for PY6 [22]. The high value of the stress exponent is attributed to cavitation during the creep process. The high value of the apparent activation energy has been discussed by Raj and Morgan [24], and is probably a consequence of the high apparent activation energy for the diffusivity of glass at the grain boundaries and the variability of solubility of silicon nitride in the glass as a function of temperature. These possibilities are discussed in reference [21]. For most data, the measured minimum creep rate lies within one order of magnitude of the least squares fit line. A major exception to this observation occurs for data collected at 1200°C. For this set of data, the least squares fit line for 1200°C (the dotted line in the figure) lies far from the data. This deviation of the fit from the data is similar to that observed by Ferber and Jenkins on PY6 [22], and has been attributed to a change in mechanism from creep cavitation to crack growth by these authors.

The data shown in Fig. 8 were compared with data collected on the same material by Ferber and Jenkins [22] at ORNL and by Khandelwal at the Allison Gas Turbine Division of the General Motors Corporation (GM) [25], Fig. 9 (This work was carried out at SoRI for GM). The lines in these figures were calculated from NIST data, equation 7. For each temperature, the data from ORNL show a higher creep rate than the data from NIST. Depending on stress, the minimum creep rate determined at ORNL plot from one to two orders of magnitude higher. By contrast, the data from GM tend to plot close to the lines predicted from the NIST data.

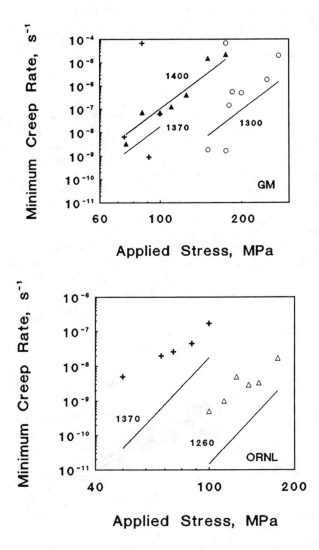

FIG. 9--Comparison of data collected at NIST with data collected at GM and at ORNL. The individual points in the upper figure were collected for GM, while in the lower figure, the data points were collected by ORNL [22]. The lines in both figures were calculated from data collected at NIST.

These results suggest that the material tested at ORNL was more susceptible to creep than that tested at NIST. Agreement between the GM and NIST data is consistent with the fact that specimens in both studies were obtained from the same powder lot and processing route. Furthermore, as the SoRI specimen is substantially larger than the one from NIST, these results suggest that specimen size does not have a strong effect on the minimum creep rate.

Creep Rupture Data

The rupture data for all three laboratories are plotted in Fig. 10 in the form of a Monkman-Grant curve, equation 4. Only the data from the creep fracture region are included in Fig. 7. All three sets of data appear to plot along the same curve; most of the data fall well within one order of magnitude of the median failure time for a given minimum creep rate. For the NIST data, the constants for equation 4 are $m=1.14$ and $\log_{10}C = -7.2$, where time is expressed in hours, and strain rate is expressed in s^{-1}. As with most Monkman-Grant plots for ceramics, the Monkman-Grant exponent, m, is slightly greater than 1. In metals, by contrast, the exponent, m, is often slightly less than 1 [9].

Based on equation 4, it is concluded that lifetime for PY6 depends only on the minimum creep rate. The faster the minimum creep rate, the shorter the lifetime, regardless of temperature or applied stress. The time to failure for a given temperature and applied stress is obtained from equation 6.

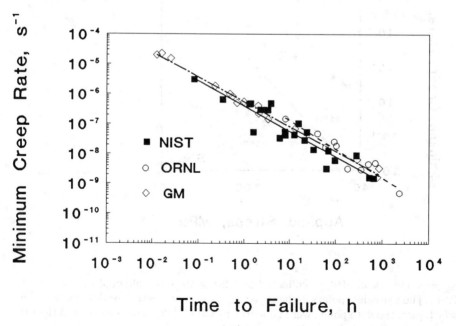

FIG. 10--Monkman-Grant curve for PY6. Data from NIST, GM and ORNL fall close to one another.

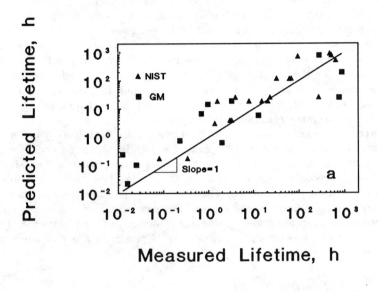

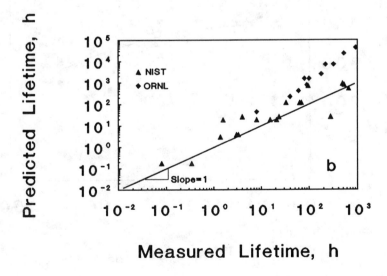

FIG. 11--Comparison of predicted and measured lifetimes: (a) GM and NIST; (b) ORNL and NIST.

Calculated and measured times to failure are compared in Fig. 11a for data from NIST and GM. The straight line in the figure represents equal measured and predicted failure times. The data tend to fall slightly above the straight line, but otherwise scatter randomly. The greatest deviations from the expected failure time are approximately one order of magnitude. The two sets of data are indistinguishable statistically. Since the Monkman-Grant plot and the creep data were the same for the two materials, this finding was expected. As the silicon nitride powder lot and processing route were the same for both sets of data, the data in Fig. 11a suggest that test geometry, by itself, has no major influence on the creep and creep rupture results for silicon nitride.

A comparison of the data from ORNL with those from NIST is given in Fig. 11b. For short failure times <100 h, the ORNL data lie on the outer fringe of the NIST data and about an order of magnitude above the predicted line. For long failure times, approximately 1000 h, the ORNL data lie about two orders of magnitude above the predicted line. The ORNL data are internally more consistent than those from the other two laboratories. Since creep rupture data from all three laboratories can be represented by the same Monkman-Grant plot, Fig. 10, the difference between the ORNL and NIST data is primarily a consequence of the difference in the creep behavior.

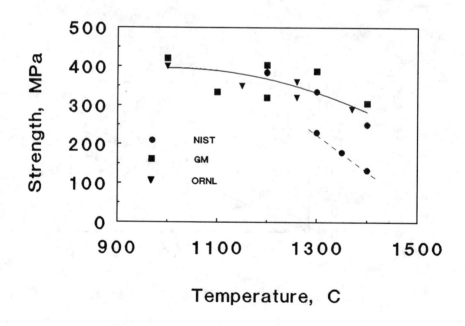

FIG. 12--Strength data collected on PY6 at three laboratories. Strength is found to depend on rate of loading. High rates of loading, >16 MPa/s, are given by the solid curve; low rates, <0.06 MPa/s are given by the dashed line.

Strength Data

Data on the tensile strength of PY6 were collected at ORNL, SoRI (for GM) and at NIST. Large, button-head tensile specimens were used at ORNL; a slightly different specimen was used at SoRI; small flat dogbone specimens were used at NIST, Fig. 6. Despite these differences in test geometry, general agreement was obtained between strengths determined at test temperatures above 1000°C. A comparison of the data collected at high loading rates, >16 MPa/s, are shown in Fig. 12[5]. Within the temperature range 900°C to 1400°C, the strength of PY6 ranged from about 400 MPa to about 300 MPa. Although there was considerable scatter of the data, most fell within 50 MPa of the mean curve. Studies at lower rates of loading, approximately 0.06 MPa/s, showed that as the temperature increased from 1000°C to 1400°C, strength depended on the loading rate during the test. At lower rates of loading, creep damage accumulation and subcritical crack growth reduce the strength.

Crack Growth Data

Data in the crack growth region for PY6 were collected only at ORNL [26]. Below temperatures of 1200°C, crack growth dominates the failure process in PY6. The change in mechanism is indicated by an increase in the stress dependence of the minimum creep rate, and a shift in the position of the data from the positions predicted from creep data taken at higher temperatures, as shown for example in Fig. 8. In addition a change in mechanism of fracture is observed [26]. At temperatures above 1260°C, fracture is intergranular and cavitation is observed on the fracture surface. At 1150°C this mode of fracture changes to transgranular fracture. Based on these observations and a set of data collected by dynamic fatigue techniques[6], a set of curves representing the crack growth region of the

---

[5] The strengths in figure 12 were not adjusted for specimen volume. If failure occurs by the weakest link mechanism, and if the flaw distribution are the same in all specimens, then the volume adjusted strengths of the NIST specimens should be lower than given in figure 12 because the NIST specimen volume is only about one-tenth that of the ORNL or SoRI specimen volumes. For a Weibull modulus, m, of 11, a reduction in strength of about 19% would be expected. This Weibull modulus was estimated from the strength data on NIST specimens.

[6] In static fatigue, the time to failure is measured as a function of stress for a constant applied stress. In dynamic fatigue, the breaking stress is measured as a function of the stressing rate. Both techniques can be used to obtain the constants in equation 3. The relationship between failure time in dynamic fatigue and failure time in static fatigue was developed first by Davidge et al. [27]: $t_s = t_d/(N+1)$, where N is the stress exponent in the crack growth equation, equation 4.

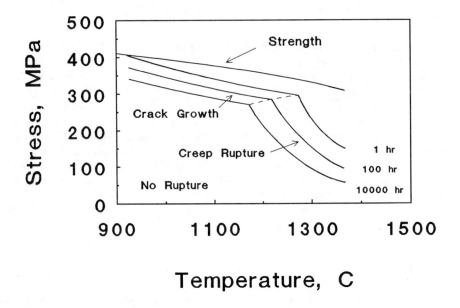

FIG. 13--Fracture mechanism map for PY6 in air in tension.

fracture mechanism map was obtained. Fitting the experimental data to equation 3, the constants were determined: $Q_{scg} \approx 760 \, kJ/mol$; $N \approx 55$; $\log_{10}(B \cdot Si_i^{N-2}) = 110.38$.

Fracture Mechanism Map for PY6

The fracture mechanism map for PY6, Fig. 13, was developed from the data discussed above. Four regions of behavior are indicated on the figure: no failure, fast fracture, slow crack growth and creep rupture. Lines of constant failure time are indicated on the figure as solid lines. The boundary for regions of slow crack growth and creep rupture, dotted line, are obtained by solving equations 3 and 6 for the given failure times. The map differs somewhat from the earlier one developed by Quinn for NC132. Because of the higher stress dependence of the crack growth data for PY6, distinct breaks are observed in the time contours. The time contours in NC132 are continuous from the creep fracture to the crack growth regions, suggesting similar stress dependencies for both mechanisms. Also the creep fracture region in PY6 goes to very much higher stresses, approximately 250 MPa. Reasons for this difference in behavior are not understood at the present time.

Maps such at that shown in Fig. 13 can be used either for materials selection, as a guide for improving the microstructure of materials, or for purposes of design. For a given application, materials selection requires the comparison of two or more

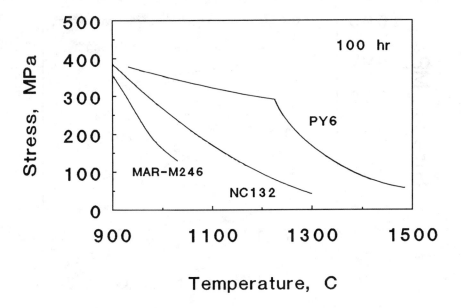

FIG. 14--A comparison of the high temperature behavior of PY6 with an older grade of silicon nitride, NC132 and a high temperature alloy, MAR-M246.

such maps. Thus, one might select a line of constant time to failure and compare it for two grades of silicon nitride, or for a grade of silicon nitride and another material to determine the best one for a given application. A comparison of this sort is given in Fig. 14 for PY6, MAR-M246 and NC132 for a total expected lifetime of 100 h[7]. The MAR-M246 is a high temperature alloy used in turbine engines; the NC132 is an older grade of silicon nitride; the PY6 is one of the newer grades of silicon nitride. At an applied stress of 200 MPa, the PY6 has an advantage of approximately 290°C over the MAR-M246 and approximately 200°C over the NC132. The temperature advantage over the NC132 represents a considerable improvement in high temperature mechanical behavior for silicon nitride as a consequence of processing research over the past 10 years.

Designing with Fracture Mechanism Maps

Fracture mechanism maps were originally intended as a qualitative means of comparing the relative fracture behavior of materials, and were not intended for the final design of components. Matsui et al. [11], however, showed that if confidence

---

[7] A 100 hour lifetime was selected because data on the alloy were easily obtained in this range [28].

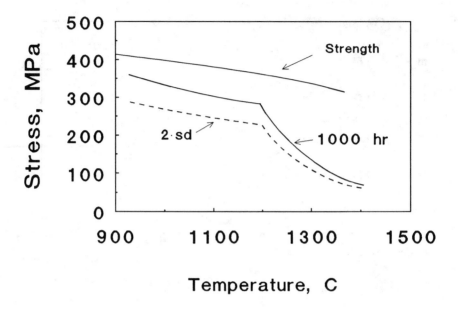

FIG. 15--Effect of measurement scatter on the design envelope.

levels are put on the different regions of the fracture mechanism map then an allowable stress-temperature envelope could be defined, within which failure would not be expected. It is possible to place such error bands on the lines of constant time-to-failure shown in Fig. 13. In both the crack growth region and the creep rupture region, these bands can be determined empirically from the actual scatter in time to failure.

This procedure is illustrated in figure 15 for the NIST creep rupture data of Fig. 10. First, the differences between the expected and measured times to failure are determined. The logarithmic mean and standard deviation of these differences are then evaluated. Since the scatter of the data about the line in Fig. 10 does not depend on the logarithm of the failure time, the standard deviation is also assumed to be independent of the logarithm of the time to failure. The standard deviation, sd, of the data in Fig. 10 was approximately 0.46. As the data are expressed in terms of $\log_{10}t$, this means that approximately 95% of the data lies between $\log_{10}(t-2sd)$ and $\log_{10}(t+2sd)$, for any given value of $\log_{10}t$. If 2sd is selected as the lower confidence limit of $\log_{10}t$, only 2.5% of the components will fail in times shorter than this value[8]. In Fig. 15, a curve of constant failure time for 1000 h is

---

[8] As only 19 specimens were tested to failure, a confidence level of greater than 2sd was not justified.

plotted on a fracture mechanism map along with a curve indicating the lower confidence bound for two standard deviations. This curve is congruent with the curve given by $\log_{10}(1000)+.92$, which equals 3.92, or 8318 h.

The same procedure was followed in the crack growth regime using data collected by Jenkins, Ferber and Lin [26]. In this case, dynamic fatigue data collected at 1150°C and 1260C° were used. The pooled standard deviation of $\log_{10}t$ from both sets of data was 2.82. This very large standard deviation is typical of the type of scatter observed in times to failure for materials that fail by crack growth. The scatter reflects the variation in severity of the life-limiting flaw that is present in the specimen and therefore is responsible for failure. The lower confidence bound for the crack growth region is thus $\log_{10}(1000)+5.64$, or 4.37×10^8 h. This curve is plotted in Fig. 15 as a dashed line.[9]

The effect of the confidence band is to limit the conditions under which the material can be used. The crack growth band effectively lowers the allowable stress by about 50 MPa over the temperature range 950°C to 1200°C. This reduction in allowed stress is consistent with the scatter in the strength reported for this material at elevated temperatures. The lower confidence band for the creep rupture region has a much smaller effect on the allowable operating conditions. Here, the reduction in strength ranges from approximately 30 MPa at 1200°C to approximately 10 MPa at 1400°C. This modest reduction in strength is consistent with the smaller scatter in the rupture time usually observed in the creep rupture region.

SUMMARY

In this paper fracture mechanism maps are suggested as a means of characterizing the structural reliability of ceramic materials. These maps can be used to generate design envelopes that define stress allowables for a given application. Lines of constant failure time and confidence levels for lifetime prediction are determined by the mechanisms that control failure in each material. In silicon nitride, three mechanisms are important: brittle fracture at low temperatures; slow crack growth at higher temperatures; creep rupture at the highest temperatures. In the creep rupture regime, lines of constant lifetime are easily determined from the minimum creep rate as a function of temperature and stress and the constants of the Monkman-Grant equation. In the slow crack growth regime, lines of constant failure time are determined either from strength

---

[9] As the scatter in failure time for crack growth data is a consequence of the scatter in initial flaw size, Weibull statistics are more appropriate for describing the lower confidence bounds for the crack growth region. It can be shown [29] that the time to failure is related to cumulative failure probability, P, through the following equation: $t=t_0\{\ln(1-P)^{-1}\}^{1/\alpha'}$, where $t_0=B\sigma^{-N} S_0^{N-2}$ and $\alpha'=\alpha/(N-2)$. The treatment given in this paper for the crack growth region is an approximation of the correct procedure.

measurements as a function of loading rate and temperature, or from time-to-failure measurements at constant stress. In the brittle fracture regime, failure probability lines are determined from the stress distribution as a function of temperature. In the course of these studies, a comparison was made between creep and creep rupture data obtained on large and small test specimens. Provided the powder lot and manufacturing process used to make the specimens was the same, specimen geometry did not appear to have a major effect on results.

ACKNOWLEDGEMENT

The authors gratefully acknowledge the support of the General Motors Corporation, and the Ceramic Technology for Advanced Heat Engines Program, U.S. Department of Energy, under Interagency Agreement No. DE-AI05-85OR21569.

REFERENCES

[1]   Smyth, J.R. and Morey, R.E. "Advanced Turbine Technology Applications Project Progress in Year Four," pp. 47-55 in Proceedings of the Annual Automotive Technology Development Contractors' Coordination Meeting 1991, P-256, Dearborn, MI, October 28-31, 1991. Society of Automotive Engineers, Inc., 400 Commonwealth Dr., Warrendale, PA 15096-0001, June 1992.

[2]   Ashby, M.F., Gandhi, C. and Taplin, D.M.R., "Fracture-Mechanism Maps and their Construction for F.C.C. Metals and Alloys," Acta Metallurgica, Vol. 27, 1979, pp 699-729.

[3]   Gandhi, C. and Ashby, M.F., "Fracture-Mechanism Maps for Materials which Cleave: F.C.C., B.C.C. and H.C.P. Metals and Ceramics," Acta Metallurgica Vol. 27, 1979, pp 1565-1602.

[4]   Quinn, G.D. "Fracture Mechanism Maps for Silicon Nitride," pp. 931-9 in Ceramic Materials and Components for Engines, W. Bunk and H. Hausner eds., Verlag Deutsche Keramische Gesellschaft, D-5340 Bad Honnef, Germany (1986).

[5]   Quinn, G.D. "Fracture Mechanism Maps for Advanced Structural Ceramics, Part 1, Methodology and Hot Pressed Silicon Nitride Results," Journal of Materials Science, Vol. 25, 1990, pp 4361-4376.

[6]   Weibull, W. "A Statistical Distribution Function of Wide Applicability," Journal of Applied Mechanics, Vol. 18, 1951, pp 293-297.

[7]   Evans, A.G. and Wiederhorn, S.M., "Crack Propagation and Failure Prediction in Silicon Nitride at Elevated Temperatures," Journal of Materials Science, Vol. 9, 1974, pp 270-278.

[8]   Ritter, J.E., Jr., "Engineering Design and Fatigue Failure of Brittle Materials," pp. 667-686 in Fracture Mechanics of Ceramics, Vol. 4, Edited by R.C. Bradt, D.P.H. Hasselman and F.F. Lange, Plenum Publ. Corp., New York (1978).

[9]   Monkman, F.C. and Grant, N.J. "An Empirical Relationship between Rupture Life and Minimum Creep Rate in Creep-Rupture Tests," Proceedings of the American Society of Testing and Materials, Vol. 56, 1956, pp 593-620.

[10]   Quinn, G.D. and Braue, W.R. "Fracture Mechanism Maps for Advanced Structural Ceramics, Part 2, Sintered Silicon Nitride," Journal of Materials Science, Vol. 25, 1990, pp 4377-92.

[11]   Matsui, M., Ishida, M., Soma, T. and Oda, I. "Ceramic Turbocharger Rotor Design Considering Long Term Durability," pp. 1043-50 in Ceramic Materials and Components for Engines, W. Bunk and H. Hausner eds., Verlag Deutsche Keramische Gesellschaft, D-5340 Bad Honnef, Germany (1986).

[12]   Carroll, D.F., Wiederhorn, S.M. and Roberts, D.E. "Technique for Tensile Creep Testing of Ceramics," Journal of the American Ceramic Society, Vol. 72, 1989, pp 1610-14.

[13]   Foley, M.R., Rossi, G.A., Sundberg, G.J., Wade J.A. and Wu, F.J., "Analytical and Experimental Evaluation of Joining Silicon Carbide to Silicon Carbide and Silicon Nitride for Advanced Heat Engine Applications, Final Report, Subcontract 86X-SBO45C, Norton Company, September 30, 1991, pp 56-86.

[14]   Tatstuki Ohji and Yukihiko Yamauchi, "Long-Term Tensile Creep Testing for Advanced Ceramics," Journal of the American Ceramic Society, Vol. 75, No. 8, August 1992, pp 2304-307.

[15]   Liu, K.C., Pih, H. and Voorhes, D.W., "Uniaxial Tensile Strain Measurement for Ceramic Testing at Elevated Temperatures: Requirements, Problems, and Solutions," International Journal of High Technology Ceramics, Vol. 4, 1988, pp 161-171.

[16]   Pears, C.D. and Digesu, F.J., "Gas-Bearing Facilities for Determining Axial Stress-Strain and Lateral Strain of Brittle Materials to 5500 F," Proceedings of the American Society of Testing Materials, Vol. 65, 1965, pp 855-73.

[17]   Gürtler, M. and Grathwohl, G., "Tensile Creep Testing of Sintered Silicon Nitride," pp. 399-408 in Proceedings of the Fourth International Conference on Creep and Fracture of Engineering Materials and Structures, Institute of Metals, 1 Carlton House Terrace, London SW1Y 5DB (1990).

[18]   M. Gürtler, A. Weddigen and G. Grathwohl, "Werkstoffprüfung von Hochleistungskeramik mit Zugproben," Mat.-wiss. u. Werkstofftech., Vol. 20, 1989, pp 291-299.

[19]   Arons, R.M. and Tien, J.K., "Creep and Strain Recovery in Hot-Pressed Silicon Nitride," Journal of Materials Science , Vol. 15, 1980, pp 2046-2058.

[20]   Hockey, B.J., Wiederhorn, S.M., Liu, W., Baldoni J.G. and Buljan, S.-T., "Tensile Creep of Whisker-Reinforced Silicon Nitride," Journal of Materials Science, Vol. 26, 1991, pp 3931-3930.

[21]   Wiederhorn, S.M., Hockey, B.J., Cranmer D.C. and Yeckley, R., "Transient Creep Behavior of Hot Isostatically Pressed Silicon Nitride," Journal of Materials Science, Vol. 28, 1993, pp 445-453.

[22]   Ferber, M.K. and Jenkins, M.G., "Evaluation of the Elevated-Temperature Mechanical Reliability of a HIP-ed Silicon Nitride," Journal of the American Ceramic Society, Vol. 75, No. 9, 1992, pp 2453-62.

[23]   Ohji, T. and Yamauchi, Y., "Tensile Creep and Creep Rupture Behavior of Monolithic and SiC Whisker Reinforced Silicon Nitride," Journal of the American Ceramic Society, to be published.

[24]   Raj, R. and Morgan, P.E.D., "Activation Energies for Densification, Creep and Grain-Boundary Sliding in Nitrogen Ceramics," Journal of the American Ceramic Society Vol. 64, 1981, pp C143-C145.

[25]   Khandelwal, P.K., "Life Prediction Methodology for Ceramic Components of Advanced Vehicular Heat Engines," pp. 176-179 in Ceramic Technology Project Bimonthly Technical Progress Report to DOE Office of Transportation Technologies, April to May 1992, D.R. Johnson, Project Manager, Ceramic Technology Project.

[26]   Jenkins, M.G., Ferber, M.K. and Lin, C.-K. J., "Apparent Enhanced Fatigue Resistance under Cyclic Tensile Loading for a HIPed Silicon Nitride," Journal of the American Ceramic Society, Vol. 76, 1993, pp 788-792.

[27]   Davidge, R.W., McLaren, J.R. and Tappin, G., Journal of Materials Science "Strength Probability Time (SPT) Relationships in Ceramics," Vol. 8, 1973, pp 1699-1709.

[28]   Hertzberg, R.W., Deformation and Fracture Mechanics of Engineering Materials, John Wiley and Sons, New York (1976).

[29]   Wiederhorn, S.M. and Fuller, E.R., Jr., "Structural Reliability of Ceramic Materials," Materials Science and Engineering, Vol. 71, 1985, pp 169-86.

Jow-Lian Ding[1], Kenneth C. Liu[2], and Charles R. Brinkman[2]

A COMPARATIVE STUDY OF EXISTING AND NEWLY PROPOSED MODELS FOR CREEP DEFORMATION AND LIFE PREDICTION OF Si₃N₄

REFERENCE: Ding, J.-L., Liu, K. C., and Brinkman, C. R., **"A Comparative Study of Existing and Newly Proposed Models for Creep Deformation and Life Prediction of Si₃N₄,"** Life Prediction Methodologies and Data for Ceramic Materials, ASTM STP 1201, C. R. Brinkman and S. F. Duffy, Eds., American Society for Testing and Materials, Philadelphia, 1994.

ABSTRACT: This paper summarizes recent experimental results, obtained at Oak Ridge National Laboratory (ORNL), on creep behavior and creep rupture of a commercial grade of Si_3N_4 ceramic in the temperature range of 1150°C to 1300°C. A uniaxial model capable of describing the behavior under general thermomechanical loading is introduced and compared with existing models. An exploratory extension of the new model to a multiaxial form is then discussed. Issues are also discussed concerning the standardization of data analysis methodology and future research needs in the area related to development of creep database and life prediction methodology for high temperature structural ceramics.

KEYWORDS; silicon nitride, creep, creep-rupture, constitutive equations, multiaxial deformation

---

Demonstrated improvements in materials processing technology have shown advanced silicon nitride ceramics to

---

[1] Associate Professor, Department of Mechanical and Materials Engineering, Washington State University, Pullman, WA 99164-2920.

[2] Senior Scientist and Group Leader, respectively, Oak Ridge National Laboratory, Metals and Ceramics Division, Oak Ridge, TN 37831.

be feasible structural materials for many engineering applications at high temperatures. Because these materials are relatively new and behave differently from metal alloys, a new database and a new constitutive model must be developed for structural design and analysis applications in order to improve the reliability of ceramic components.

Creep properties of ceramics have been studied predominantly using flexure tests. Information on the creep behavior of ceramics under pure uniaxial loading condition is limited. In a recent study [1], the creep and creep rupture properties of a commercial grade of Si_3N_4 ceramic were systematically characterized in the temperature range of 1150°C to 1300°C. Based on the uniaxial data, a constitutive model capable of describing creep deformation and life prediction was proposed for ceramic materials subjected to general thermomechanical loading conditions [2]. This paper summarized key results obtained in [1, 2]. The model was expressed in the form of scalar equations. An exploratory study was made to extend the model to a tensor equation, employing the techniques used in the theory of viscoplasticity. Finally, issues concerning existing methods of creep data analysis and potential future research needs in the area related to database development and life prediction methodology for high temperature structural ceramics were discussed.

MATERIAL, SPECIMEN, AND EXPERIMENTAL APPARATUS

The material used in the recent study [1] was a commercial grade of hot-isostatically-pressed (HIP) Si_3N_4, known as GN-10[3], containing small additions of Y_2O_3 and SrO as densification aids.

Buttonhead tensile specimens having a 6.3-mm diameter and a uniform gage length of 26 mm were used. Creep tests were performed on specimens with dead weight using four standard lever-arm creep testing machines. To maintain a uniform tensile stress in the test specimen, self-aligning gripping fixtures [3], developed at ORNL, were incorporated in the load train. A low-profile, two-zone-controlled resistance furnace capable of achieving temperatures to 1600°C was used to heat the specimen. The temperature along the gage section was measured with three thermocouples, one placed near the middle and one each near both ends of the gage length. The temperature gradient between the center and the end of the gage section was less than 1% of the maximum temperature at the center. Creep strain was measured with a mechanical extensometer, also developed at ORNL [4]. The extensometer has a resolution of 5 microstrain

---

[3]Garrett Ceramic Components Division of Allied-Signal Aerospace Company, Torrance, California.

and an absolute accuracy better than 100 microstrain for long term creep testing.

EXPERIMENTAL DATA

Creep and Creep Rupture Behavior of the As-HIPed Material

The test matrix including test parameters and creep rupture times is shown below (Table 1). Resultant creep curves are illustrated (Figs. 1 to 4). The symbols used in these figures represent experimental data and the solid lines are predictions of the proposed model to be discussed later.

Two characteristic features were observed from the overall creep behavior of GN-10 Si_3N_4 (Figs. 1 to 4), showing the lack of tertiary creep and the strong nonlinearity of the creep behavior with respect to both stress and temperature. Some curves showed a slight hint of tertiary creep but not conclusively. The latter can be seen from the sharp transitional changes in both creep behavior and rupture time relative to the same incremental change of stress or temperature when a combined state of stress and temperature exceeded a critical level. For example, testing at 1300°C showed a sharp change in the creep behavior when

TABLE 1--Matrix for creep tests of GN-10 Si_3N_4

Temp. (°C)	Stress (MPa)	Time to Rupture (h)	Temp. (°C)	Stress (MPa)	Time to Rupture (h)
1150	250	733.8	1250	75	>2238[c]
...	300	365.4	...	100	>1030[c]
			...	125	2996
			...	150	135.9
			...	175	25.5
1200	125	>1031[a]	1300	50	>650[d]
...	150	1204	...	75	>936.9[e]
...	175	>3405[b]	...	100	1721
...	200	203.1	...	125	15.2
...	225	96.3	...	150	0.2
...	250	7.5			

Notes:
   a. Stress increased to 225 MPa after 1031 h of testing.
   b. Buttonhead failure due to power outage.
   c. Fracture at shank.
   d. Test ongoing.
   e. Specimen was cooled and unloaded, and then reheated and reloaded at the end of 936.9 h.

the applied stress was increased from 100 to 125 MPa (Fig. 4). The sharp transition was discerned to have occurred between 125 and 150 MPa at 1250°C (Fig. 3), and between 175 and 200 MPa at 1200°C (Fig. 2). Because of the limited information with only two 1150°C creep curves generated at high stresses, the transition stress level could not be determined but was believed to fall on somewhere between 150 and 250 MPa. A comparison of the creep curves at 1250°C and 1300°C with an applied stress of 125 MPa revealed that a 50°C temperature difference can make significant influences to both the creep rate and rupture time.

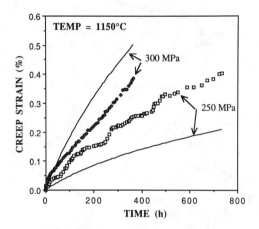

FIG. 1--Creep curves of GN-10 Si₃N₄ tested at 1150°C. (Symbols are experimental data, and solid lines are predictions of the proposed model. The legend is used throughout this paper unless specified otherwise).

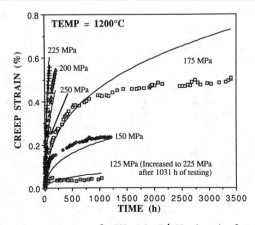

FIG. 2--Creep curves of GN-10 Si₃N₄ tested at 1200°C.

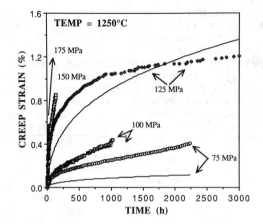

FIG. 3--Creep curves of GN-10 Si₃N₄ tested at 1250°C.

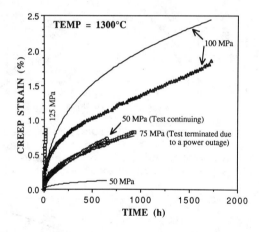

FIG. 4--Creep curves of GN-10 Si₃N₄ tested at 1300°C.

An analysis of the creep rupture behavior indicates that the creep rupture time, t_r, generally decreases with an increase in stress except in a case where the value of t_r at 1200°C with an applied stresses of 150 MPa was lower than that at 175 MPa. Anomalies similar to the above were observed in the situations at 1250°C. However, it should be noted that the specimens tested at the low stresses (75 and 100 MPa) fractured outside the gage section at specimen shank. Therefore these two tests are considered to be

incomplete. The premature fractures are believed to be due
to some preexisting defects. In general, scatter in t_r
values as a function of stress is not unusual for most
engineering materials, especially for brittle materials such
as ceramics.

Effects of Prior Creep and Annealing (Aging) on Subsequent Creep Behavior

In metals, a decrease in creep rate during primary
creep is usually attributed to an increase in dislocation
density and proliferation of dislocation pile-ups, known as
strain hardening. In ceramics, dislocation activity
occurring inside the matrix grains does not contribute
significantly to creep deformation [5, 6]. The major
strengthening mechanism in GN-10 Si$_3$N$_4$ was found to be
devitrification of the grain boundary phase by high
temperature annealing. Creep behavior of the as-HIPed
material tested at 1200°C with an applied stress of 225 MPa
is compared with that of a specimen precrept at 125 MPa for
1031 h prior to the application of additional load to 225
MPa (Fig. 5). To facilitate comparison, the zero time for
the precrept specimen was reset at the beginning of the
second loading to 225 MPa. Although the precept specimen
fractured prematurely at the buttonhead, another case of
incomplete test, the enhancement in creep resistance
exhibited by the precept specimen is clearly discernable.
Since the total creep strain accumulated during low stress
creep was only 0.044%, the strengthening was attributed to
thermal annealing instead of strain hardening.

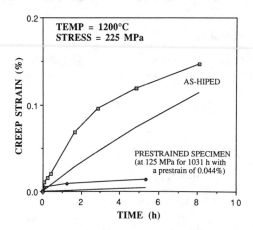

FIG. 5--Comparison of initial transient creep behavior
of a specimen tested in the as-HIPed condition and that of a
specimen precrept at 125 MPa for 1031 h. Both specimens
were tested at 1200°C and 225 MPa.

To understand the strengthening due to thermal annealing, four specimens were annealed at 1370°C in air without applied stress for 150 hours. In general, thermal annealing enhanced both creep resistance and creep rupture life of GN-10 Si_3N_4 (Fig. 6). An anomaly was noted at 1150°C where the annealed specimen showed rupture life substantially lower than that of the as-HIPed specimen.

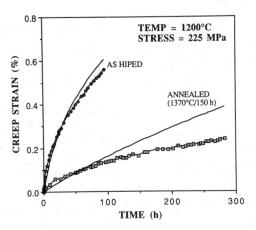

FIG. 6--Comparison of creep behavior of annealed and the as-HIPed specimens tested at 1200°C under 225 MPa.

CURRENT PRACTICES IN ANALYSIS OF CREEP AND CREEP RUPTURE DATA

Analysis of Creep Data

Most reported creep data on ceramics have been analyzed with the Norton power-law relation [7] described by

$$\dot{\epsilon} = A\sigma^n e^{-\frac{Q}{RT}} \tag{1}$$

where
    $\dot{\epsilon}$ = steady state creep rate, h^{-1},
    σ = stress, MPa,
    Q = activation energy, kJ/mole,
    R = gas constant,
    T = absolute temperature, °K,
    n = stress exponent, and
    A = material constant.

Since creep curves of ceramic materials may exhibit neither steady-state creep nor tertiary creep, as are the cases shown above for GN-10 Si_3N_4, the minimum creep rate at the end of the extensive primary creep is often used synonymously as the steady-state rate in Eq. 1. Because the semantic change does not alter the basic characteristics of Eq. 1, it remains as an ineffective model to describe the intrinsic transient creep behavior unless some complementary rules are additionally introduced. In reality, however, the most serious deficiency of Eq. 1 is its inability to represent the details of creep behavior in practical situations where both stress and temperature vary in time.

In [1], the minimum creep rates were determined by taking the time derivative of the following equation which was the basic form of the proposed creep deformation model:

$$\varepsilon = x_1[(1+x_2t)^{x_3}-1], \qquad (2)$$

where
ε = creep strain,
t = time, h,
and x_1, x_2, and x_3 are coefficients to be determined by curve fitting.

Generally, Eq. 2 fits the creep curves well individually except in those cases where the primary creep is less pronounced. Minimum creep rates calculated from Eq. 2 (Fig. 7 and 8) indicate that the data are polarized in two groups (identified by open and closed symbols) with respect to the aforementioned transition region. For tests that were interrupted or incomplete due to non-gage section fracture, a downward arrow was attached to the corresponding data point to indicate the actual creep rate might be lower than the indicated one.

The above creep rate data were analyzed to determine the coefficients of Eq. 1. Two analyses were exercised. In the first analysis, the entire set of the data points, excluding the ongoing-test point at 1300°C/50 MPa, was analyzed as a group, using a multivariate regression analysis. Results of the analysis yielded

$$\dot{\varepsilon} = e^{52.1}\sigma^{8.4}e^{-\frac{1\,326}{RT}} \qquad (3)$$

with n = 8.4 and Q = 1 326 kJ/mole. The creep rate data are well represented by Eq. 3 (Fig. 7).

The polarization of the creep rate data in two groups may suggest that creep behavior was controlled by two different deformation mechanisms [8]. Therefore, in the second exercise, the data were analyzed separately in groups. Results are given below.

$$\dot{\varepsilon} = e^{50.5}\sigma^{8.3}e^{-\frac{1\,292}{RT}}; \qquad (4a)$$

$$\dot{\varepsilon} = e^{30.4}\sigma^{1.6}e^{-\frac{649}{RT}}. \qquad (4b)$$

Equation 4a represents the open-symbol group with n = 8.3 and Q = 1 292 kJ/mole, and Eq. 4b for the closed-symbol group with n = 1.6 and Q = 649 kJ/mole (Fig. 8). The transition points were determined in terms of $(T, \sigma, \dot{\epsilon})$ as a set to be (1150°C, 164 MPa, 8.355×10^{-8}/h), (1200°C, 125 MPa, 3.453×10^{-7}/h), (1250°C, 96 MPa, 1.300×10^{-6}/h), and (1300°C, 76 MPa, 4.501×10^{-6}/h).

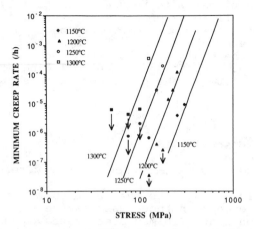

FIG. 7--Comparison of the minimum creep rate data with Eq. 3 represented by lines.

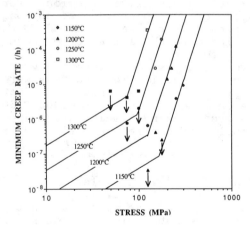

FIG. 8--Comparisons of the minimum creep rate data in open symbols with Eq. 4a, and those in closed symbols with Eq. 4b.

Since the data points in closed symbols fell in the proximity of the intersections of Eqs. 4a and 4b, Eqs. 3 and 4a are about the same. Although direct comparisons can not be made, the values of n and Q in Eqs. 3 and 4a are comparable to the data reported for other comparable Si_3N_4

ceramics such as PY-6 [9] and NT-154 [10]. A high n value
indicates that creep deformation is a stress-controlled
process such as slow crack growth.
    The values of n and Q in Eq. 4b are not firm due to the
lack of data in the low creep-rate range, and will be
revised as more data become available in the future.
However, it is interesting to point out that n = 1.6 given
by Eq. 4b suggests that the mechanism may be a diffusion
type, i.e. n = 1; and Q = 649 kJ/mole given by Eq. 4b agrees
well with that reported for viscous flow of the secondary
phase in Si₃N₄ [11].

Analysis of Creep Rupture Data

    To predict creep rupture life under isothermal and
constant stress conditions, two models, namely, the
Larson-Miller model (LM) [12] and the minimum commitment
method (MC) [13] have been widely used.
    The Larson-Miller model is described by the following
equation:

$$\log t_r = B_o + \frac{B_1}{T} + \frac{B_2}{T} \log \sigma,$$ (5)

where B_0, B_1, and B_2 are constants. The creep lifetime data
(Table 1) were analyzed to determine the values of the
constants as B_0 = -58.28, B_1 = 136 600 and B_2 = -20 510,
using the multivariate regression technique. Predictions
and experimental results are compared (Fig. 9). Note that
data affixed with a horizontal arrow representing the
interrupted and premature tests were excluded in the above
analysis for the reason of incomplete information.

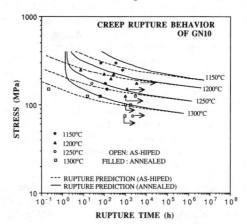

FIG. 9--Comparisons of experimental data and
predictions of models. Data for the interrupted and
incomplete tests are affixed with an arrow to indicate the
actual rupture time would be longer than the indicated one.

Although Eq. 1 is severely limited in capability of describing the pronounced primary creep feature, it remains widely in use. The virtue of Eq. 1 is the simplicity of its form that can be used to predict t_r in association with the Monkman-Grant relationship [14]:

$$t_r = G \cdot \dot{\varepsilon}^h , \tag{6}$$

where G and h are constants. Substituting Eq. 1 into Eq. 6 leads to the same expression as Eq. 5.

The minimum-commitment method is described by the following equation:

$$\log t_r + [R_1 (T-T_m) + R_2 (\frac{1}{T} - \frac{1}{T_m})] = B + C \log \sigma + D\sigma + E\sigma^2 \tag{7}$$

where T_m is the mid point of the temperature range used in the tests, or 1498 °K in the present analysis. The model constants were determined to be: $R_1 = 0.1731$, $R_2 = 303\ 700$, $B = 87.24$, $C = -45.69$, $D = 0.1156$, and $E = -0.9733 \times 10^{-4}$. Predictions and experimental data were compared (Fig. 9).

Generally speaking, both models fit the experimental data reasonably well. Results of the regression analyses indicate that the minimum commitment model fits the data slightly better than the Larson-Miller model. The reason is that the MC model has five independent variables whereas the LM model has only two.

A NEW MODEL FOR CREEP DEFORMATION AND LIFE PREDICTION

In view of the shortcomings in the existing models discussed above, the need of a new model is recognized. Fracture mechanisms of advanced Si_3N_4 ceramics are complex and the situations in the creep range are further complicated by temperature. Therefore, an accurate description of the mechanisms may not be known until more studies are completed. However, a simplified and reasonable model can be contemplated based on behavioral features observed from experimental results in association with intelligent deduction from the information gleaned from micrographs of scanning electron microscopy (SEM) and transmission electron microscopy (TEM) [1, 15]. The new model is assumed to be an aggregate of rigid Si_3N_4 grains which are bonded by a grain boundary phase material with some imperfections in the forms of surface microcracks and internal voids which are preexisting defects in the bulk. The proposed model was formulated on the basis of the following assumptions:

1.   Time dependent behavior of the material is attributed to the viscosity of residual amorphous phase in the grain boundary region, and creep rate is a nonlinear function of both stress and temperature.

2.   Matrix grains are assumed to be rigid, and there is little or no intragranular dislocations occurring inside the grains. "Hardening", which describes the enhancement of creep resistance, is therefore attributed to devitrification that occurs progressively in the residual amorphous phase as temperature and soaking time increase.
3.   Although cavity nucleation and growth occur concurrently (especially in the annealed specimen) during creep, creep fracture is assumed to have occurred only when preexisting macrocracks propagate to a critical size.
4.   No interaction is assumed between creep deformation and damage induced either by cavity nucleation and growth or by propagation of macrocracks. Therefore, the damage accumulation is assumed to inflict no accelerated or tertiary creep.

More specifically, the proposed model is of the following mathematical form.

$$\dot{\varepsilon} = \frac{d\varepsilon}{dt} = \frac{f\left(\frac{\sigma}{\sigma_0}\right)e^{-\frac{Q_\varepsilon}{RT}}}{\delta} = \frac{\alpha}{\delta}, \tag{8a}$$

$$\dot{\delta} = \frac{d\delta}{dt} = \frac{\delta_0 e^{-\frac{Q_\delta}{RT}}}{\delta^m} = \frac{\beta}{\delta^m}, \tag{8b}$$

$$\dot{\omega} = \frac{d\omega}{dt} = \frac{\dot{\omega}_0\left(\frac{\sigma}{\sigma_0}\right)^\nu e^{-\frac{Q_\omega}{RT}}}{\delta(1-\omega)} = \frac{\gamma}{\delta(1-\omega)}, \tag{8c}$$

where

$$\alpha = f\left(\frac{\sigma}{\sigma_0}\right)e^{-\frac{Q_\varepsilon}{RT}}, \tag{8d}$$

$$\beta = \delta_0 e^{-\frac{Q_\delta}{RT}}, \tag{8e}$$

$$\gamma = \dot{\omega}_0\left(\frac{\sigma}{\sigma_0}\right)^\nu e^{-\frac{Q_\omega}{RT}}, \tag{8f}$$

δ = hardening variable,
ω = damage variable,
σ_0 = reference stress = 100 MPa,
and Q_ε, Q_δ, Q_ω, δ_0, m, $\dot{\omega}_0$ and ν are constants.

The upper dots symbolize the rate with respect to time. Function f indicates the initial creep rate is stress dependent and strongly nonlinear based on assumption 1. The hardening variable, δ, takes a value of unity at the initial state and then increases as the amorphous phase in the grain boundary region crystallizes progressively. Since the initial value of δ for the as-received specimen is 1, α is actually the initial creep rate at the inception of creep loading. Equation 8b indicates that the rate of hardening is a function of soaking temperature but decreases as the value of δ increases. At the inception of creep loading, no cracks are assumed to have developed from the initial defects, which are represented by $\omega = 0$. The rate of damage, $\dot{\omega}$, is always positive and approaches infinity as ω approaches unity. The condition of $\omega = 1$ signifies the prevailing crack has propagated to the critical size and specimen failure. Based on assumptions 2, 3, and 4, Eqs. 8a and 8c are individually coupled with Eq. 8b but are independent of each other. When Eqs. 8a and 8c are coupled to each other, they are often used to describe the transition behavior from secondary to tertiary creep of metals [16]. With ε being independent of ω, Eq. 8a can dispose of the tertiary creep term.

Characterization of Proposed Model

Details of the model characterization were described in [2]. The function f and the material constants in Eqs. 8a and 8b are determined from the creep curves by curve fitting, and those in Eq. 8c from the creep rupture life data for both the as-HIPed and annealed specimens. The function f was determined to be of the following form :

$$f\left(\frac{\sigma}{\sigma_0}\right) = \dot{\varepsilon}_0 \sinh\left(\frac{\sigma - \sigma_{th}}{100}\right)^2 \qquad (9)$$

where

$\sigma_{th} = 2.8342 \times 10^{-6} \times 10^{11\,069/T}$,
$\dot{\varepsilon}_0 = e^{62.259}$ (for T≤1200°C); $e^{36.455}$ (for T>1200°C),
$Q_\varepsilon = 711.34$ kJ/mole (for T≤1200°C);
$\quad = 395.51$ kJ/mole (for T>1200°C),
$\dot{\delta}_0 = e^{78.8}$ (for T≤1200°C); $e^{2.62}$ (for T>1200°C),
$Q_\delta = 997.12$ kJ/mole (for T≤1200°C);
$\quad = 64.65$ kJ/mole (for T>1200°C),
$\dot{\omega}_0 = e^{110.74}$,
$\nu = 10.86$
$Q_\omega = 1\ 510$ kJ/mole, and
$m = 0.43$.

EVALUATION OF PROPOSED MODEL

Prediction of Creep Deformation

Isothermal, constant-stress creep--Under isothermal and constant stress conditions, a closed-form solution can be obtained from Eq. 8a for the creep strain as a function of time as described by the following equation:

$$\varepsilon = \frac{\alpha}{m\beta}\{[1 + (1+m)\,\beta t]^{\frac{m}{1+m}} - 1\}. \qquad (10)$$

Creep curves predicted by Eq. 10 are compared with experimental data (Figs. 1 to 4). Examinations of these figures indicate that the essential features of creep behavior are reasonably well described in each case with consideration of the temperature and stress ranges that were covered. Noted that Eqs. 2 and 10 are the same.

<u>Effects of annealing on subsequent creep behavior</u>--
For the annealed specimens, a closed-form solution for the creep strain can be derived as:

$$\varepsilon = \frac{\alpha}{m\beta}\{[(1+m)\,\beta\,t+\delta^{m+1}]^{\frac{m}{1+m}}-\overline{\delta}^{m}\}. \tag{11}$$

where $\overline{\delta}$ is the value of δ at the end of annealing given by

$$\overline{\delta} = [1+(1+m)\,\beta_{1370^{\circ}}\cdot 150]^{\frac{1}{1+m}}. \tag{12}$$

In general, calculated creep curves overpredict the experimental creep curves at T < 1250°C and underpredict at T > 1250°C. However, the creep rates for both curves (Fig. 6) agree well in the post primary creep range.

<u>Creep under stepwise-varied loading</u>--Since the model assumes no strain hardening, the creep strain at each loading step can also be calculated using Eq. 11 with $\overline{\delta}$ updated at the beginning of each loading step. The total creep strain is accumulated from each loading step. Comparison of calculated and experimental creep curves for a specimen subjected to a stepped load (Fig. 5) shows predicted creep rate is in agreement with the experimental result.

A case of stepped loading at 1300°C was studied (Fig. 10). The specimen was initially loaded at 75 MPa.

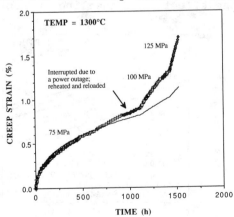

Fig. 10--Comparison of experimental and predicted creep curves for a specimen subjected to stepwise-varied loading.

The test was interrupted at t = 937 h due to a power
failure. Although the specimen cooled down to ambient
temperature, the subsequent creep behavior did not appear to
have been altered when testing resumed. The load was
increased to 100 MPa after completing 1125 h of testing, and
further increased to 125 MPa at t = 1437 h, until specimen
fracture occurred at t = 1529 h. The solid line indicates
the predicted creep curve which agrees well with the first
segment of the experimental data, but progressively
underestimates the remaining data as the applied stress
increases in steps. One possible reason for the
discrepancies may be omission of the interaction between
creep deformation and damage accumulation postulated in
assumption 4.

Prediction of Creep Rupture

Isothermal, constant-stress creep rupture--Under
isothermal and constant-stress condition, the model leads to
the following equation for predicting rupture time for both
as-HIPed and annealed specimens:

$$t_r = \frac{1}{(1+m)\,\beta}\left(\left(\frac{m\beta}{2\gamma}+k^{m}\right)^{\frac{m+1}{m}}-k^{m+1}\right),$$ (13)

where k is the initial value of δ at the inception of creep
loading. Predicted rupture-time curves, solid lines for the
annealed and dashed for the as-HIPed specimens (Fig. 11),
compare quite well with experimental data. It is interesting
to note that both curves in each temperature pair merge
together as rupture time increases. This observation
implies that the effects of annealing on rupture time

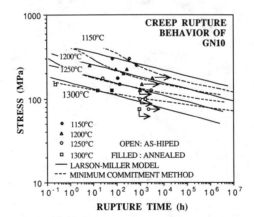

Fig. 11--Comparison between experimental creep rupture
times of both the as-HIPed and annealed specimens and
predictions of the proposed model.

diminish if tests are run at low stresses over a long period
of time. In the low rupture-time range, the solid curves
for 1250 and 1300°C turn sharply upward, indicating fast
fracture as in short-term tensile tests. The sharp
transitions of the rupture curves in dashed lines will occur
outside the plot (Fig. 11) in the range below 10^{-1} h. This
is physically plausible from the view point that the stress
corresponding to the sharp transition point may fall close
to the tensile strength of the material at that
temperature. Therefore, any stresses above the transition
point bear no physical significance but indicate instant
specimen rupture.

Creep rupture under stepwise varied loading--Under
stepwise-varied loading conditions, it can be shown [2] that
ω can be calculated for each loading step from the following
expression

$$\frac{1}{2}(1-\omega)^2 = \frac{\gamma}{\beta m}\left[(1-m)\beta t_i + k^{1+m}\right]^{\frac{m}{1+m}} - \frac{1}{2}(1-\overline{\omega})^2 - \frac{\gamma}{m\beta}k^m, \quad (14)$$

where t_i is the time expended at the "i"th step load, and $\overline{\omega}$
is the initial value of ω when the step load was applied to
the specimen. The rupture time of the specimen tested in
steps (Fig. 10) was estimated, using Eq. 14. Values of ω =
0.031, 0.033, and 0.103 were obtained at t = 937, 1125, and
1437 h, respectively. Under the last step of loading at 125
MPa, the model predicts a rupture time of 239 h, which
overpredicts the actual rupture time of 92 h by 147 h.

MULTIAXIAL MODEL

Although the model described above is capable of
predicting the creep and creep rupture behavior of ceramic
materials under general thermal mechanical loadings, its
present form in scalar expression is applicable only to
uniaxial stress states such as simple tension or
compression. Since nearly all engineering components are
subjected to complicated stress states in practical
applications, the model must be extended to the tensor form
in order to be truly useful to mechanical reliability
analysis. This requires a multiaxial database, which is
virtually nonexistent. In such an environment, a matured
multiaxial model will not be possible until relevant
information becomes available. Nonetheless, an extension of
the scalar model to a multiaxial model can be contemplated
based on the underlying assumptions.
    As mentioned earlier, creep deformation is assumed to
be mainly attributed to viscous flow of the amorphous phase
in the grain boundary, driven by shear stress. Therefore,
it is reasonable to apply a concept usually employed in the
viscoplasticity theory to extend the creep flow equation
(Eq. 8a) from the scalar form to tensor form through the

relationship of effective stress and stress deviators, which is equivalent to a generic shear stress. Since creep fracture is dominantly caused by growth of macrocracks or voids, damage is expected to be controlled by the maximum principal tensile stress. This is a realistic assumption due to the fact that the fracture surface is always perpendicular to the tensile direction. On the basis of the above arguments, a multiaxial model having the general characteristics of the uniaxial model is proposed as follows:

$$\dot{\varepsilon}_{ij} = \frac{3}{2} \frac{\dot{\varepsilon}_0 \sinh\left(\dfrac{\overline{\sigma}-\sigma_{th}}{100}\right)^2 e^{-\frac{Q_\varepsilon}{RT}}}{\delta} \frac{\sigma'_{ij}}{\overline{\sigma}} , \tag{15a}$$

$$\delta = \frac{\delta_0}{\delta^m} e^{-\frac{Q_\varepsilon}{RT}} , \tag{15b}$$

$$\dot{\omega} = \frac{\dot{\omega}_0 \left(\dfrac{\sigma_{max}}{\sigma_0}\right)^v e^{-\frac{Q_\omega}{RT}}}{\delta(1-\omega)} , \tag{15c}$$

where

$\sigma'_{ij} = \sigma_{ij}-(1/3)\sigma_{kk}\delta_{ij}$ = stress deviator,
σ_{ij} = stress tensor,
δ_{ij} = the Kronecker delta,
σ_{max} = maximum principal stress, and
$\overline{\sigma}$ = effective stress defined by

$$\overline{\sigma} = \sqrt{\frac{3}{2}\sigma'_{ij}\sigma'_{ij}} . \tag{15d}$$

Since both the effective stress and the maximum principal stresses are reduced to the applied stress in uniaxial loading condition, Eq. 15a reduces to Eq. 8a.

DISCUSSION

Until recently, one major issue in the aspect of high temperature engineering design with ceramics has been the lack of experimental data. With the advance of material testing techniques, a reliable database is gradually growing. However, in addition to the need for continuous expansion of the database, several other important issues need to be addressed concerning the refinement of existing or new models for engineering analysis and design. Several points are raised here for discussions.

Current Practices in Creep and Creep Rupture Data Analysis

Nearly all creep test data reported in the literature have been analyzed with the Norton power-law relation, which was widely used in the analysis of metal creep. The norton

equation can approximate creep curves with a short primary
creep range relative to a long steady-state creep range.
Since the second-stage creep dominates the creep lifetime,
both creep rate and creep rupture time can be defined rather
unambiguously. Contrasting to the well defined three-stage
creep curves exhibited by many metal alloys, creep curves of
ceramic materials such as GN-10 Si$_3$N$_4$ may exhibit only
extensive first-stage creep with little or no second and
third stage creep. In the case of ceramics, a serious
difficulty may arise in confirming whether the specimen
rupture is genuinely the end of the creep life or a
premature failure due to slow crack growth initiating from
large defects preexisting in the bulk. The uncertainty is
obviously a factor that contributes the diversity of creep
data for ceramic materials. Because of the pronounced
primary creep feature, the Norton equation becomes less
applicable to creep curves of ceramic materials.
Nevertheless, the stress exponent and activation energy
terms used in the equation remain to be meaningful indices
to characterize creep properties of different ceramics.
Therefore, it is imperative that a unified method must be
established to determine the values of creep rate and creep
rupture life that are consistent with existing models
discussed in earlier sections.

Multiaxiality and Other Essential Features of Model

The importance of multiaxiality in the proposed model has
been discussed previously and needs not further emphasis.
While the mutiaxiality feature is desirable in any model,
the asymmetric feature of creep behavior in tension and in
compression is an immediate concern. Since the proposed
model is exploratory in nature, the asymmetry is not built
in the present form but should be addressed in the refined
version.
     In the proposed model, the creep rate equation (Eq. 8a)
is assumed to be independent of the damage variable based on
the results of microscopic observations, which show no
significant formation of cavities for the as-HIPed specimens
during creep [1, 15]. This assumption also renders the
scheme of closed form solution possible for cases involving
stepwise-varied loading conditions. For ceramics showing
cavity formation as a prevailing factor of creep deformation
[17-20], Eq. 8a must include the damage variable. A model
formulated with such variables is discussed in a paper
included in this publication [21].
     As the distribution of fracture initiating defects in
ceramics are random, an ultimate model must take into
account the volume effect as well as the stochastic nature
of the creep rupture behavior. For example, in a
deterministic approach of modeling as described in the
proposed model, fracture is assumed to initiate at the
defect under the highest principal stress. However, in

reality, the occurrence of fracture depends on the distribution, size, and shape of defects, and results could be highly stochastic in nature. Therefore, provisions to account for the statistical diversity of creep and creep rupture behavior must be considered in the future model to make reliability analysis practical. To this end, a much wider data base is needed and discussed in the following section.

Experimental Data Under Variable Multiaxial Loadings

With the increasing maturity of the uniaxial testing technique, future research efforts in material testing should focus on the study of multiaxial creep behavior of ceramic materials. These data are essential for model validation and refinement work. Some biaxial studies of ceramic materials at room temperature have been reported [22, 23]. But no multiaxial creep data are currently available for evaluating the proposed multiaxial model.

It should also be noted that constant stress and temperature are ideal laboratory test condition. Experimental data of this type are of fundamental importance to material characterization. However, stress and temperature usually vary with time in practical applications [24]. Experimental data simulating the operating condition are also desirable in order to gain the insight of material behavior and to evaluate the theoretical model. Since not all field conditions can be reproduced in laboratory, data may be obtained under somewhat modified conditions. Nevertheless, information such as that obtained under stepwise-varied loading condition can be extremely useful.

CONCLUSION

This paper examined the strength and limitation of existing models used to predict creep deformation and creep rupture behavior based on a comprehensive set of experimental data of a commercial grade of Si_3N_4. The proposed model was shown to be effective and capable of describing the essential features of uniaxial creep and creep rupture behavior of the material under both constant stress and stepwise-varied loading conditions. Introduction of the hardening and damage variables in the model has further enhanced its ability to predict the effects of annealing on creep behavior and creep rupture lifetime. Although exploratory in nature, a cursory multiaxial model was proposed based on the theory of viscoplasticity. Discussions were also given concerning the unified approach of creep data analysis for ceramic materials, data needs for model refinement to include features such as multiaxiality, asymmetry of creep in tension and compression, and stochastic nature of defects inherent to ceramic materials.

ACKNOWLEDGEMENTS

The authors thank Drs. A. E. Pasto and J. H. Schneibel for reviewing the manuscript. This research was sponsored by the U.S. Department of Energy (USDOE), Assistant Secretary for Conservation and Renewable Energy, Office of Transportation Technologies, as part of Ceramic Technology Project of Materials Development Program, under contract DE-AC05-84OR21400 with Martin Marietta Energy Systems, Inc.

J. L. Ding would also like to acknowledge the partial support provided by the USDOE Faculty Research Participation Program administered by Oak Ridge Associated Universities.

REFERENCES

[1] Ding, J. L., Liu, K. C., More, K. L., and Brinkman, C. R., " Creep and Creep Rupture of An Advanced Silicon Nitride Ceramic," submitted to Journal of the American Ceramic Society, 1993.

[2] Ding, J. L., Liu, K. C., and Brinkman, C. R., "Development of a Constitutive Model for Creep and Life Prediction of Advanced Silicon Nitride Ceramics," Proceedings of the Annual Automotive Technology Development Contractors' coordination Meeting, Dearborn, Michigan, November 2-5, 1992, Society of Automotive Engineers, Inc., Warrendale, Pa, 1993 (in press).

[3] Liu, K. C. and Brinkman, C. R., "Tensile Cyclic Fatigue of Structural Ceramics," Proceedings of the 23rd Automotive Technology Development Contractors' Coordination Meeting, Dearborn, Michigan, October 21-24, 1985, P-165, Society of Automotive Engineers, Inc., Warrendale, Pa, March 1986, pp 279-283.

[4] Liu, K. C. and Ding, J. L., "A Mechanical Extensometer for High-Temperature Tensile Testing of Ceramics," Journal of Testing and Evaluation, American Society for Testing and Materials, September, 1993 (in press).

[5] Evans, A. G. and Sharp, J. V., "Microstructural Studies on Silicon Nitride," Journal of Materials Science, Vol. 6, 1971, pp 1292-1302.

[6]. Kossowsky, R., "The Microstructure of Hot-Pressed Silicon Nitride," Journal of Materials Science, Vol. 8, 1973, pp 1603-1615.

[7] Norton, F. H., "The Creep of Steel at High Temperatures," McGraw-Hill, 1929.

[8] Frost, H. J. and Ashby, M. F., "Deformation Mechanism
     Maps - The Plasticity and Creep of Metals and
     Ceramics," Pergamon Press, 1982.

[9] Ferber, M. K. and Jenkins, M. G., "Empirical Evaluation
     of Tensile Creep and Creep Rupture in a HIPed Silicon
     Nitride," Creep : Characterization, Damage and Life
     Assessment, Woodford, D. A., Townley, C. H. A., and
     Ohnami, M., Eds., ASM International, 1992, pp 81-90.

[10] Cranmer, D. C., Hockey, B. J., and Wiederhorn, S. M.,
     "Creep and Creep-Rupture of HIP-ed Si_3N_4," Proceedings
     of Ceramic Engineering Science, 1991, (in press).

[11] More, K., Davis, R. F. and Carter, C. H., Jr., "A
     Review of Creep in Silicon Nitride and Silicon
     Carbide," Advanced Ceramics, Saito, S., Ed., Oxford
     University Press and Ohmsha Ltd., 1988, pp 95-125.

[12] Larson, F. R. and Miller, J., "Time-Temperature
     Relationship for Rupture and Creep Stress,"
     Transactions of the American Society of Mechanical
     engineers, Vol. 74, 1952, pp 765-771.

[13] Manson, S. S. and Muralidharan, U., "Analysis of Creep
     Rupture Data for Five Multi-heat Alloys by the Minimum
     Commitment Method Using Double Heat Term Centering,"
     Progressing Analysis of Fatigue and Stress Rupture,
     MPC-23, American Society of Mechanical engineers, 1984,
     pp 1-46.

[14] Monkman, F. C. and Grant, N. J., "An empirical
     Relationship Between Rupture Life and Minimum Creep
     Rate in Creep-rupture Tests," Proceedings of Society of
     Testing and Materials, Vol. 56, 1956, pp 593-620.

[15] More, K. L., Ding, J. L., Liu, K. C., and Brinkman, C.
     R., "Microstructural Evolution During Creep and Creep
     Rupture of an Advanced Silicon Nitride Ceramic," (in
     preparation).

[16] Rides, M., Cooks, A. C. F., and Hayhurst, D. R., "The
     Elastic Response of Creep Damaged Materials," Journal
     of Applied Mechanics, Vol. 56, 1989, pp 493-498.

[17] Chuang, T.-J., Wiederhorn, S. M., "Damage-Enhanced
     Creep in a Siliconized Silicon Carbide: Mechanics of
     Deformation," Journal of the American Ceramic Society,
     Vol. 71, No. 7, 1988, pp 595-601.

[18] Wiederhorn, S. M., Roberts, D. E., Chuang, T.-J., and
     Chuck, L., "Damage-Enhanced Creep in a Siliconized
     Silicon Carbide: Phenomenology," Journal of the

American Ceramic Society, Vol. 71, No. 7, 1988, pp 602-608.

[19] Chen, C.-F., Wiederhorn, S. M., and Chuang, T.-J., "Cavitation Damage during Flexural Creep of SiALON-YAG Ceramics," Journal of the American Ceramic Society, Vol. 74, No. 7, 1991, pp 1658-1662.

[20] Luecke, W., Wiederhorn, S. M., Hocky, B. J., and Long, G. G., "Cavity evolution during Tensile Creep of Si$_3$N$_4$," Proceedings of Scientific and Technological Advances, Material Research Society, 1992.

[21] Chuang, T-J. and Duffy, S. F., "A Methodology to Predict Creep Life for Advanced Ceramics Using continuum Damage Mechanics concepts," Life Prediction Methodologies and Data for Ceramic Materials, American Society for Testing and Materials, STP 1201, Brinkman, C. R. and Duffy, S. F. Eds., 1993

[22] Chao, L. Y. and Shetty, D. K., "Reliability Analysis of Structural Ceramics Subjected to Biaxial Flexure," Journal of the American Ceramic Society, Vol. 74, No. 2, 1991, pp 333-344.

[23] Kim, K. T. and Suh, J., "Fracture of Alumina Tube Under Combined Tension/Torsion," Journal of the American Ceramic Society, Vol. 75, No. 4, 1992, pp 896-902.

[24] Fang, H. T., Cuccio, J. S., Wade, J. C., and Seybold, K. G., "Progress in Life Prediction Methodology for Ceramic Components of Advanced Heat Engines," Proceedings of the Annual Automotive Technology Development Contractors' coordination Meeting, Dearborn, Michigan, October 28-31, 1991, P-256, Society of Automotive Engineers, Inc., Warrendale, Pa, 1992, pp 261-272.

Jonathan A. Salem[1] and Sung R. Choi[2]

CREEP BEHAVIOR OF SILICON NITRIDE DETERMINED FROM CURVATURE AND NEUTRAL AXIS SHIFT MEASUREMENTS IN FLEXURE TESTS

REFERENCE: Salem, J. A. and Choi, S. R., "Creep Behavior of Silicon Nitride Determined from Curvature and Neutral Axis Shift Measurements in Flexural Tests," Life Prediction Methodologies and Data for Ceramic Materials, ASTM STP 1201, C. R. Brinkman and S. F. Duffy, Eds., American Society for Testing and Materials, Philadelphia, 1994.

Abstract: The creep behavior of a hot-pressed silicon nitride was determined in flexure in air at 1200 and 1300°C by monitoring the creep deflection, the specimen curvature and the position of the neutral axis. The resulting data was used to evaluate the steady-state creep rate from the conventional elastic solution, curvature-moment relations and a model accounting for neutral axis shift. Fractography and measurements of specimen compliance before and after testing indicated bulk cracking and a loss of stiffness. The validity of flexural data to determine creep life parameters was considered.

KEYWORDS: silicon nitride, creep, tension, compression, bending, flexure, neutral axis, cracking, stiffness

---

INTRODUCTION

Accurate measurement of stresses, strains and empirical parameters associated with creep of advanced ceramics is critical to component life prediction. Frequently, flexure or compression tests are used because of the small material

---

[1] Materials Research Engineer, Structural Integrity Branch, NASA LeRC, Cleveland, OH, 44135.

[2] Resident Research Associate, Cleveland State University, Cleveland, OH, 44115.

volume required, the simple specimen configuration and the
ease of testing (flexure only). The small volume is
especially convenient when new materials are being
developed.

In metallic materials the creep rate in tension
typically equals that in compression, so that compression or
flexural tests can be used. However, in ceramic materials
which contain a hard granular phase (e.g. Si_3N_4 or Al_2O_3)
surrounded by a continuous glassy or semicrystalline phase,
creep rates in tension and compression are different,
leading to two creep equations and four parameters: $\dot{\epsilon} = A_t\sigma^{Nt}$
and $\dot{\epsilon} = A_c\sigma^{Nc}$, where subscripts c and t indicate parameters
derived from uniaxial compression and tension. This
difference can be traced to the nature of creep in a
multiphase material, in which the intergranular phases flow
around the hard grains until grain-to-grain contact occurs
in the compression region and void formation occurs in the
tensile region.

Creep measurements in flexure and tension have been
made on the same ceramic [1, 2] and models to determine the
tension and compression parameters from a flexure test have
been developed [3, 4]. These results indicate that creep in
pure tension is far greater than creep in flexure or
compression. Thus, parameters derived from flexural tests
using the conventional theory of simple beam bending in
steady state creep, which assumes the neutral axis to be
fixed at the beam center, can be misleading.

Data from flexural creep tests can be analyzed by
several methods: (1) application of the conventional elastic
solution [5], (2) application of specimen curvature -
displacement or moment relations assuming that the neutral
axis does not shift [5, 6] and (3) application of models
that account for the neutral axis shift in determining the
creep parameters [3, 4]. These three approaches make
various assumptions that may not hold. To compare the
applicability and differences between these models, creep
data was generated with silicon nitride in four-point
flexure.

EXPERIMENTAL PROCEDURES AND ANALYSES

The material used in this study was a hot-pressed
Si_3N_4[3] with 1% MgO. Average grain size was one and five μm
with the largest grains ranging from 12 to 17 μm. X-ray
diffractometer scans of as-received material and material
heated for 24 h at 1300°C indicated little detectable
amorphous phase and little or no change in the quantity
detectable. Typical mechanical properties of the test
material are presented in Table 1. This material was chosen
because it exhibits fatigue and creep susceptibilities at

---

[3] Ceralloy 147A, Ceradyne Inc.

Table 1 -- Mechanical properties of Ceralloy silicon nitride at room temperature

Toughness[1] K_c (MPa.√m)	Hardness[2] H (GPa)	Young's Modulus[3] E (GPa)	Density[4] (g/cm³)
5.78 (0.10)[5]	14.7 (0.9)	316 (1)	3.210 (0.007)

1. By the SEPB method, ref [8], three tests.
2. By Vickers indenter, 10 kg, $H = P/2d^2$, four tests.
3. By strain gaging a 4-point bend specimen, three tests.
4. By the buoyancy method, three test specimens.
5. The value in parenthesis is one standard deviation.

high temperature [7], enabling the comparison of life prediction parameters from various testing and analysis methods.
        Creep testing was conducted in ambient air at 1200 and 1300°C using a SiC four-point bend fixture. The nominal dimensions of the test bars were 3.2 x 4.2 x 50 mm. Two inner and outer test spans were used: one set with 20 and 40 mm spans, respectively, and another with 10 and 22 mm spans. The test specimens were preloaded with 20 N to maintain good alignment relative to the test fixture, and held at the test temperature for 20 min prior to testing in ambient air. The heating rate was 1200°C/h.
        The specimens with larger spans (20 and 40 mm) were tested at initial stress levels (elastic solution) of 120 to 320 MPa, and 80-260 MPa, respectively, at 1200 and 1300°C. During the testing, the deflection at the middle of the inner span of each specimen was monitored with an LVDT system. Also, specimen curvature was monitored by periodically interrupting the test. It was assumed that the interruptions had no effect on the creep behavior.
        The specimens with short spans were tested at nominal initial stresses of 100 and 130 MPa at 1200°C, and 60 and 85 MPa at 1300°C in air. The creep strain across the specimen height was monitored to determine the location of the neutral axis during creeping. This was done by scribing three scratch lines (≈2 mm between adjacent lines) within the inner span on a polished side surface of the specimen, as illustrated in Fig. 1. The scratch marks were made with a Vickers microhardness indenter. A similar method was used by Chen and Chuang [4] to measure neutral axis shift by

placing rows of Vickers impression marks ≈2 mm apart.
However, the post-creep visibility was found to be better
for line scratches than for impression marks. After some
time interval under steady-state creep conditions, the
specimen was unloaded, the furnace cooled (≈1 h) and the
specimen removed. The specimen was then immersed into a
20%HF-20%H_2SO_4-60%H_2O solution for ≈15 minutes to remove the
oxide layer and delineate the scratch lines. The distances
L_1' and L_2' between two adjacent lines were measured along
the specimen height at 0.2 mm spacing using a machinist
microscope. The corresponding strains, ϵ_1 and ϵ_2, were
calculated and the average strain at each point along the
specimen height was obtained by averaging ϵ_1 and ϵ_2 ($\epsilon =$
$(\epsilon_1+\epsilon_2)/2$). During this test interruption, the curvature of
the crept specimens was measured lengthwise using a
machinist optical microscope with a travelling stage.

    Also, for verification of the technique, the neutral
axis position of type 304 stainless steel, which has similar
creep behaviors in tension and compression, was monitored as
described above by flexural testing at 700°C in air with
initial stresses of 150 to 350 MPa. The neutral axes of the
stainless steel test specimens remained almost unchanged, as
shown in Fig. 2, indicating that tensile and compressive
creep were identical to each other, as expected [9].

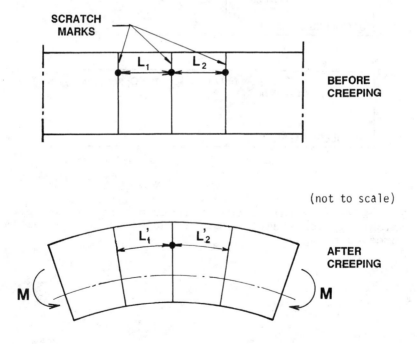

FIG. 1 -- Schematic of scratch lines made on a flexure creep
specimen.

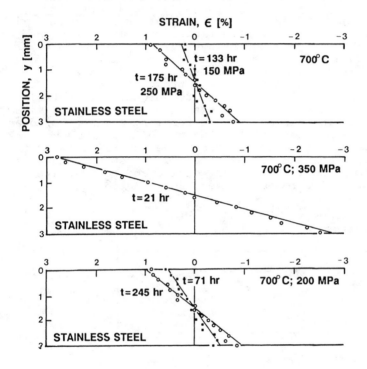

FIG. 2 -- Flexural creep strain as a function of specimen height (y) for stainless steel subjected to the initial stresses shown.

RESULTS AND DISCUSSION

The steady state creep deformation of many advanced ceramics is usually expressed by Norton's law

$$\dot{\epsilon} = A \, \sigma^N \tag{1}$$

where A is a constant associated with creep compliance, σ is the initial maximum applied stress, and N is the stress exponent in steady state.

Elastic Solution

Eq 1 is typically used to analyze bend beam data with the assumptions that creep in tension is identical to creep in compression, no neutral axis shift takes place during deformation [5], and macroscopic cracking does nor occur. Further, assumption of a constant radius of curvature between the inner loading points allows calculation of strains in Eq (1) from

$$\epsilon_{max} = 4hd/L_i^2 \tag{2}$$

where ϵ_{max} is the maximum strain in the outer fiber, d is the relative deflection of the bar center with respect to the inner load points, h is the specimen height and L_i is the inner span [5].

Observation of the strain-time curves for this material indicated that most of the specimens underwent steady-state creep within the range of applied stresses, and that the specimen span size had no effect on the strain-time curves. The results of the creep strain-rate measurements as a function of initial applied stress, based on the conventional theory (Eqs 1 and 2) are summarized in Fig. 3. The stress exponent (N) was determined to be N = 6.0 and 5.0, respectively, for 1200 and 1300°C. For a given initial applied stress, the creep rate at 1300°C is about 10 times greater than that at 1200°C. The high N indicates that creep was associated with cavitational damage or crack formation [2].

Direct Curvature Measurement

If the stress and strain distributions in the beam are assumed unknown or not directly determinable from deflection, the parameters can be determined from measurement of the actual curvature and the applied moment, which both can be accurately determined. The relation between curvature and applied bending moment can be derived by combining Eq (1) with the elastic strain-curvature relation [5] to give

$$1/\rho = A \, [M/\Phi]^{N} \tag{3}$$

where $1/\rho$ is the curvature, M is the applied bending moment expressed in M = Px for 0 < x < 10 mm with P being the reaction force at the outer loading point of the specimen and x being the position along the beam as measured from the outer reaction point toward the inner load point. For 10 < x > 30 mm the moment is constant at M= $P(L_o - L_i)/2$. Φ is a constant associated with A, N and specimen geometry. Substituting M = Px into Eq (3) and taking the logarithm of both sides yields

$$\log (1/\rho) = N \log x + B \tag{4}$$

where B = $\log [A(P/\Phi)^{N}]$. Equations (3) and (4) indicate that on a logarithmic plot the curvature is a linear function of x with a slope of N in regions between the inner span and the outer reaction points (i.e. for 0 < x <10 mm and 30 < x < 40 mm); whereas, the curvature in the inner span is constant because of the constant moment therein.

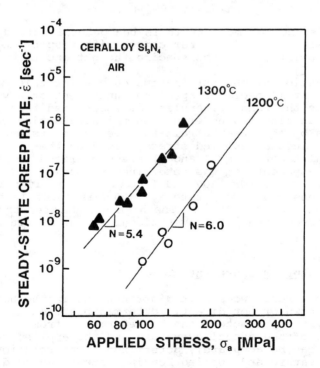

FIG. 3 -- Flexural steady-state creep rate as a function of initial applied stress at 1200 and 1300°C in air. The stress exponent (N) is based on the conventional solution.

The advantage of this solution over Eq (2) is that strain
is related to actual curvature along the beam length instead
of the load point displacements and assumed shapes. The
disadvantage is that the actual neutral axis position is
still assumed (at center) and thus strains based on
curvature are an average value, lower than the maximum
because of the axis shift.
    Curvature of the specimens was determined from optical
measurements of specimen deflection as a function of
position from an outer load point, and differentiation of
polynomial fits to the resulting data. It was found that a
fifth order polynomial equation was sufficient to fit
deformation curves with a correlation coefficient $r^2 \geq 0.99$.
The deflection curves $y(x)$ were used to calculate the
specimen curvature as a function of specimens position (x)
by applying the equation from differential calculus

$$1/\rho = (d^2y/dx^2)/[1 + (dy/dx)^2]^{3/2} \qquad (5)$$

where $1/\rho$ is the curvature. A similar method was used
previously by Jakus and Wiederhorn to calculate the creep
parameter N of alumina and glass [6].
    Some typical creep deformation curves obtained at
initial applied stresses of $\sigma = 80$ to 170 MPa are shown in
Fig. 4, where x is the distance from the outer load point
and y is the permanent deflection due to creep. This figure
indicates that for a given applied stress and elapsed time,
appreciable creep deformation occurred.
    The curvatures obtained from Eq (5) and the data shown
in Fig. 4 are presented in Fig. 5, where $\log (1/\rho)$ is
plotted as a function of $\log (x)$. The curves in Fig. 5 show
that $\log (1/\rho)$ is nearly, though not exactly, a linear
function of $\log (x)$ in the region between the outer reaction
point and the inner load point (i.e. for $0 < x < 10$ mm and $30 < x < 40$ mm). The $\log (1/\rho)$ was reasonably constant within
the inner span (i.e. for $10 < x < 30$ mm), consistent with
Eqs (3). Assuming that Eqs (3) and (4) are fully
applicable, the creep stress exponent N was evaluated. The
average N parameters are N = 3 to 3.5 at 1200°C and 2.7 at
1300°C. These values of N are about one-half of those (N =
6.0 and N = 5.4 for 1200°C and 1300°C, respectively)
obtained from the conventional solution. This indicates
that the tensile creep rate did not equal the compressive
creep rate and that the neutral axis shifted to accommodate
the moment equilibrium condition. Further, the creep
parameter N is not unity, which is the case for symmetric
creep deformation between the tension and compression sides,
so that neutral axis shift is expected. The trend of a
lower value N via Eq. (4) as compared to that determined
directly from Eq. (2) is in agreement with the results of
Jakus and Wiederhorn [6].

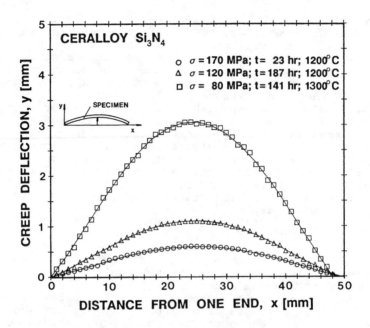

FIG. 4 -- Creep deflection as function of distance from an outer load point for the initial stresses shown.

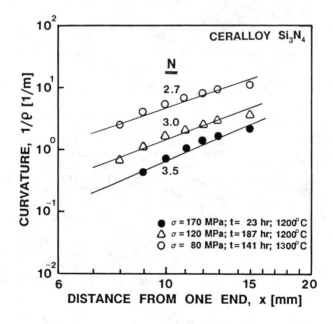

FIG. 5 -- Curvature as a function of distance from an outer load point for the initial stresses shown.

Neutral Axis Shift Measurements

   The results of the neutral axis determinations, based
on the scratch mark technique described in the Experimental
Procedure section, are presented in Fig. 6. The strain as a
function of location along the y axis is presented for
different conditions of initial applied stress, sustained
time and temperature.
   The results in Fig. 6 clearly show that the neutral
axis rapidly shifted to the compression side and stopped
shifting for all the test conditions. These results suggest
again that the stress exponent (N) as well as any of the
other creep or fatigue parameters determined in flexure
should be accepted with some caution and that uniaxial
tensile and compressive creep tests should be performed, as
emphasized previously (for example, [6]).
   Application of the analysis of Chen and Chuang [4] to
predict uniaxial tension and compression creep parameters
from the scratch mark data resulted in the parameters listed
in Table 2. The results for 1200°C indicate a much larger
tension parameter (N_t = 9.5) than that determined
conventionally (N = 6.0). The results for 1300°C indicate a
tension value, N_t, between the N determined by the
conventional theory and the curvature moment relation, and a
compression value, N_c, less than unity. With the exception
of the 1200°C N_t value, the results are in reasonable
agreement with the results of Ferber et al. [2] and Chen and
Chuang [4] in which the N parameter for uniaxial tension is
equal to or greater than that determined from flexure, and
the compressive value is near unity.
   The success of the scratch mark technique depends on
the degree of oxidation a given material exhibits and the
ability to remove it by polishing or etching. Although
etching was effective, it may affect mechanisms that are
controlled by the surface layer. However, no such effects
were noted between tests with and without etching.
   Fig. 7 shows the polished side surface of a specimen
crept with an initial applied stress of 80 MPa at 1300°C for
141 hours. Cracks measuring 0.5 mm in depth developed along
the tensile side of the inner span at roughly 1 mm
intervals. Such large scale cracking should lower the
specimen stiffness. Measurements of Young's modulus of the
specimen (via strain gages and assumption of simple beam
theory: σ = Mh/2I) after creeping indicated a modulus of 264
± 3 GPa on the tension side of the beam, and a modulus of
272 ± 7 on the compression side. The cracking lowered the
apparent modulus of the material by roughly 15%.

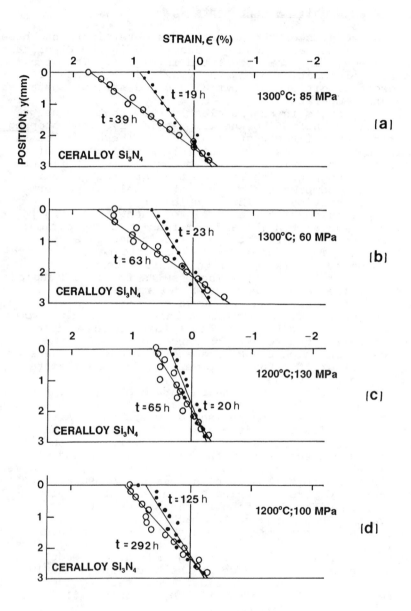

FIG. 6 -- Flexural creep strain as a function of specimen height (y) for silicon nitride subjected to the initial stresses shown at 1200 and 1300°C in air.

Table 2 -- <u>Summary of creep parameters</u>

Method	Temperature (°C)	N	
Conventional[1]	1200	6.0	(directly from flexure)
	1300	5.4	
Curvature[2] measurement	1200	3.0 - 3.5	(directly from flexure)
	1300	2.7	
Scratch [3] technique	1200	9.6	(tension, extracted from flexure)
		≈ 0	(compression, extracted from flexure)
	1300	3.9	(tension, extracted from flexure)
		0.6	(compression, extracted from flexure)

1. Direct application of Eq (2).
2. Application of Eqs (4) through (6).
3. Analysis of Chen and Chuang, ref 4.

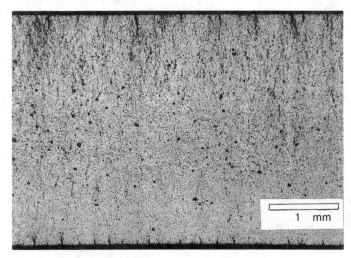

FIG. 7 -- Side surfaces of a specimen subjected an initial stress of σ = 80 MPa for 141 hours at 1300°C.

CONCLUSIONS

Although the creep behavior of a ceramic in flexure appeared to follow Norton's law, neutral-axis shift occurred for all test conditions. The neutral axis shift occured rapidly, within 20 h, and appeared to nearly stop thereafter. As a result, different methods of analyzing flexural creep data resulted in different creep parameters. Thus, the experimental and analysis methods used are critical to the evaluation of a creep parameter from flexural data. Complete characterization of creep behavior requires, at a minimum, that the neutral axis position be monitored and the strain rate and stress values be modified accordingly, or that individual uniaxial tension and compression tests be used. Verification of the models used to extract tension and compression creep parameters from flexural data will require further testing.

ACKNOWLEDGEMENTS

This research was sponsored in part by the U.S. Department of Energy as part of the Ceramic Technology Project, DOE Office of Transportation Technologies, under contract DE-AC05-84OR21400 with Martin Marietta Energy Systems, Inc. The authors are grateful to R. Pawlik for his experimental and SEM work.

REFERENCES

[1] Carroll, D. F., Chuang, T. J., and Wiederhorn, S. M., "A Comparison of Creep Rupture Behavior in Tension and Bending" Ceramic Engineering and Science Proceedings, Vol. 4, No.7-8, 1988, pp. 635-649.

[2] Ferber, M. K., Jenkins, M. G., and Nolan, T. A. "Creep-Fatigue Response of Structural Ceramics: I, Comparison of Flexure, Tension, and Compression Testing," Proceedings of the 37th Sagamore Army Materials Research Conference, D.J. Viechnicki, Ed., 1990, pp. 343-351.

[3] Chuang, T. J., "Estimation of Power-Law Creep Parameters From Bend Test Data," Journal of Materials Science, Vol. 21, 1986, pp 165-175.

[4] Chen, C. F., and Chuang, T.J., "Improved Analysis for Flexural Creep with Application to Sialon Ceramics," Journal of the American Ceramic Society, Vol. 73, No. 8, 1990, pp. 2366-73.

[5] Hollenberg, G. W., Terwilliger, G. R., and Gordon, R. S., "Calculation of Stress and Strain in Four-Point Bending Creep Tests," Journal of the American Ceramic Society, Vol. 54, No. 4, 1971, pp. 196-199.

[6] Jakus, K., and Wiederhorn, S. M., "Creep Deformation of Ceramics in Four-Point Bending, "Journal of the American Ceramic Society, Vol. 71, No. 10, 1988, pp. 832-836.

[7] Salem, J. A., and Choi, S. R., Ceramic Technology Project Semiannual Progress Report for April 1991 through September 1991, pp 305-319, ORNL/TM-11984, Oak Ridge National Laboratory, 1992.

[8] T. Nose and T. Fujii, "Evaluation of Fracture Toughness for Ceramic Materials by the Single-Edge-Precracked-Beam Method," Journal of the American Ceramic Society, Vol. 71, No. 5, 1988, pp. 328-333.

[9] Metals Handbook, ninth edition, Vol. 8, 1985, pp. 306.

Sung R. Choi[1], Jonathan A. Salem[2], and Joseph L. Palko[1]

COMPARISON OF TENSION AND FLEXURE TO DETERMINE FATIGUE
LIFE PREDICTION PARAMETERS AT ELEVATED TEMPERATURES

REFERENCE: Choi, S. R., Salem, J. A., and Palko, J. L., "Comparison of
Tension and Flexure to Determine Fatigue Life Prediction Parameters at
Elevated Temperatures," Life Prediction Methodologies and Data for
Ceramic Materials, ASTM STP 1201, C. R. Brinkman and S. F. Duffy, Eds.,
American Society for Testing and Materials, Philadelphia, 1994.

ABSTRACT: High temperature slow crack growth of a hot-pressed silicon
nitride (NCX 34) was determined at temperatures of 1200 and 1300°C in
air. Three different testing methods were utilized: dynamic and static
fatigue with bend specimens, and static fatigue with dog-bone-shaped
tensile specimens. Good agreement exists between the dynamic and static
fatigue results under bending. However, fatigue susceptibility in
uniaxial tensile loading was greater than in bending. This result
suggests that high temperature fatigue behavior should be measured with
a variety of specimen configuration and loading cycles so that adequate
lifetime prediction parameters are obtained.

KEYWORDS: silicon nitride, slow crack growth, dynamic fatigue, static
fatigue, tensile fatigue, lifetime prediction

Silicon nitride ceramics are candidate materials for high
temperature structural applications in advanced heat engines and heat
recovery systems. The major limitation of this material in high
temperature applications is fatigue-associated failure, where slow crack
growth of inherent defects or flaws can occur until a critical size for
catastrophic failure is reached. Therefore, it is very important to
evaluate fatigue behavior with specified loading condition so that
accurate lifetime prediction of ceramic components is ensured.

There are several ways of determining fatigue parameters.
Dynamic, static or cyclic fatigue loading can be applied to smooth
specimens with inherent flaws or to precracked fracture mechanics
specimens in which the crack velocity measurements are made directly. A
considerable number of studies have been carried out to characterize
fatigue behavior of silicon nitride ceramics using the testing methods

---

[1] Research Associates, Cleveland State University, Cleveland, OH
44115; Resident Research Associates, NASA Lewis Research Center,
Cleveland, OH 44135.
[2] Research Engineer, NASA Lewis Research Center, Cleveland, OH
44135.

mentioned above [1-7]. Although the reported results agree to some
degree, there remains uncertainty and disagreement among the testing
methodologies, depending on test materials and even on researchers.

        In this study, high temperature fatigue behavior of a hot-pressed
silicon nitride was determined at 1200 and 1300°C in air using three
different loading conditions: dynamic and static loading for flexure
beam specimens and static loading for dog-bone-shaped uniaxial tensile
specimens. Finite element analysis was carried out for the tensile
specimens to obtain stress distributions and to assure the
appropriateness of the specimen configurations designed. The material
was chosen because it exhibited moderate fatigue susceptibilities at
high temperatures, enabling the comparison of fatigue lifetime
prediction results from various testing methods. This material has been
previously used under bending loading to study high-temperature
structural reliability [8], long term environmental exposure [9] and
effects of oxidation on strength distribution [10].

EXPERIMENTAL METHOD

Material

        The material used in this study was a hot-pressed silicon nitride
containing 8% Y_2O_3*. The room temperature basic physical and mechanical
properties of the material are shown in Table 1. The material exhibited
a slightly bimodal grain structure of large elongated and fine equiaxed
grains. This bimodal grain structure resulted in a high fracture
toughness ($K_{IC} \approx 7$ MPa√m) as well as a rising R-curve [11], typical to
most in-situ toughened silicon nitrides with elongated grain structure.

Test Procedures

        Dynamic and static fatigue testing for the as-machined flexure
beam specimens was conducted in ambient air at 1200 and 1300°C using a
SiC four-point bend fixture in an electromechanical testing machine.
The inner and outer span of the test fixture were 10 mm and 30 mm,
respectively. The nominal dimensions of the rectangular test specimens
were 3 mm by 7 mm by 35 mm, respectively, in height, width, and length.
Four loading rates of 4.2 to 4200 N/min were used in the dynamic fatigue

TABLE 1--Physical and mechanical properties of NCX 34
          silicon nitride at room temperature [11,12]

Young's[1] Modulus, E (GPa) [11]	Hardness[2] H(GPa) [11]	Density[3] (g/cm³) [11]	Fracture Toughness[4] K_{IC}(MPa√m) [11]	RT Strength[5] (MPa) [12]
296	14.5±0.6	3.37	6.90±0.56	805±50

Notes:
1. By strain gaging; 2. By Vickers microhardness indenter;
3. By buoyancy method; 4. By SEPB method; 5. With specimens of 6.35 mm
(width) by 3.17 mm (height) using a four-point fixture with 9.525/19.05
mm- inner and outer spans [12].

NCX 34, fabricated in 1979, Norton Co. Northboro, MA.

testing, resulting in the corresponding stressing rates of 2 to 2000 MPa/min. Stress levels applied in the static fatigue tests were 250 to 500 MPa at 1200°C, and 75 to 400 MPa at 1300°C, respectively. The number of the test specimens in the dynamic fatigue testing was four at each loading rate per temperature; whereas, the total number of the test specimens in the static fatigue testing was 14 and 21, respectively, at 1200 and 1300°C. Each test specimen was preloaded with 20 N to maintain good alignment relative to the test fixture, and held at the test temperature for 20 min prior to testing.

The tensile fatigue behavior was investigated at temperatures of 1200 and 1300°C in air. The dog-bone-shaped tensile test specimens, similar to those used in creep testing measurements by Wiederhorn et al. [13], were utilized for this testing. A test specimen with strain gages attached is shown in Fig. 1. The dimensions of the test specimens were 2.5 mm by 2.5 mm by 20 mm in cross section and gage length, respectively. To minimize the degree of misalignment of the tensile test specimen, the loading pin holes of each test specimen were tapered toward the center so that load was applied to the center of the specimen, as suggested by Carroll et al [14]. With this tapered pin hole configuration and careful specimen mounting, it was possible to achieve less than two percent misalignment at a stress of 150 MPa.

The tensile specimens were preloaded with 35 N at room temperature and heated to the test temperature. Each test specimen was kept at the test temperature for about 20 min prior to applying the full test load. The testing was conducted in dead weight creep machines. A total of 14 test specimens were used at 1200°C with a nominal applied stress range of 80 to 200 MPa; whereas, at 1300°C a total of 15 were used with applied stresses from 50 to 100 MPa. A finite element analysis of the test specimen has shown that high stress, similar in magnitude to those occurring in the gage section, occurred around the loading pin hole due to the stress concentration. Hence, particular care was taken to minimize the possibility of machining-induced damage around the tapered pin-hole. Every pin hole was carefully diamond polished with a specially designed hand tool. Also, all the surfaces and edges of each as-machined test specimen were carefully hand-sanded to minimize any machining damage.

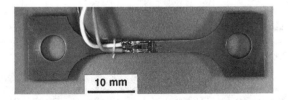

FIG. 1--A dog-bone-shaped tensile test specimen with strain gages attached.

Finite Element Analysis of Tensile Specimens

The finite element model for the tensile test specimen consisted of 1040 HEX/20 elements with a total of 5747 nodes. The model was analyzed using the MSC/NASTRAN finite element package. Linear static analysis was employed, with one eighth symmetry for simplicity. The specimen was loaded such that a specified uniform uniaxial tensile stress was present in the gage section. Three FORCE cards were used to apply the load. The load was applied to the one side of one element through the thickness. Since HEX/20 elements were used for the analysis, the load was divided into four parts. The outer nodes of the element received one part of the load each and the center node received two parts of the load. This load scheme was used to remain consistent with the element shape function formulations employed for the 20 node element [15].

Typical stress contours thus obtained are shown in Fig. 2, where σ_{zz} (principal stress along the specimen length) is plotted for one eighth symmetry of the specimen. This figure indicates that the maximum stress is present both at the gage section and at the pin hole (in the nine o'clock direction) due to the stress concentration. Note that the stresses in the neck region are always lower than that occurring in the gage section. Although the maximum stress occurs at the pin hole as well as at the gage section, the probability of failure is considerably higher at the gage section than at the pin hole since the volume or surface area stressed under the maximum stress is much greater at the gage section than at the pin hole. However, every attempt was made to minimize machining-induced damage.

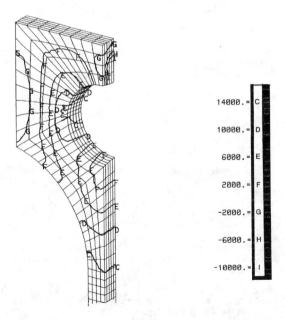

FIG. 2--Contours of principal stress along specimen length obtained from finite element analysis (one eighth symmetry).

RESULTS AND DISCUSSION

For most ceramics and glasses, slow crack growth can be expressed by the empirical relation

$$v = A \ [K_I/K_{IC}]^n \tag{1}$$

where A and n are the fatigue parameters associated with material and environment, K_I is the mode I stress intensity factor, and K_{IC} is fracture toughness. For dynamic fatigue testing which employs constant loading rate ($\dot{P}$) or constant stressing rate ($\dot{\sigma}$), the corresponding fatigue strength, σ_f, is expressed [16]

$$\sigma_f = [B \ (n+1) \ S_i^{n-2}]^{1/n+1} \ \dot{\sigma}^{1/n+1} \tag{2}$$

where $B = 2/[AY^2(n-2)K_{IC}^{n-2}]$ with Y being the crack geometry factor and S_i is the inert strength. The fatigue constants n, B and A can be obtained from the intercept and slope, respectively, of the linear fit of Log σ_f versus Log $\dot{\sigma}$. In the same way, for static fatigue testing where constant stress is applied, the time to failure (t_{fs}) can be derived easily in terms of applied stress (σ) as follows [1]

$$t_{fs} = [B \ S_i^{n-2}] \ \sigma^{-n} \tag{3}$$

Likewise, static fatigue parameters B and n can be evaluated by a linear regression analysis of the static fatigue curve when time to failure (Log t_{fs}) is plotted against applied stress (Log σ). However, it should be noted that there are several statistical approaches to estimate the fatigue parameters from dynamic and static fatigue data [17].

The relationship in fatigue life between dynamic and static fatigue is [18]

$$t_{fs} = t_{fd}/(n+1) \tag{4}$$

where t_{fd} is the time to failure in dynamic fatigue, which corresponds to $t_{fd} = \sigma_f/\dot{\sigma}$. By substituting t_{fs} in Eq. (4) into Eq. (3) with $\sigma = \sigma_f$, the dynamic fatigue curve can be converted into an equivalent static fatigue curve as follows:

$$t_{fd} = \alpha \ \sigma_f^{-n} \tag{5}$$

where α is the value associated with B, n and S_i.

Dynamic and Static Fatigue in Bending

A summary of the dynamic fatigue results at 1200 and 1300°C in air is presented in Fig. 3. The solid lines in the figure represent the best-fit lines based on Eq. (2). The decrease in fatigue strength with decreasing stressing rate, which represents fatigue susceptibility, was evident at both temperatures. The fatigue parameter (n) was determined to be n = 16.0 and 15.0 at 1200 and 1300°C, respectively, from a linear regression of Log σ_f versus Log $\dot{\sigma}$. Fractographic analysis of the failure surfaces revealed the presence of slow crack growth zones at the lower stressing rates, while no appreciable slow crack growth was obtained at higher stress rates.

The results obtained from the static fatigue tests at 1200 and 1300°C in air are shown in Fig. 4. The arrow marks in the figure represent the specimens that did not break before about 600 hr. Also, the solid lines in the figure represent the best-fit lines based on Eq. (3). The fatigue constant n was evaluated to be n = 20.7 and 15.0 at 1200 and 1300°C, respectively. The fatigue constant n = 15-21 thus

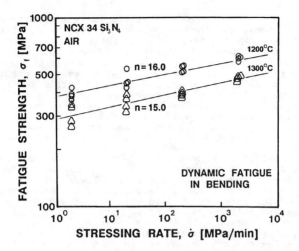

FIG. 3--Results of dynamic fatigue testing in bending of NCX 34 silicon nitride at 1200 and 1300°C in air.

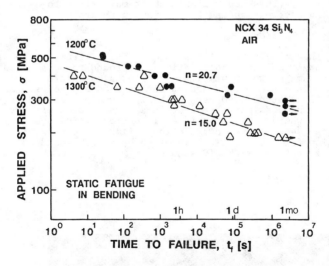

FIG. 4--Results of static fatigue testing in bending of NCX 34 silicon nitride at 1200 and 1300°C in air.

obtained agrees reasonably well with the value of n = 15-16 determined from the dynamic fatigue testing. A value of n = 12 was reported previously, obtained from static fatigue with MOR bars at 1400°C in air [19].

It is interesting to compare the static fatigue results obtained in this study with those obtained by previous researchers. Particularly at 1200°C, the stress rupture data of Quinn [9] is very different from those reported in this study. Similarly, Wiederhorn and Tighe [8] observed almost no failure at 1200°C at 400 MPa, even with specimen containing Knoop indents. The data in Fig. 4, however, indicates that they all should have failed as about 1000 seconds. This is due to the fact that there are billet to billet variations in the material, as pointed out previously [8]. Similarly, Quinn [9] reported bands of nonuniform material which cracked and crept differently. Thus, NCX 34 silicon nitride seems to be a such a material that some subtle chemistry or microstructural variation leads to radically different behavior.

It is important to note that appreciable creep deformation occurred for the specimens subjected at 1300°C to the lowest stressing rate in the dynamic fatigue testing and to the lowest applied stress in the static fatigue testing. Fracture surfaces of the specimens tested at different stressing rates or applied stresses showed that slow crack growth zones dominate failure as stressing rate or applied stress decreases. One complication evident from the fractography is the shape of the crack developed, especially in the specimens subjected to long time to failure. The cracks, though initially half-pennies in configuration, develop into corner and straight-through cracks as the crack size approaches the specimen size, as shown in Fig. 5. This may affect the values of the measured fatigue parameters, due to changes in crack geometry and net section stress. Further, enhanced creep at high temperature can result in neutral axis shift attributed to asymmetric creep behavior between the compression and tension sides of a flexure beam specimen [20]. This neutral axis shift may affect the stress distribution and possibly change the fatigue parameters.

Static Fatigue in Tension

Some of the uniaxial tensile specimens failed from the loading pin holes. This undesirable pin-hole failure was found to be associated with machining damage, which was minimized later by careful hand polishing with diamond compound around the tapered pin holes. Another undesirable failure associated with machining damage occurred at the intersection of the straight gage section and the radius of curvature of the neck region, where a small surface discontinuity (damage) existed. The typical gage section and intersection failures are shown in Fig. 6. This neck region failure was also minimized by careful hand polishing around the intersections with SiC sand paper.

FIG. 5--Fracture surface of a specimen subjected to static fatigue in bending at 1300°C; σ = 190 MPa and t_f = 456 hr.

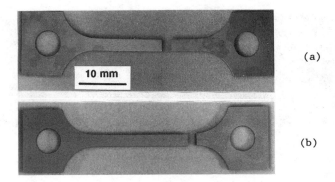

FIG. 6--Fractured uniaxial tensile test specimens: (a) desirable gage
section failure; (b) undesirable neck region failure.

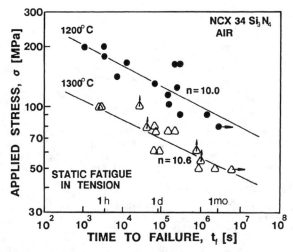

FIG. 7--Results of static fatigue testing in uniaxial tension of NCX 34
silicon nitride at 1200 and 1300°C in air.

        A summary of the uniaxial tensile fatigue results obtained at 1200
and 1300°C in air is presented in Fig. 7. The solid lines in the figure
represent the best-fit lines obtained by a linear regression analysis of
Log t_f versus Log σ. It should be noted that specimens failed from the
pin holes and neck region (marked with vertical arrows) were excluded in
the regression analysis. The horizontal arrow marks in the figure
represent the test specimens that did not break before 600 hr. The
fatigue parameter n was determined to be n = 10.0 and 10.6 at 1200 and
1300°C, respectively. This fatigue parameter of n $\approx$ 10 is somewhat
lower than that (n $\approx$ 15-21) from the dynamic and static fatigue testing
with the flexure beam specimens. By contrast, the parameter of n $\approx$ 10

(a)	(b)

Fig. 8--Fracture surfaces of specimens subjected to static fatigue in
uniaxial tension at 1300°C: (a) σ = 100 MPa; t_f = 42 min;
(b) σ = 50 MPa; t_f = 682 h.

determined for the smooth (as-machined) specimens is higher than the
value of n = 5.2 obtained for the uniaxial tensile specimens with Knoop
indent cracks by Henager and Jones [21].

At high applied stress, failure was usually (but not always
clearly) associated with slow crack growth; whereas, at lower applied
stress creep-induced failure was dominant at both temperatures.
Multiple creep crack formation in the gage section was typically
observed for the specimens failed at lower stresses. Fig. 8 shows the
fracture surfaces of specimens failed at 1300°C at two different applied
stresses of σ = 100 and 50 MPa. The formation of a dominant crack is
evident for the specimen failed at 100 MPa with t_f = 42 min (volume
failure). However, a clear fracture origin was not readily discernable
from the specimen failed at 50 MPa with t_f = 682 hr, suggesting that
crack coalescence associated with creep damage caused specimen failure.

<u>Comparison of Fatigue Parameters</u>

Comparison of dynamic and static fatigue behavior in bending can
be evaluated by converting the dynamic fatigue data into a corresponding
static fatigue curve via Eqs. (4) and (5). The resulting plots are
shown in Fig. 9, where the converted dynamic fatigue data are compared
with the static fatigue data. Also included in the figure are the data
obtained from the uniaxial tensile fatigue testing.

Overall agreement between dynamic and static fatigue in bending is
reasonably good, notwithstanding a little variation, especially at
1200°C. Therefore, based on these results it is possible to obtain slow
crack growth parameters from either static or dynamic fatigue testing
techniques, as observed previously for Ceralloy 147A silicon nitride
[22]. The dynamic fatigue testing is preferable since the time to
failure is shorter in dynamic fatigue than in static fatigue. Care
should be taken when an extrapolation based on the dynamic fatigue data
is made to predict slow crack growth behavior in the low applied stress
regime.

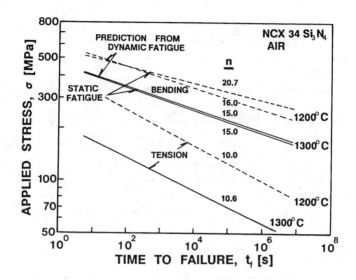

FIG. 9--Comparison of fatigue life curves obtained from dynamic and static bend fatigue and tensile static fatigue for NCX 34 silicon nitride at 1200 and 1300°C in air.

The difference between the tensile static fatigue and the dynamic or static bend fatigue is much larger as seen in Fig. 9. The difference in fatigue strength between the tensile and bend loading is probably due to the difference in effective volume or surface area between the two specimen configurations (dog-bone-shaped tensile specimens and flexure beam specimens). The ratio of bending strength to tensile strength can be calculated using the following equation

$$\sigma_T/\sigma_B = [A_{effB}/A_{effT}]^{1/m} \quad \text{or}$$

$$= [V_{effB}/V_{effT}]^{1/m} \tag{6}$$

where A_{eff} and V_{eff} are effective surface area and effective volume, respectively, and m is the Weibull modulus. Since the Weibull modulus at 1200 and 1300°C were not known, a value of m ≈ 10 was arbitrarily chosen. By assuming that failure is controlled by volume-associated failure, and taking $V_{effB} = 11.28$ mm^3 and $V_{effT} = 62.5$ mm^3 based on the fixture and specimen geometry, a value of $\sigma_T/\sigma_B = 0.786$ is obtained. This ratio of 0.786 is a reasonable estimate at 1200°C, but a poor one at 1300°C since the ratios of the tensile fatigue strength to the bend fatigue strength (dynamic or static) at $t_f = 1$ s (assuming little fatigue) are 0.7 and 0.4, respectively.

Using the experimental data and the appropriate equation and assuming $K_{IC} \approx 5$ MPa√m at 1200 to 1300°C, the fatigue parameter A was calculated and tabulated in Table 2. By using the A value, the fatigue life curve (Fig. 9) was converted into a crack velocity curve, Fig. 10, where crack velocity is plotted, based on Eq. (1), as a function of normalized stress intensity factor K_I/K_{IC}. The crack velocity curve obtained at 1300°C from the uniaxial tensile fatigue with Knoop indent crack [18] was also included in the figure for comparison. As already

TABLE   2--Summary of fatigue parameters of NCX 34 silicon nitride at
1200 and 1300°C in air under various loading.

Loading and geometry	Temp (°C)	n	B (MPa^{2}min)	A (m/min)
Dynamic in Bending	1200	16.0	14.702	0.194
	1300	15.0	10.269	0.300
Static in Bending	1200	20.7	11.148	0.191
	1300	15.0	11.697	0.262
Static in Tension	1200	10.0	16.242	0.319
	1300	10.6	11.126	0.417
Static in Tension with Indents [21]	1300	5.2	-	-

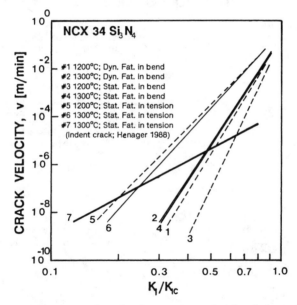

FIG. 10--Summary of crack velocity curves obtained from different
testing methods for NCX 34 silicon nitride at 1200 and 1300°C in air.

shown in the fatigue life curve (Fig. 9), there is reasonably good
agreement between the dynamic and static fatigue in bending, but a large
discrepancy between the uniaxial tension and bend fatigue data.   Fig. 10
also shows that in uniaxial tensile loading the fatigue susceptibility
of the indent cracks was much greater than that of the smooth (as-
machined) specimens, as observed previously for the GN-10 silicon
nitride material at 1200°C [5].

The difference in fatigue susceptibility of smooth and Knoop precracked specimens may be due to the nature of the cracks used or the method of parameter estimation: The dynamic and static methods determine parameters empirically from stress and time-to-failure data that results from inherent defects that develop into cracks, grow and cause failure; whereas fracture mechanics methods, such as the Knoop indent, attempt to directly observe and track the length of an initially sharp, well developed crack.

These results indicate that there is no unique fatigue testing methodology, implying that a variety of fatigue loading cycles, specimen configurations and flaw systems should be used to thoroughly characterize fatigue behavior of ceramic components that will have multiaxial stresses. An important result obtained from this fatigue testing study is that the dog-bone-shaped tensile specimens that have been used primarily in creep studies of ceramics can be applied to high-temperature tensile fatigue life (stress rupture) testing.

CONCLUSIONS

(1). The high-temperature fatigue parameters for this material determined from dynamic and static fatigue bend data are in good agreement. Fatigue parameters were $n = 15$ to 20.

(2). A discrepancy exists between bend (dynamic or static) fatigue and uniaxial tensile fatigue, resulting in more fatigue susceptibility for uniaxial tension. The discrepancy is presumably due to creep associated mechanism, different in bending and tension primarily attributed to neutral axis shift occurring in the bend specimens.

(3). Creep-associated failure became dominant as applied stress or stressing rate decreased. In this case, neutral axis shift via asymmetric creep deformation may have affected the stress distribution of flexure specimens, and presumably changed the fatigue parameters, which were based on an elastic stress solution. The use of tensile specimens is thus strongly preferable in this case.

(4). Fatigue behavior should be evaluated with a variety of stress states, loading cycles and flaw (inherent or artificial) configurations to ensure accurate life prediction parameters.

ACKNOWLEDGEMENTS

This research was sponsored in part by the Ceramic Technology Project, DOE Office of Transportation Technologies, under contract DE-AC05-84OR21400 with Martin Marietta Energy Systems, Inc. The authors are grateful to R. Pawlik at NASA Lewis for his experimental and SEM work.

REFERENCES

[1] Ritter, J. E., "Engineering Design and Fatigue Failure of Brittle Materials," pp 661-686 in Fracture Mechanics of Ceramics, Vol. 4, Bradt, R. C., Hasselman, D. P. H., and Lange, F. F. Eds., Plenum Publishing Co., NY, 1978.

[2] Trantina, G. G., "Strength and Life Prediction for Hot-Pressed Silicon Nitride," Journal of American Ceramic Society, Vol. 62, 1979, pp 377-380.

[3] Govila, R. K., "Uniaxial Tensile and Flexural Stress Rupture

Strength of Hot-Pressed Si₃N₄," _Journal of American Ceramic Society_, Vol. 65, 1982, pp 15-21.

[4] Quinn, G. D., and Quinn, J. B., "Slow Crack Growth in Hot-Pressed Silicon Nitride," pp 603-636 in _Fracture Mechanics of Ceramics_, Vol. 6, Bradt, R. C., et al. Eds., Plenum Press, NY, 1983.

[5] Choi, S. R., and Salem, J. A., "Comparison of Dynamic Fatigue Behavior Between SiC Whisker-Reinforced Composite and Monolithic Silicon Nitrides," NASA TM 103707, NASA Lewis Research Center, Cleveland, OH, 1991.

[6] Chuck, L., McCullum, D. E., Hecht, N. L., and Goodrich, S. M., "High-Temperature Tension-Tension Cyclic Fatigue for a Hipped Silicon Nitride," _Ceramic Engineering and Science Proceedings_, Vol. 12, 1991, pp 1509-1523.

[7] Evans, A. G., and Wiederhorn, S. M., "Crack Propagation and Failure Prediction in Silicon Nitride at Elevated Temperatures," _Journal of Material Science_ Vol. 9, 1974, pp 270-278.

[8] Wiederhorn, S. M., and Tighe, N. J., "Structural Reliability of Yttria Doped Hot Pressed Silicon Nitride at Elevated Temperatures," _Journal of American Ceramic Society_, Vol. 66, 1983, pp 884-889.

[9] Quinn, G. D., "Characterization of Turbine Ceramics After Long Term Environmental Exposure," U.S. Army Tech. Report TR 80-15, U.S Army Materials and Mechanics Research Center, Watertown, MA, 1980.

[10] Easler, T.E., Bradt, R. C., and Tressler, R. E., "Effect of Oxidation and Oxidation Under Load on Strength Distribution of Silicon Nitride," _Journal of American Ceramic Society_, Vol. 65, 1982, pp 317-319.

[11] Choi, S. R. and Salem, J. A., Bimonthly Progress Report, Ceramic Technology for Advanced Heat Engines Project, W.B.S. Element 3.2.1.7., November-December/1991, Oak Ridge National Laboratory, Oak Ridge, TN, 1991.

[12] Sanders, W. A., unpublished work, NASA Lewis Research Center.

[13] Wiederhorn, S. M., Roberts, D. E., Chuang, T.-Z., and Chuck, L., "Damage-Enhanced Creep in a Siliconized Silicon Carbide: Phenomenology," _Journal of American Ceramic Society_, Vol. 71, 1988, pp 602-608.

[14] Carroll, D. F., Wiederhorn, S. M., and Roberts, D. E., "Technique for Tensile Creep Testing of Ceramics," _Journal of American Ceramic Society_, Vol. 72, 1989, pp 1610-1614.

[15] Bathe, K. J., _Finite Element Procedures in Engineering Analysis_, Prentice-Hall, Englewood Cliffs, NJ, 1982.

[16] Evans, A. G., "Slow Crack Growth in Brittle Materials under Dynamic Loading Conditions," _International Journal of Fracture_, Vol. 10, 1974, pp 1699-1705.

[17] Jakus, K., Coyne, D. C., and Ritter, J. E., "Analysis of Fatigue Data for Lifetime Prediction for Ceramic Materials," _Journal of Materials Science_, Vol. 13, 1978, pp 2071-2080.

[18] Davidge, R. W., McLaren, J. R., and Tappin, G., "Strength-Probability-Time (SPT) Relationship in Ceramics," _Journal of_

Materials Science, Vol. 8, 1973, pp 1699-1705.

[19] Larsen, D. C., and Adams, J. W., "Property Screening and Evaluation of Ceramic Turbine Materials," AFWAL-TR- 83-4141, Wright Aeronautical Laboratories, Wright-Patterson AFB, OH, 1984.

[20] Chuang, T.-J., and Wiederhorn, S. M., "Damage-Enhanced Creep in a Siliconized Silicon Carbide: Mechanics of Deformation," Journal of American Ceramic Society, Vol. 71, 1988, pp 595-601.

[21] Henager, C. H., and Jones, R. H., "Environmental Effects on Slow Crack Growth in Silicon Nitride," Ceramic Engineering and Science Proceedings, Vol. 9, 1988, pp 1525-1530.

[22] Salem, J. A., and Choi, S. R., Ceramic Technology Project Semiannual Progress Report for April 1991 through September 1991, pp 305-319, ORNL/TM-11984, Oak Ridge National Laboratory, Oak Ridge, TN, 1992.

François HILD,[1] Didier MARQUIS[2]

MONOTONIC AND CYCLIC RUPTURE OF A SILICON NITRIDE CERAMIC

REFERENCE: Hild, F. and Marquis, D., "Monotonic and Cyclic Rupture of a Silicon Nitride Ceramic," Life Prediction Methodologies and Data for Ceramic Materials, ASTM STP 1201, C. R. Brinkman and S. F. Duffy, Eds., American Society for Testing and Materials, Philadelphia, 1994.

ABSTRACT: A testing system was developed to perform tensile tests on monolithic ceramics. Tensile tests under monotonic and cyclic conditions were carried out on a silicon nitride ceramic. Fractographic observations were made to identify the causes of failure. The monotonic tensile tests are compared with three different kinds of monotonic flexural tests. A size effect analysis is performed using a Weibull-type of modeling. An expression of the cumulative failure probability, which takes account of the initial flaw distribution, is derived in the framework of the weakest link assumption and the independent events hypothesis.

KEYWORDS: silicon nitride, failure properties, initial flaws, statistical analysis, size effect, monotonic rupture, cyclic rupture, cumulative failure probability.

A large scatter of failure stress is a common feature of all brittle materials. The strength and data scatter are due to initial flaw size and distribution within the material. These flaws are either intrinsic (porosities, inclusions) from the manufacturing process of the ceramic (sintering, pressing), or extrinsic from grinding of specimen surface.

In this paper, tensile tests are performed in order to analyze the initial flaws under the influence of homogeneous loading conditions. The experiments are carried out at room temperature on specimens made of silicon nitride under monotonic and cyclic loading conditions. Fractographic observations are made to characterize the causes of failure. The monotonic tensile tests are compared with experiments where the stress field profile within the specimen is heterogeneous. A size effect analysis is performed to study the correlation between mean failure stress and effective volume.

[1]Postgraduate researcher, Laboratoire de Mécanique et Technologie, 61 avenue du Président Wilson, F–94235 Cachan Cedex, France.
[2]Professor, Laboratoire de Mécanique et Technologie, 61 avenue du Président Wilson, F–94235 Cachan Cedex, France.

An expression of the cumulative failure probability is given under monotonic and cyclic conditions. It allows a unified approach of the failure probability considering the flaw distribution and its evolution in the case of cyclic loading.

TENSILE TEST SYSTEM

The advantage of a tensile test is the homogeneous macroscopic stress field within the specimen. This test is more complicated to perform on brittle materials than on ductile materials. This is because the failure strains of brittle materials are very small (typically 0.1 %). Therefore, if no particular caution is taken, spurious flexural strains induced by the testing grips can easily reach the magnitude of tensile failure strains.

To avoid such problems, a new testing system is designed [1]. It is composed of four main parts (Fig. 1). The role of the elastic joints (4) is to avoid flexural strains. To obtain an equivalent Cardan joint, the elastic joints are constituted of thin elastic bonds in two orthogonal directions. Because the specimen shape is axisymmetric, the grips (2) are entirely axisymmetric-like. Two half-shells (3) maintain the specimen in the grip and evenly distribute the load onto the cone-shaped surfaces of the specimen. To avoid stress concentration, a smooth axisymmetric specimen (1) without singularities is designed. It is composed of a cylindrical testing volume (1000 mm^3), cone-shaped heads and tore-like connecting parts in the same material.

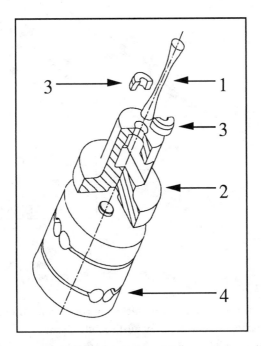

FIG. 1—The gripping system for tensile tests [1]
(1- specimen, 2- grip, 3- half shells, 4- elastic joint).

Some specimens have been tested with four strain gauges on the central section to verify the stress alignment: the maximum difference between the four gauges is less than 1.3×10^{-4} (i.e. 7.4 % of the tensile strain), and we conclude that the flexural strains are small (at maximum about 11%) compared with the tensile strains. As it is shown in the following, the scatter in failure stress induced by the flexural strains is small in comparison with the scatter induced by initial flaws.

MONOTONIC TENSILE TESTS ON SILICON NITRIDE

Eighteen specimens were tested in tension. They were made of isostatically pressed silicon nitride (SN 220M, Kyocera, Japan), density of 3,200 kg/m³, and cylindrically machined from rods (18 mm in diameter). The value of the effective surface roughness (R_a) of the specimens was less than 0.6 µm. The loading rate was equal to 1.5 MPa/s.

First, the silicon nitride stress-strain behavior is purely linear elastic with brittle failure. Second, the failure stress varied from 374 to 635 MPa , with a mean value of 526 MPa (Fig. 2).

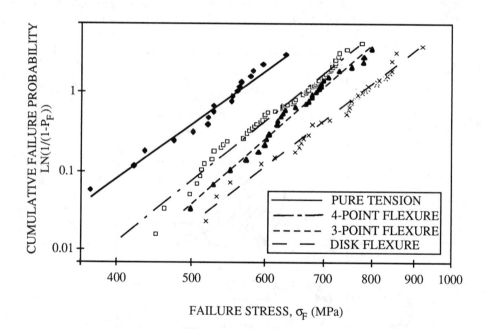

FIG. 2—An experimental data-base for the rupture of silicon nitride specimens subjected to different loading patterns (P_F : cumulative failure probability, σ_F : failure stress).
Tensile tests (♦) and Weibull correlation (————),
Four-point flexural tests (□) and Weibull correlation (—— · ——),
Three-point flexural tests (▲) and Weibull correlation (- - - - - - -),
Disk flexural tests (×) and Weibull correlation (—— —— ——).

These experimental results can be fitted (with least squares) to a two-parameter Weibull model [2] (Eqn. (1)) with a value of the Weibull modulus m equal to 8.0, the effective volume (V_{eff}) and the reference volume (V_0) are equal to 1000 mm^3 and 1 mm^3, respectively, and the Weibull stress S_0 equals 1360 MPa.

$$\ln\left(\ln\left(\frac{1}{1-P_F}\right)\right) = m \ln\left(\sigma_F\right) - m \ln\left(S_0\right) + \ln\left(\frac{V_{eff}}{V_0}\right) \tag{1}$$

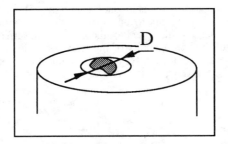

FIG. 3.1—Definition of the flaw size.

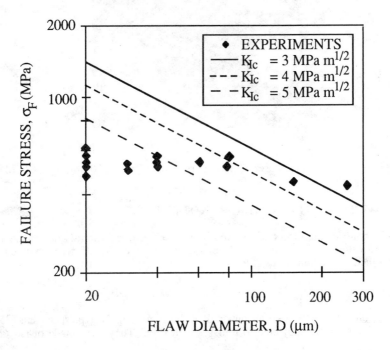

FIG. 3.2—Correlation between failure stress and flaw diameter in pure tension.

The initiation sites observed on the fracture surfaces of the specimens has flaws which size lies between 20 and 250 μm (Fig. 3.1–2). In Fig. 3.2, the experimental failure stresses are compared with failure stresses derived from a calculation where the initial flaws are assumed to be penny-shaped cracks loaded in mode I. For large flaws (70 ≤ D ≤ 250 μm) a correlation given by a Linear Elastic Fracture Mechanics approach with a critical stress intensity factor K_{Ic} on the order of 4–5 MPa$\sqrt{m}$ seems to correlate the experimental data (For SN 220M the value of K_{Ic} is known to be on the order of 5 MPa$\sqrt{m}$). Conversely, for small flaws (D ≤ 70 μm) a correlation given by a constant failure stress on the order of 550–600 MPa seems better.

Seventeen out of eighteen failures initiated within the volume of the specimen (Fig. 4) and not at the surface. This seems to be an *a posteriori* proof of the quality of the experimental setup that has been designed for these tests: the surface flaws do not play any leading role because of the homogeneity of the stress field that is induced by this setup. Moreover, because of this homogeneity, it is likely that the initiation flaw observed is, whatever its size, the largest flaw, i.e. the weakest link [3], [4], within the specimen.

FIG. 4—Fractographic observation in pure tension (bar = 100 μm, σ_F = 425 MPa).

COMPARISON WITH OTHER MONOTONIC TESTS

To analyze the effect of the stress field pattern, the monotonic tensile tests is compared with three different kind of flexural tests (Fig. 2). An effective volume analysis is performed on the four series of tests.

Flexural Tests

4–Point flexural tests—One hundred machined rectangular-type specimens (35 mm x 4 mm x 3 mm) were subjected to four-point flexure. They were made of isostatically pressed SN 220M and the R_a value was less than 1 μm. The load and support spans were equal to 12.5 and 24 mm, respectively.

The load-displacement curve has the same linear aspect as the curve obtained in pure tension. The failure stress (defined as the maximum stress over the structure and calculated by beam theory) lies between 448 MPa and 780 MPa (Fig. 2). The value of the Weibull modulus m (using a least squares method) is 8.9, the effective volume is equal to 8 mm^3, and the Weibull stress S_0 equals 875 MPa.

Compared to the pure tension case, the size of the initiation flaws is on average smaller (15 to 110 μm) [5], and the mean failure stress is higher (629 MPa instead of 526 MPa). Moreover, the initiation sites are located near the surface of the specimens that is subjected to tensile stresses (97% were in the volume). These observations lead to two conclusions that are valid for this type of loading. First, the highest tensile stress levels play a leading role in the failure process. Second, the initiation flaws located near the surface in tension are not in general the largest flaws within the material.

3–Point flexural tests—Fifty machined rectangle-type specimens (70 mm x 4 mm x 3 mm) were subjected to three-point flexure. The material and the R_a value were the same as for the four-point flexure specimens. The support span was 60 mm. The load-displacement curve has the same linear aspect as the curve obtained in pure tension.

The failure stress (defined as the maximum stress over the structure and calculated by beam theory) lies between 491 MPa and 801 MPa (Fig. 2). The value of m (using a least squares method) is 9.6, the effective volume is equal to 3.2 mm^3, and the Weibull stress S_0 is 780 MPa.

Compared to the four-point flexural case, the size of the initiation flaws is on average similar [5] (90% were located in the volume), and the mean failure stress is higher (660 MPa instead of 629 MPa).

Disk flexural tests—To understand the behavior of silicon nitride when subjected to a multiaxial stress field, a set of fifty disks was subjected to biaxial stress loading using ring on ring. The disk specimens were 30 mm in diameter and 3 mm in thickness. They were made of uniaxially pressed SN 220M and the value, after machining, of R_a was less than 1 μm. The disk-flexural testing setup consisted of a loading and a supporting ring 2.5 mm in diameter and 24 mm in diameter, respectively.

The failure stress, defined as the maximum principal stress over the structure and derived from a finite element analysis, lies between 513 MPa and 918 MPa (Fig. 2). The value of m (using a least squares method) is 8.4, of the effective volume is 0.5 mm^3 (a maximum principal stress criterion is used to compute the effective volume), and of S_0 is 720 MPa.

Compared to the three-point flexure case, the size of the initiation flaws is on average similar [5] (70% were located in the volume), the mean failure stress is higher (721 MPa instead of 660 MPa).

Effective Volume Analysis

Using the results of Davies [6] an effective volume analysis can be performed to correlate mean failure stress and effective volume. It is worth noting that in the framework of a Weibull model [2], the expression of the effective volume [6] can be related to the Weibull stress heterogeneity factor H_m and to the total volume V of a structure Ω by

$$V_{eff} = V H_m \qquad (2)$$

where H_m is defined as [7]

$$H_m = \left\{ \frac{1}{V} \int_{\Omega} \|\sigma\|^m \, dV \right\} / \sigma_F^m \qquad (3)$$

where $\|\sigma\|$ is the equivalent stress (we used the maximum principal stress), and σ_F is the failure stress corresponding to the maximum equivalent stress in Ω, $\sigma_F \equiv \underset{\Omega}{Max} \|\sigma\|$. The mean failure stress, $\overline{\sigma}_F$, is then related to the effective volume by

$$\overline{\sigma}_F = S_0 \left(\frac{V_0}{V_{eff}} \right)^{1/m} \Gamma\left(1 + \frac{1}{m}\right) \qquad (4)$$

where Γ is the Euler function of the second kind. Therefore, a Weibull model leads to a linear correlation between mean failure stress and effective volume in a log-log plot with a slope equal to $-1/m$. The correlation between the two quantities (Fig. 5) leads to a slope equal to -0.04 ($=-1/25$) with a coefficient of correlation ρ^2 equal to 0.995. The value of the slope does not correspond to the experimental Weibull modulus.

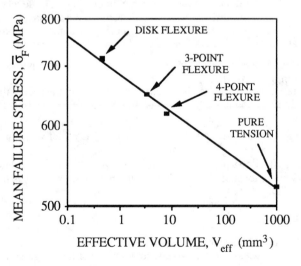

FIG. 5—Correlation between mean failure stress ($\overline{\sigma}_F$) and effective volume (V_{eff}).

Contrary to Katamaya and Hattori [8], for this set of experiments, a correlation in terms of effective volume is not satisfactory. Also, a correlation in terms of effective surface is not satisfactory since it leads to a slope equal to $-1/17$ with ρ^2 equal to 0.983.

In summary, the values of the scale parameter S_0 were found to be different. They lead to a value of the shape parameter m, derived from an effective volume analysis (on the order of 25), different from the values observed experimentally (on the order of 9). The batches of specimens were made by using various processing techniques: therefore the flaw distribution was not the same for the four sets of specimens. It is worth noting that the data plotted in Fig. 2 correspond to the same material but do not correspond to a unique initial flaw distribution.

CYCLIC TENSILE TEST ON SILICON NITRIDE

Testing Procedure

The tensile testing system is also used to perform cyclic tests. These tests were carried out to study the influence of cyclic loadings at different stress levels corresponding to known values experimentally obtained of cumulative failure probabilities under monotonic conditions. The minimum stress is equal to about 20 MPa. The maximum stress is also controlled during a test. The test is divided into series of 10,000 cycles. The first stress level (422 MPa) corresponds to a cumulative failure probability under monotonic conditions equal to 10%, the second stress level (470 MPa) corresponds to a cumulative failure probability under monotonic conditions equal to 20% , and so on. The frequency of the first cycle is equal to 5 x 10^{-3} Hz and corresponds to a monotonic test. The 9,999 remaining cycles have a frequency equal to 10 Hz. The test stops when the specimen breaks.

Experimental Results

Seventeen specimens were tested from a second batch. Seven out of seventeen failed at a stress level less than 422 MPa (i.e. the number of cycles to rupture is equal to unity). This shows the advantage of having a first cycle of each series with a very low frequency. For one of the specimens, a stiffness analysis is performed each 2,000th cycle. Fifteen analyses were carried out (the number of cycles to rupture was equal to 28,600, and the corresponding failure stress is equal to 507 MPa). The variation of the Young's modulus was less than 1% of the initial value of the Young's modulus. The stress alignment was again checked: the maximum difference between the four gauges was less than 11% (value observed under monotonic conditions). The number of cycles to rupture is plotted against the failure stress in Fig. 6. It is worth noting that all the ruptures do not occur when the load level is increased. The highest number of cycles is equal to 45,000 and corresponds to a failure stress of 545 MPa (i.e. a cumulative failure probability under monotonic conditions equal to 50%).

Fractographic analyses were made and led to observations similar to those found in the case of monotonic tension. In particular the mirror zone is still observed and the initiation sites are within the volume for the tests with a number of cycles to rupture greater than unity. When the number of cycles to rupture is equal to unity, the initiation sites are all at the surface. This may indicate that the grinding of the batch of specimen tested in cyclic tension is of a poorer quality than the batch of specimens tested in monotonic tension. The initiation sites observed on the fracture surfaces of the specimens are flaws which size is between 60 and 150 µm in diameter.

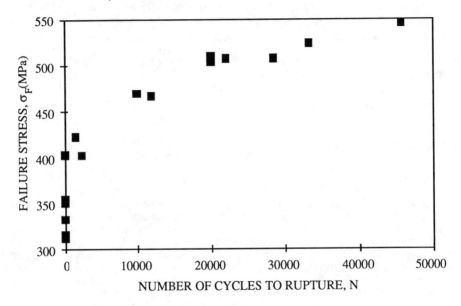

FIG. 6—Number of cycles to rupture versus failure stress.

In Fig. 7, the experimental failure stresses are compared with failure stresses derived from a calculation where the initial flaws are assumed to be penny-shaped cracks loaded in mode I. With this assumption, a Linear Elastic Fracture Mechanics approach indicates that the critical stress intensity factor K_{Ic} is again on the order of 3–5 MPa√m.

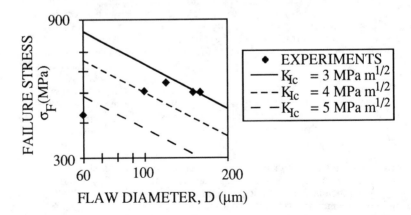

FIG. 7—Correlation between failure stress and flaw diameter in pure tension.

RELIABILITY ANALYSIS UNDER MONOTONIC AND CYCLIC LOADING CONDITIONS

The aim of this section is to analyze the reliability of structures, subjected to monotonic and cyclic loading, by considering initial flaw distributions. The study is conducted in the framework of the weakest link theory. An independent events hypothesis is made.

Expression of the Cumulative Failure Probability

The initial flaws are characterized by an initial flaw distribution density, f_0, which depends only upon their size, a (=D/2). Let us denote a_c the critical flaw size at a given stress level under monotonic and cyclic loading. Its expression is directly related to the type of modeling [9]. The cumulative failure probability P_{F0} is the probability of finding a flaw which size is larger than the critical flaw size a_c

$$P_{F0} = \int_{a_c}^{+\infty} f_0(a)\, da \tag{5}$$

In the case of cyclic loading, the flaws grow with the number of cycles, N (from $a(0)$ to $a(N)$), and the flaw distribution evolves from f_0 to f_N. The cumulative failure probability P_{F0} is then related to the flaw distribution f_N

$$P_{F0} = \int_{a_c}^{+\infty} f_N(a)\, da \tag{6}$$

It is useful to introduce a function ψ such that $a(0)=\psi(a(N))$ given by the evolution law of the flaws. If no new flaw nucleates, f_0 can be related to f_N by [10]

$$f_N(a) = f_0(\psi(a)) \frac{\partial \psi}{\partial a} \tag{7}$$

where the coefficient $(\partial\psi/\partial a)$ comes from the change of measure (from da to $d\psi(a)$). Therefore for monotonic and cyclic loading we have

$$P_{F0} = \int_{\psi(a_c)}^{+\infty} f_0(a)\, da \tag{8}$$

where $\psi(a_c)$ denotes the initial flaw size that, after N cycles of loading with a maximum equivalent stress over a cycle of period T, $\sigma = \underset{0 \le t \le T}{\mathrm{Max}} \, \|\sigma\|$, reaches the critical flaw size a_c.

Eqn. (8) constitutes a unified expression of the cumulative failure probability in the case of monotonic and cyclic loading.

On the structural level, the failure occurs when one flaw within the volume becomes critical. It means that if this flaw reaches the critical flaw size, a_c, failure occurs. Determining the failure at a structural level is equivalent to finding the 'weakest link' of the structure. When the interaction between flaws can be neglected, an hypothesis of independent events can be made. We consider a structure Ω of volume V, subjected to any stress field. It can be divided into a large number of elements of volume V_0 (i.e. a representative volume element subjected to a uniform remote stress field characterized by an equivalent stress). The cumulative failure probability, P_F, of a structure Ω can be related to the cumulative failure probability, P_{F0}, of a link by

$$P_F = 1 - \exp\left\{\frac{1}{V_0} \int_\Omega \ln\left(1 - P_{F0}\right) dV\right\} \tag{9}$$

By means of expressions (8) and (9), a general relationship between the initial flaw distribution and the cumulative failure probability of a structure Ω can be derived

$$P_F = 1 - \exp\left\{\frac{1}{V_0} \int_\Omega \ln\left(1 - \int_{\psi(a_c)}^{+\infty} f_0(a)\, da\right) dV\right\} \tag{10}$$

For monotonic loading, the same equation holds with the substitution $\psi(a_c)$ into a_c.

<u>Correlation with a Weibull Law</u>

In the case of monotonic loading, a correlation can be obtained between the Weibull parameters and the parameters of the flaw size distribution. If we assume that the initial flaws are modeled by cracks, the value of the critical flaw size, a_c, is related to the equivalent applied stress, σ, by

$$Y\sigma\sqrt{a_c} = K_{Ic} \tag{11}$$

where Y is a dimensionless parameter characterizing the shape of the flaw, and K_{Ic} is the critical stress intensity factor. If we assume that the maximum flaw size is bounded by a_M, the flaw size distribution f can be given by a beta distribution [11]

$$f_0(a) = \frac{a_M^{-1-\alpha-\beta}}{B_{\alpha\beta}} a^\alpha (a_M - a)^\beta \qquad 0 < a < a_M \quad, \quad \alpha > -1, \beta > -1 \tag{12}$$

where α and β are the parameters of the beta function, and $B_{\alpha\beta}$ is equal to $B(\alpha+1,\beta+1)$, where B is the Euler function of the first kind. There exists a threshold stress, S_u, defined as the lowest value of the stress, under which the failure probability has a zero value

$$S_u = \frac{K_{Ic}}{Y\sqrt{a_M}} \tag{13}$$

Assuming that a_c is close to a_M, Eqns. (8) and (12) lead to the following approximation of the cumulative failure probability of a single link, P_{F0}

$$P_{F0} \cong \left\{ \frac{<\|\sigma\| - S_u>}{S_0} \right\}^{\beta+1} \quad \text{with} \quad S_0 = \frac{S_u}{2}\left\{ (\beta+1)B_{\alpha\beta} \right\}^{1/(\beta+1)} \tag{14}$$

Using Eqns. (9) and (14), we obtain a Weibull expression where m is related to β (one of the parameters of the flaw distribution f_0) by

$$m = \beta + 1 \tag{15}$$

The β parameter gives the trend of the defect distribution for the large sizes (i.e. a being close to a_M). In that case, large defects are more critical for failure. Eqn. (14) gives a relationship between a parameter of the flaw distribution (i.e. a physical parameter) and m (i.e. a mechanical parameter) that can be obtained through an analysis of a set of macroscopic failure tests.

Analysis of the Monotonic Tests

For the four sets of tests, the identification was performed on the basis of a two parameter Weibull law. A similar correlation as the previous one can be derived [9] in the case of an unbounded flaw distribution. Since the values of m and S_0 are different for the four series of tests, we can conclude that the flaw distribution for small failure stresses is different.

This can also be shown by a direct identification of Eqns. (10) and (12). A good set of values of the parameters is given by $\alpha = -0.81$, $\beta = 7.11$, when $K_{Ic} = 4$ MPa$\sqrt{m}$ ($Y \cong 1$), $a_M = 125$ μm, and $V_0 = 1$ mm^3. Fig. 8 shows the computations compared with the experiments. The difference between experimental and theoretical data in the three other sets of experiments again indicates that the flaw distribution is different. This is due to the difference in processing techniques used to make the various specimen batches.

In summary, we have used two different approaches to show that the initial flaw distribution is different for the four sets of experiments. Another way to prove this difference is to perform direct measurements of the flaw distribution. This work is more tedious and the probability of finding a large flaw is very small.

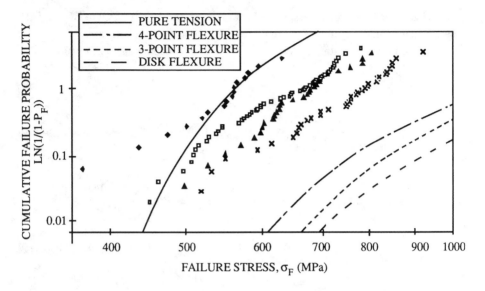

FIG. 8—Comparison between experiments and predictions for the rupture
of silicon nitride specimens subjected to different loading patterns.
Experiment ($\blacklozenge$) and model (——————————) in tension,
Experiment ($\square$) and model (——— · ———) in four-point flexure,
Experiment ($\blacktriangle$) and model (– – – – – –) in three-point flexure,
Experiment ($\times$) and model (—— —— ——) in disk flexure.

Analysis of the Cyclic Tests

Eqn. (10) enables a computation of the cumulative failure probability in the case of monotonic and cyclic loading. We can therefore compute the cumulative failure probability when the number of cycles to failure is equal to 1, 10,000, 20,000, and so on. We assume that the critical stress intensity factor K_{Ic} on the order of 4 MPa$\sqrt{m}$ can model the fracture behavior as it is suggested in the fractographic analysis. The evolution law is assumed to be given by the power law regime of a Paris law

$$\frac{da}{dN} = C\,K_{max}^{n} \tag{16}$$

where the constants C and n are taken to be 6.23×10^{-24} m $(MPa\sqrt{m})^{-n}$ and 21.1, respectively [12]. For a given load level we can compute the value of $\psi(a_c)$ as a function of the number of cycles N

$$\psi(a_c)\,/\,a_M = \left(\frac{\sigma}{S_u}\right)^{-2} \left\{ 1 + \frac{n-2}{2}\,C^*N\left(\frac{\sigma}{S_u}\right)^2 \right\}^{2/(2-n)} \tag{17}$$

where $C^* = C K_{Ic}^n / a_M = 2.52 \times 10^{-7}$. When the quantity $\dfrac{n-2}{2} C^* N \left(\dfrac{\sigma}{S_u}\right)^2 \ll 1$, the previous expression can be simplified to become

$$\psi(a_c) / a_M \cong \left(\frac{S_u}{\sigma}\right)^2 - C^* N \qquad (18)$$

In the considered cases, the value of S_u is 360 MPa ($Y \cong 1$) and $C^* N \ll 1$ therefore the cumulative failure probability does not evolve very much during the series of cyclic loading. However it is worth remembering that failures occurred during sequences of constant load levels. This shows that a stable degradation can occur and can lead to a catastrophic failure. Moreover, the fractographic analyses did not spot any micro cracking around initial flaws. This is consistent with the fact that initial flaws grew only very little to reach the critical size. Therefore initial and critical flaws are almost identical and the correlation between initial flaw size and failure stress of Fig. 7 is relevant.

Furthermore, it is possible to compute the stress drop for the same cumulative failure probability after N Cycles. After N cycles, we get the same cumulative failure probability by writing [10]

$$\psi(a_c) / a_M = A \qquad (19)$$

where the constant A depends upon the details of the flaw distribution. Eqn. (19) enables us to get the relationship between the number of cycles to failure and the stress level to give the same cumulative failure probability.

For instance, in the case of the experiments in monotonic tension, a failure stress equal to 515 MPa corresponds to a cumulative failure probability equal to 0.25. Using the material parameters defined earlier, a drop of 10% of the applied stress corresponds to a number of cycles to rupture greater than 1.8 million of cycles.

CONCLUSION

A series of tensile tests was carried out on silicon nitride specimens. The mean strength in tension is less than the mean strength in four-point flexure, in three point flexure and in disk flexure. The Weibull modulus in tension is on the same order as in the different types of flexure. However, the value of the scale parameter changes. A correlation between the size of the flaw leading to rupture and the failure stress is given. In tension, seventeen out of eighteen failures initiated in the volume of the specimen.

The size effect analysis performed for the present results leads to a good correlation in a log-log plot. Yet the slope does not correspond to the experimentally determined Weibull modulus.

In the case of cyclic tension, a stable propagation was observed. Seven out of seventeen failures occurred for number of cycles less than unity. A correlation between the size of the flaw leading to rupture and the failure stress is given.

An expression of the cumulative failure probability taking account of the initial flaw distribution has been derived in the case of monotonic and cyclic loading. It enables us to correlate the reliability of a structure to the initial flaw distribution, which corresponds to the cause of rupture. The cumulative failure probability in the case of cyclic tension is shown to be on the same order as the cumulative failure probability in the case of monotonic tension.

In this study, it is shown that the failure properties of a structure made of ceramics not only depend upon the volume of structure and the stress field profile, but also upon the flaw distribution. It is therefore important to specify these three quantities to be able to fully describe the failure conditions.

ACKNOWLEDGEMENTS

The authors gratefully acknowledge the financial support of Renault through contract ENSC/24 (H5–25–511) with the Laboratoire de Mécanique et Technologie, Ecole Normale Supérieure de Cachan.

REFERENCES

[1]   Hild, F., "Dispositif de traction-compression d'une éprouvette," French Patent, Patent number 90 06848, 1990.

[2]   Weibull, W., "A Statistical Theory of the Strength of Materials," Ing. Vetenskap Akad., 153, 1939.

[3]   Weibull, W., "A Statistical Distribution Function of Wide Applicability," J. Appl. Mech., Vol. 18, No. 3, 1951, pp 293-297.

[4]   Freudenthal, A. M., "Statistical Approach to Brittle Fracture," Fracture (an Advanced Treatrise), Vol. 2, 1968, pp 591-619.

[5]   Amar, E., Gauthier, F. and Lamon, J., "Reliability Analysis of a Si_3N_4 Ceramic Piston Pin for Automotive Engines," 3rd Int. Symp. Ceramic Materials and Components for Engines, Las Vegas, Nevada, V. J. Tennery (Edt.), American Ceramic Society, 1989, pp 1334-1356.

[6]   Davies, D. G. S., "The Statistical Approach to Engineering Design in Ceramics," Proceeding Brit. Ceram. Soc., Vol. 22, 1973, pp 429-452.

[7]   Hild, F., Billardon, R. and Marquis, D., "Stress Heterogeneity versus Failure of Brittle Materials," C. R. Acad. Sci. Paris, tome 315, série II, 1992, pp 1293-1298.

[8]   Katamaya, Y. and Hattori, Y., "Effects of Specimen Size on Strength of Sintered Silicon Nitride," J. Am. Ceram. Soc., 1982, pp C-164/C-165.

[9]   Hild, F. and Marquis, D., "A Statistical Approach to the Rupture of Brittle Materials," Eur. J. Mech. A/Solids, Vol. 11, No. 6, 1992, pp 753-765.

[10]   Hild, F. and Roux, S., "Fatigue Initiation in Heterogeneous Brittle Materials," Mech. Res. Comm., Vol. 18, No. 6, 1991, pp 409-414.

[11]   Spanier, J. and Oldham, K. B., An Atlas of Functions, Springer Verlag, 1987.

[12]   Hoshide, T., Ohara, T. and Yamada, T., "Fatigue Crack Growth from Indentation Flaw in Ceramics," Int. J. Fract., Vol. 37, 1988, pp 47-59.

Pramod K. Khandelwal[1]

MECHANICAL PROPERTIES AND NDE OF A HIP'ED SILICON NITRIDE[*]

REFERENCE: Khandelwal, P. K., **"Mechanical Properties and NDE of a Hip'ed Silicon Nitride,"** Life Prediction Methodologies and Data for Ceramic Materials, ASTM STP 1201, C. R. Brinkman and S. F. Duffy, Eds., American Society for Testing and Materials, Philadelphia, 1994.

ABSTRACT: Life prediction analytical methodology for ceramic gas turbine engine components requires the incorporation of time-temperature-stress dependent behavior and properties of the structural material. Injection molded and hot isostatic pressed PY6 silicon nitride with 6% yttria was used for this study. The room temperature fast fracture modulus-of-rupture (MOR) strength of 812.9 MPa degraded at 1400°C to 465.6 MPa. The monotonic tensile strength at room temperature was 407.9 MPa, which degraded to 303.3 MPa at 1400°C. Scanning electron microscopy (SEM) indicated that MOR specimens predominantly failed from surface flaws at room temperature, whereas the tensile specimens failed from volume defects. At higher temperatures, however, the failures occurred from surface and volume flaws in both the MOR bars and tensile specimens. Tensile creep data between 1260°C and 1400°C resulted in a stress exponent (n) of 7.8 ± 1.0 with an apparent activation energy of 957 ± 149 kJ/mole. SEM revealed that the material between 1300°C and 1400°C failed due to the formation, linking, and growth of cavities that caused the material to creep. At lower temperatures, failure was controlled both by slow crack growth and creep.

Microfocus X-ray detected 50 micron laser drilled surface holes in flat and button-head cylindrical specimens with a thickness sensitivity of 1% to 2%. Surface wave acoustic microscopy detected 10 micron surface holes in button-head tensile specimens.

KEYWORDS: silicon nitride; fast fracture, slow crack growth, creep, NDE

INTRODUCTION

[1]Staff Research Scientist, Allison Gas Turbine Division, General Motors Corporation, Indianapolis, IN 46206-0420

[*]This research was sponsored by the U.S. Department of Energy, Assistant Secretary for Energy Efficiency and Renewable Energy, Office of Transportation Technologies, as part of the Ceramic Technology Project of the Materials Development Program, under contract DE-AC05-84OR21400 with Martin Marietta Energy Systems, Inc.

The design requirements for structural ceramic components in an
automotive gas turbine engine include a turbine inlet temperature up to
2500°F and 3500 hr service life encompassing both cyclic and steady-
state operation. The resulting stress state in the ceramic components
will not only be multiaxial, but will be thermomechanically driven. As
the ceramic gas engine technology advances, a primary challenge is to
develop and validate the methodology to predict the time dependent
failure modes of structural ceramics. Allison is currently developing a
life prediction methodology (LPM) under a program funded by the
Department of Energy (DOE) and managed by Oak Ridge National
Laboratories (ORNL). Allison selected GTE injection-molded and hot
isostatic pressed (HIP) PY6 silicon nitride material with 6% yttria for
this program since it was a leading candidate for the fabrication of the
Advanced Turbine Technology Applications Program (ATTAP) gasifier rotor—
a requirement of the program (Ref 1). Allison has selected NASA-Lewis
Research Center (LeRC) CARES program as the centerpiece of our LPM
development efforts (Ref 2). The post processor analytical code is
available in the public domain and has been successfully used by many
industrial organizations for designing ceramic components. The purpose
of the DOE funded program is to incorporate four failure controlling
mechanisms into the CARES code (i.e., fast fracture, slow crack growth,
oxidation, and creep rupture). Another objective of the program is to
develop nondestructive evaluation (NDE) technology to detect failure
controlling flaws. In this paper, we will review the results obtained
to date using MOR and tensile specimens, two failure modes—fast
fracture, and creep along with some NDE development efforts.

MATERIAL AND SPECIMENS

Injection molded and HIP processed PY6 silicon nitride material with 6%
yttria was used to manufacture all the specimens by GTE Laboratories. A
large quantity of starting powders (coarse and fine) was acquired and
isolated in the beginning of the program to assure batch to batch
consistency of the material. MOR bars were individually injection
molded to near net shape and longitudinally machined to fabricate 3 x 4
x 50 mm Type-B specimens. Button-head tensile specimens (Figure 1) were
machined from cylindrical rods. Inspection of representative MOR bars
and tensile specimens indicated that the surface finish was between 0.38
to 0.51 microns (15 to 20 micro inches) for both type of specimens.

Reference standard specimens with surface laser drilled holes of 10 to
100 micron diameter were fabricated for both the MOR bars and
cylindrical button-head specimens for optimizing NDE inspection
parameters.

EXPERIMENTAL METHODS

Fast fracture MOR or flexural strength from room temperature (RT) to
1400°C was measured using the Type-B bars in 4-point bending at a cross-
head speed of 0.508 mm/min. The inner loading span was 20 mm while the
outer loading span was 40 mm.

Monotonic tensile strength from RT to 1400°C was measured at Southern
Research Institute (SoRI), Birmingham, Alabama, at a loading rate of
4137 MPa/min using a button-head specimen as shown in Figure 1. Fast
fracture tensile strength of advanced ceramics is commonly measured in

stressing rate mode (Ref. 3), and therefore it was used in the present
work. The button-head specimen geometry was empirically designed and
experimentally evaluated to assure that gage failures were accomplished.
The specimen was 137.2 mm long with a gage section of 55.9 mm length and
4.76 mm diameter. The tensile evaluations were performed in a gas-
bearing tensile system, a schematic of which is shown in Figure 2. This
system uses gas-bearing universals in the load linkage to prevent
introduction of unknown bending moments in the load train from the
cross-head motion or eccentricity and allows monitoring of the
straightness of the load train. Emphasis in the design of the load
train was placed on large length to diameter ratios at each connection,
close sliding fits of all the mating surfaces, elimination of threaded
connections, and other factors essential to minimize bending stresses.
All the components were machined true and concentric to within 12.5
microns. The entire load train was regularly checked by spinning a
circular alignment specimen between the lower and upper loading rods.
The precision machined tensile grip uses a three-piece split collet
assembly. The deformation of the specimen gage length from RT to 1400°C
is measured using two-clip-on extensometers 180 deg apart. The
electrical outputs of the two extensometers were averaged. The clip-ons
were mounted on special flags clamped to the specimen using specially
machined gage block with a specified initial separation. The clip-ons
were calibrated over the expected range of deformation by means of a
supermicrometer. The load and deformation data were recorded on a
calibrated X-Y chart recorder.

Concentricity and run out of each tensile specimen was checked at SoRI

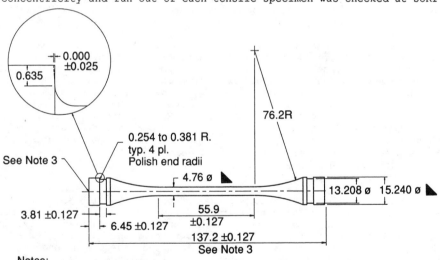

Notes:
1. Diameters true and concentric to ±0.0127 mm
2. Do not under cut radius at tangent points
3. Both ends to be flat and perpendicular to ±0.0127 mm
4. Axial grind gage and radius together with 152.4 mm ø wheel
5. All dimensions are in millimeters
6. Tolerances unless otherwise noted: decimals ±0.0254 mm,
   on diameters ±0.0508 mm, on lengths ±0.0508 mm

TE92-3969

Figure 1. Schematic of button-head tensile specimen.

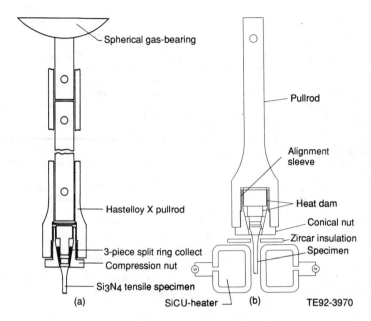

Spherical gas-bearing

Pullrod

Alignment
sleeve

Hastelloy X pullrod

Heat dam

Conical nut

Zircar insulation

3-piece split ring collect

Specimen

Compression nut

Si3N4 tensile specimen

(a)

SiCU-heater

(b)

TE92-3970

Figure 2.   (a) Schematic of the gas bearing tensile facility;
(b) close-up view of pull rods with the button-head specimen and heating
elements.

using the loading rods, collet, and specimen assembly in a specially
designed rig. The total length of the assembly is around 457.2 mm.
While mounted vertically, one end of the assembly is spun while the
runout is measured on the other end using a dial indicator.
At high temperatures, the gage section described earlier for both the
monotonic and creep rupture specimens was resistively heated using
silicon carbide elements developed at SoRI (Ref. 4) and has been
successfully used up to 1650°C in air. The temperature gradient from
the center of the gage section to its two ends, ±27.95 mm, was within
±5°C in the worst case at 1400°C. The loading pull rods were also
heated to about 850°C to minimize the temperature gradient across the
gage length of the specimen.

Creep and creep rupture behavior of the button-head tensile specimens
was measured using dead-weight loading. Standard creep frames were used
with the aforementioned gripping, heating, and strain measurement system
used for fast fracture tensile characterization of the material. The
output of the clip-on extensometers was digitized and time-dependent
strain data were continuously acquired using a commercially available
IBM-PC based data acquisition system. Calibrated X-Y chart recorders
were also used to collect real time analog data as a back-up to the
automatic data acquisition system.

NDE of specimens was conducted using microfocus X-ray and acoustic
microscopy to detect process and machining induced defects. Feinfocus
microfocus 160 kV and 2 mA system was used to optimize the flaw
detection sensitivity. This system has five degrees of freedom, which
include linear translation in x,y,z directions and two angular motions
(0 and Ø) around the Z-axis. In addition. the linear translation of the

imaging plane lends itself to provide the sixth degree of freedom. Kodak-M type film was used for recording the images.

A 100 MHz acoustic microscopy system has been developed and implemented at Allison to characterize flat and button-head cylindrical specimens. This system utilizes Panametrics pulser/receiver, peak detector, transducers, and computer control data acquisition and imaging software. The scanner has x,y,z, and q axis of scanning. All the axes are programmable and computer controlled. The data is acquired and stored on the disk in real time. It can be presented on the screen or hard copy generated.

RESULTS AND DISCUSSION

Flexural Strength

The four-point fast fracture flexural strength of the material was measured from room temperature to 1400°C, as shown in Figure 3. The number of specimens tested at each temperature are also shown in Figure 3. The two parameter Weibull modulus (m) and the characteristic strength (s_0) were estimated as a function of temperature using the maximum likelihood method using suspended item analysis. The average room temperature strength of the material was 812.9 MPa with m = 6.34 and s_0 = 652.3 MPa. The average strength decreased to 734.3 MPa at 1000°C with m = 10.7 and s_0 = 659.4 MPa, and to 603.7 MPa at 1100°C with m = 15.2 and s_0 = 536.5 MPa. The decrease in strength at 1100°C is believed to be due to the softening of the grain boundary intergranular glassy phase. The calculated value of m and s_0 at 1200°C were 9.5 and 530.7 MPa, respectively. At 1300°C, the estimated value of m = 16.7 and s_0 = 529.2 MPa. The average strength further reduced to 490.9 MPa at 1400°C with m = 12.6 and s_0 = 409.1 MPa. Fractographic analysis revealed that the failure at room temperature was predominantly controlled by surface flaws, as shown in Figure 4(a). Above 1000°C, the failure initiated both from surface and volume flaws, Figure 4(b) and 4(c). Furthermore, X-ray energy dispersive analysis (XEDA) indicated that failure controlling inclusions contained iron. Erosion of steel components of the injection molding equipment by hard silicon nitride powder may have contaminated the material with iron.

Uniaxial Tensile Testing

Uniaxial or monotonic tensile strength of the PY6 material was measured from RT to 1400°C at a loading rate of 4137 MPa/min (Figure 5) in air. The fast loading rate minimized any oxidation or time dependent effect on the strength of the material. Dimensional inspection of each specimen was conducted prior to testing. Concentricity of each specimen and loading rod assembly, 4.572 cm long, was 75 micron, which was within the 250 micron allowable over the entire length of the assembly. In other words, the bending stresses in the loading system were very low. The two parameter Weibull parameters (m and s_0) at each temperature were calculated using the maximum likelihood method. The average room temperature tensile strength was measured to be 407.9 MPa with m = 7.4 and s_0 = 306.7 MPa for volume flaws. The Weibull constants at 1000°C were measured to be m = 4.4 and s_0 = 250.4 MPa. The average strength at 1100°C degraded to 340 MPa with m = 9.6 and s_0 = 284.9 MPa. The trough at 1100°C in Figure 5 is hypothesized primarily to be due to softening

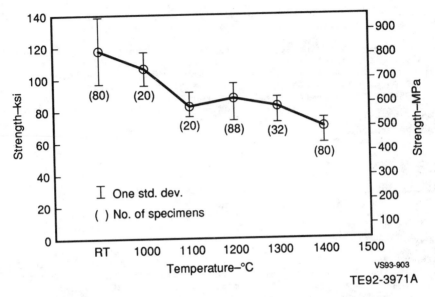

Figure 3.   MOR fast fracture strength from RT to 1400°C.

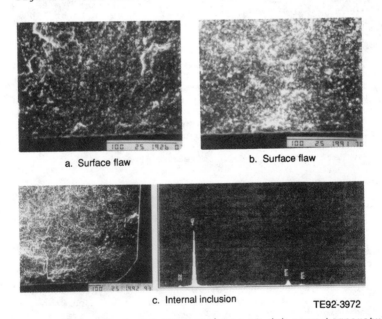

a. Surface flaw

b. Surface flaw

c. Internal inclusion

TE92-3972

Figure 4.   SEM fractography of MOR specimens at (a) room temperature and (b) and (c) at 1400°C.

of the grain boundary phase. In addition, the low Weibull modulus of the material also makes the average very sensitive to the sample population. The value of m and s_0 were estimated to be 16.1 and 367.1

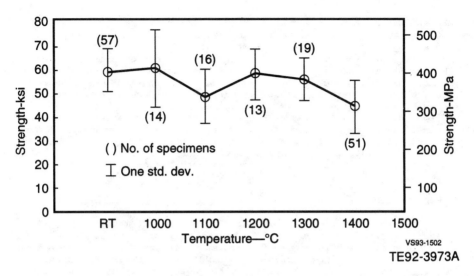

VS93-1502
TE92-3973A

Figure 5.   Uniaxial tensile strength of PY6 material from room
temperature to 1400°C.

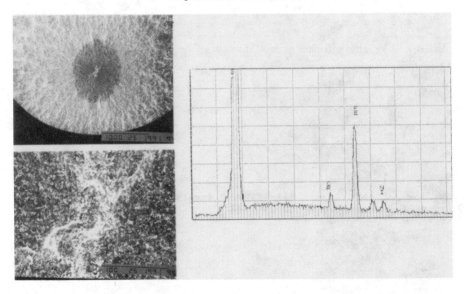

VS91-1607
TE92-428

Figure 6.   SEM fractography of room temperature tested tensile specimen
failed from internal iron inclusion containing nickel and chrome.

MPa at 1200°C, whereas m = 11.8 and σ_O = 323.9 MPa were calculated for
1300°C. At 1400°C, the average strength degraded to 317.9 MPa with m =
5.8 and s_O = 225.8 MPa. Failure at RT initiated from internal
inclusions of iron, which also showed the presence of nickel and chrome
(Figure 6), indicating abrasion of process equipment components by hard
silicon nitride powder. The majority of failures at 1400°C occurred
from volume inclusions (Figure 7). In addition, the stress-strain
behavior was nonlinear at 1400°C (Figure 8), even though the specimen
failed in less than 10 sec. Further fractography revealed that grain
boundary cavities were formed (Figure 8), during the short duration of
testing. This clearly indicates that this material may be unsuitable
for usage at 1400°C where the component undergoes severe
thermomechanical loads during transient conditions in gas turbine
engines.

Creep Rupture

Tensile creep and creep rupture behavior of the PY6 material was
measured between 1200°C to 1400°C using the button-head specimen. The
data are summarized in Figure 9. Minimum creep rate for each specimen
was determined from the strain-time curves. The stress dependence of
the creep behavior was calculated by the Norton well-known power law
equation shown as follows:

$$\dot{\varepsilon} = A\,\sigma^n \qquad\qquad (1)$$

where $\dot{\varepsilon}$ - is the minimum creep rate, s - is the applied stress, n is

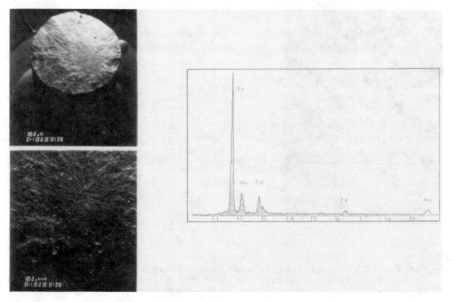

Figure 7. SEM fractography of a monotonic tensile specimen tested at
1400°C which failed from an internal iron inclusion.

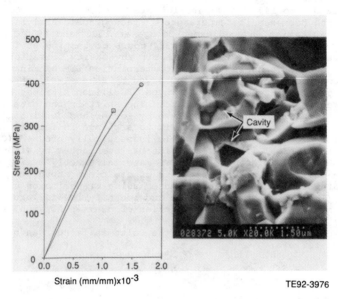

TE92-3976

Figure 8.   Fast fracture uniaxial testing at 1400°C showing non-linear
stress-strain behavior and grain boundary cavitation.

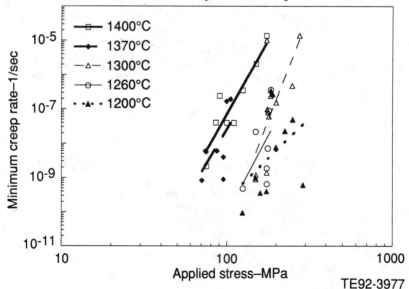

TE92-3977

Figure 9.   Tensile creep rupture behavior of PY6 injection molded and
HIPed Si₃N₄.

the creep exponent, and A is a constant. It is evident from Figure 9
that from 1260°C to 1400°C the material creeps with an expected shift to
the right with decrease in temperature. Fractographic analysis, Figure
10, showed cavitation induced creep damage at 1300° and 1400°C. At
1260°C, cavities were observed after 364 hours of testing at 175 MPa.
Specimens tested at 1200°C did not exhibit any cavity formation or any
other creep damage mechanism under high resolution SEM examination. The
typical slow crack growth features observed by Khandelwal, et.al. (Ref
5) and Chang, et.al (Ref 6) in other type of silicon nitrides have not
been observed in this material to date. The shift in the data fit line,
however, is similar to the one observed by Ferber and Jenkins (Ref. 7).
These authors have attributed the shift to a change in failure mechanism
from cavitation to crack growth in cold isostatically pressed (CIPed)
and HIPed PY6 material. A similar shift has been noted by Wiederhorn,
et.al (Ref 8) in the vintage of the PY6 material exactly similar to the
one used in our work. Further analysis of the creep rupture behavior of
PY6 material is currently being conducted using transmission electron
microscopy (TEM) to better understand and identify the failure modes
under the on-going Ceramic Life Prediction program at Allison.

The Norton equation when combined by the Boltzman factor can be
rewritten as follows:

$$\dot{\varepsilon} = A\,\sigma^n e^{(-\Delta H/RT)} \qquad\qquad (2)$$

where D H is the activation energy and R is a constant. A least squares
fit of the data using equation (2) between 1260°C and 1400°C resulted in
a stress exponent (n) of 7.8 $\pm$1.0 with an apparent activation energy of
957 $\pm$149 kJ/mole. These numbers are in good agreement with the values
of n = 8.5 $\pm$1.1 and 1350 kJ/mole obtained by Wiederhorn (Ref 8) of
National Institute of Standards and Technology (NIST). The NIST data
were acquired using flat dog-bone specimens. Allison button-head
specimens were machined from rods whereas NIST specimens were machined
from rectangular billets both of which were fabricated from the same lot
of starting powder.

Creep rupture data can also be represented by a Monkman-Grant
relationship (Ref 9),

$$t_f = c \cdot \dot{\varepsilon}^{\,m} \qquad\qquad (3)$$

where t_f is the time to failure, $\dot{\varepsilon}$ is the minimum creep rate, and C is
a constant. The data between 1300°C to 1400°C are plotted in Figure 11.
Data for all the temperatures fall around one line with a slope of -1.1,
which is in excellent agreement with Wiederhorn slope of about -1 [Ref
6]. This demonstrates that when specimens are fabricated from the same
lot of the starting powder the two types of specimens result in similar
creep rupture data within scatter.

Nondestructive Evaluation

Microfocus detection of surface holes in MOR bars was optimized by

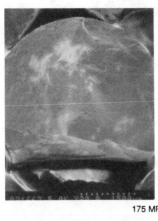 

175 MPa, >473 hrs
(a)

 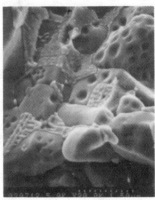

75 MPa, 844 hrs
(b)   TE92-3996

Figure 10. Fracture surfaces from creep specimens demonstrate that the failure at both 1300°C and 1400°C was controlled by cavitation induced damage accumulation.

varying X-ray tube voltage, current, exposure time, and magnification of the image. It was experimentally established that 78 kV and 0.2 mA resulted in optimum detection of surface holes. X-ray detected 50 micron diameter and 50 micron deep holes with a thickness sensitivity of 1.69% in MOR bars. Microscopy revealed that the subsurface quality of the laser-drilled holes was very irregular, they were shallower in depth than planned, were drilled at an angle, and not flat at the bottom at all. The 25 micron diameter holes were not detected because of their shallow depth and lower than anticipated effective volume. Khandelwal [Ref 10] has demonstrated detection of 25 micron surface holes and internally seeded voids using the same experimental system and Kodak-M film. The radiographic detection of the holes in button-head specimen depends on their radial orientation and location with respect to the X-ray beam because their irregular shape, size, and cumulative geometrical effects. Only 100 micron nominal diameter and depth holes were detected with a thickness sensitivity of 2.13%.

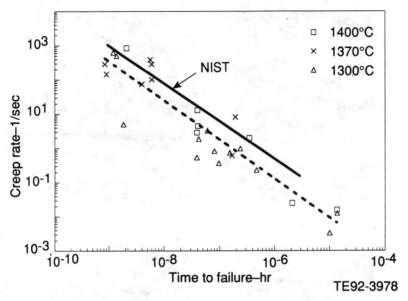

Figure 11.   Creep rupture data plotted using Monkman-Grant formulation.

Acoustic microscopy examination of the MOR bars and tensile specimens with surface holes was conducted.   Holes of 25 micron diameter and depth

were readily detected in MOR bars using a 50 MHz transducer with a focal length (F) of 12.7 mm and a diameter (d) of 6.35 mm with a water path of 12.7 mm.   Smaller holes of 10 micron nominal diameter could not be discerned from the background microstructure.   Again, the holes were imperfectly drilled as described previously.   Button-head tensile specimens with 10 to 50 micron surface holes were also evaluated by acoustic microscopy using a 50 MHz transducer with focal length of 5.08 mm and crystal diameter of 6.35 mm.   All the 10 micron diameter holes were readily detected (Figure 12).

CONCLUSIONS

The PY6 injection-molded and HIP processed silicon nitride has MOR fast fracture strength of 812.9 MPa at RT which degraded about 43% at 1400°C. The failure of MOR specimens was predominantly controlled by surface defects with failures in volume at higher temperatures.   The RT Weibull parameter (m) for MOR specimens was estimated to be 6.34, which increased in the range of 9.5 to 16.7 at high temperatures.   The characteristic strength (s_0) at RT was estimated to be 652.3 MPa, which changed between 659.4 MPa to 409.8 MPa at temperatures between 1000°C to 1400°C.   The RT tensile strength was 407.9 MPa, which decreased by 25% at 1400°C.   Tensile specimens generally failed from volume inclusions with some failures from surface flaws.   The RT Weibull parameter (m) for fast fracture tensile specimens was calculated to be 7.4, which changed to 4.4 to 16.1 at high temperatures.   The characteristic strength (s_0) at RT was estimated to be 306.7 MPa, which changed between 250.4 MPa to 367.1 MPa at elevated temperatures.   The material failed from cavitation

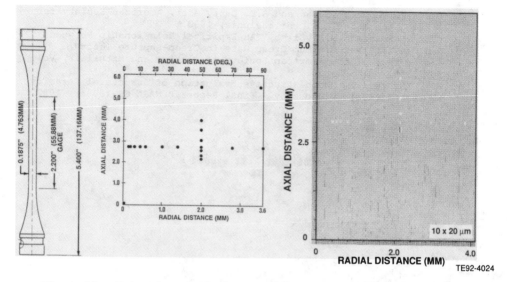

Figure 12.   Detection of 10 micron surface holes by acoustic surface wave microscopy.

during creep between 1300°C to 1400°C. At lower temperatures, however, the failure apparently occurs both from slow crack growth and creep. Microfocus X-ray detected 50-100 micron flaws whereas acoustic microscopy detected 10 micron surface voids.

REFERENCES

1.   "Advanced Turbine Technology Applications Project", NASA CR-185133, 1988 Annual Report.
2.   N.N. Nemeth, J.M. Manderscheid, J.P. Gyekenyesi, "Ceramics Analysis and Reliability Evaluation of Structures (CARES) - Users and Programmers Manual", NASA Technical Paper 2916, 1990.
3.   M.K Ferber and M.G. Jenkins, "Rotor Data Base Generation in Ceramic Technology for Advanced Heat Engine Project", Semiannual Progress Report, October 1990-March 1991, ORNL/TM-11859.
4.   Pears, C. D., et al.,"Test Methods for High Temperature Materials Characterization," AFML TR-79-4002, Air Force Materials Laboratory, Wright-Patterson Air Force Base, Ohio, February 1979.
5.   P.K. Khandelwal, J. Chang, and P.W. Heitman, "Slow Crack Growth in Silicon Nitride", Fracture Mechanics of Ceramics, Vol. 8, R.C. Brandt, A.G. Evans, D.P.H. Hasselman, and F.F. Lange, Plenum Publishing Corporation, 1986.
6.   J. Chang, P.K. Khandelwal, and P.W. Heitman, "Dynamic and Static Fatigue Behavior of Sintered Silicon Nitrides", Ceramic Engineering and Science Proceedings, Vol. 8, No. 7-8, July-August, 1987.
7.   M.K. Ferber and M.G. Jenkins, "Evaluation of the Elevated Temperature Mechanical Reliability of a HIP-ed Silicon Nitride," *Journal of American Ceramic Society*, *75*[9], 2453-62, 1992.
8.   S.M. Wiederhorn, G.D. Quinn, and R.F. Krause, "Fracture Mechanisms Maps: Their Applicability to Silicon Nitride", to be published in Life Prediction Methodologies and Data for Ceramic Materials, ASTM

STP 1201, C.R. Brinkman and S.F. Duffy eds., American Society for Testing and Materials, Philadelphia, 1993.

9.   F.C. Monkman and N.J. Grant, "An Empirical Relationship Between Rupture Life and Minimum Creep Rate in Creep-Rupture Tests", Proceedings of the American Society of Testing and Materials, Vol. 56, 1956, pp. 593-620.

10. P.K. Khandelwal, "Nondestructive Evaluation of Structural Ceramics by Photoacoustic Microscopy - Final Report," NASA CR-180858, 1987.

Life Prediction Methodologies

John Smart[1] and Siu L. Fok[1]

THE NUMERICAL EVALUATION OF FAILURE THEORIES FOR BRITTLE MATERIALS

REFERENCE: Smart, J. and Fok, S. L., "The Numerical Evaluation of Failure Theories for Brittle Materials," Life Prediction Methodologies and Data for Ceramic Materials, ASTM STP 1201, C. R. Brinkman and S. F. Duffy, Eds., American Society for Testing and Materials, Philadelphia, 1994.

ABSTRACT: Many laws have been proposed for the failure of brittle materials. These divide into two main categories - volume and surface integrals. To evaluate these for a complex body requires an initial finite element analysis followed by postprocessing to evaluate the integrals. How this can be achieved efficiently and effectively so that the errors are similar to those associated with the initial finite element analysis is discussed.

KEYWORDS: failure theories, brittle materials, finite elements, Gaussian quadrature, postprocessor, Weibull probability

INTRODUCTION

In order to make life predictions for ceramic materials three things are necessary. These are (i) basic material data, (ii) a material failure law and (iii) a method of accurately incorporating the material data into the failure law and then evaluating for a component which may not have a simple shape and is probably subject to a complex loading system. In this paper the third requirement will be examined.

Most failure laws for brittle materials are based on the Weibull probability equations with the laws differing because of the relative contribution of the stresses and whether the stresses throughout the body or just those on the surface are considered. To determine the stresses in the component a finite element analysis is performed and then the results of this analysis are used in a postprocessor which determines the failure probability of the component. There are many postprocessors which can perform these calculations such as that of the authors, BRITPOST[1,2], and CARES[3]. To ensure that these postprocessors are giving the correct answers, benchmarks have been formulated by WELFEP (Weakest-Link Failure Probability Prediction by Finite Element Postprocessors)[4].

In this paper, various ways in which life predictions can be made will be reviewed using different failure laws that have been proposed in the literature.

These postprocessors numerically evaluate the failure law for the component and incorporate integrals either over the surface of the body

[1]Lecturer and research student respectively, Department of Engineering, University of Manchester, Manchester, M13 9PL, UK

or over the volume of the body using data provided by the finite element analysis. Thus, there will be errors in the final failure predictions because of the approximate nature of the finite element analysis and because of the methods used in the postprocessing even if the material data and failure law are fully representative of the behaviour of the material. It is important that any analysis minimises these errors. Whilst this can be done by using many elements in the finite element analysis, and also by using a high integration order in the postprocessor, this is not very efficient.

THEORY

Brief Review of Failure Theories

Various theories for the failure of ceramics have been proposed in the literature and a representative sample has been chosen for analysis. The theories divide into those that assume that cracks on the surface cause failure and those that assume that the laws must be evaluated over the volume. Both are examined in this paper.

Within each of these groups, most of the failure criteria are based on the assumption that the natural flaws in ceramics act as cracks and then using different fracture criteria for deciding whether or not a crack fails. Thiemeier et al[5] have considered 6 such criteria. The difference between these criteria depends on the relative contribution to the stress intensity factor of the normal and shear stresses acting on the crack. We have chosen for analysis the normal stress averaging (NSA) model[6] in which there is no contribution from the shear stresses and the maximum non-coplanar strain energy release rate (MSE) model [7] which has the greatest shear stress contribution. We have also analysed the principle of independent actions (PIA)[8]. This is empirically based and assumes that each principal stress contributes independently to the failure.

For volume flaws the probability of failure can be written as

$$P_f = 1 - \exp\left\{ - \left(\frac{\sigma_{NOM}}{\sigma_o} \right)^m \Sigma_v \right\} \tag{1}$$

where Σ_v is the stress integral and given by

$$\Sigma_v = \int_v \frac{1}{4\pi} \int_0^{2\pi} \int_0^{\pi} \left(\frac{\sigma_e}{\sigma_{NOM}} \right)^m \cos\phi \, d\phi \, d\psi \, dv \tag{2}$$

with
    i)   for the normal stress averaging model (NSA)

$$\sigma_e = \sigma_n \tag{3}$$

    ii)   for the maximum non-coplanar strain energy (MSE) release rate criteria

$$\sigma_e^4 = \sigma_n^4 + 6\sigma_n^2 \tau^2 + \tau^4 \tag{4}$$

where σ_n and τ are the stress components acting on the crack as shown in Fig. 1.

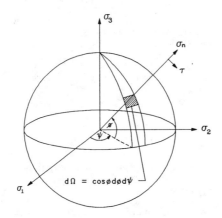

$$\sigma_n = \sum_{i=1}^{3} \sigma_i \, l_i^2 \qquad\qquad l_1 = \cos \phi \, \cos \psi$$

$$l_2 = \cos \phi \, \sin \psi$$

$$\tau^2 = \sum_{i=1}^{3} \left(\sigma_i \, l_i \right)^2 - \sigma_n^2 \qquad l_3 = \sin \phi$$

FIG. 1--Normal and shear stress components acting on a crack

For the principle of independent actions (PIA) the stress integral is given by

$$\Sigma_v = \int_v \left[\left(\frac{\sigma_1}{\sigma_{NOM}} \right)^m + \left(\frac{\sigma_2}{\sigma_{NOM}} \right)^m + \left(\frac{\sigma_3}{\sigma_{NOM}} \right)^m \right] dv \qquad (5)$$

where σ_1, σ_2 and σ_3 are the principal stresses. In all cases σ_{NOM} is a normalising stress which has no effect on the final failure probability. The maximum stress in the component is a sensible value to use for σ_{NOM}. The material properties of the brittle material define σ_0 which is a measure of the strength of the material and m which is a measure of the spread in the failure loads for nominally identical tests.

Similar equations can be written for surface failure theories.

Numerical Evaluation

An examination of equations (1) to (5) shows that the stresses have to be evaluated over the volume of the body or, if surface integrals are used, over the surface of the body. Also, the integral of equation (2) has to be integrated over a unit sphere. The question arises as to how this can be done both efficiently and accurately.

Two methods of numerical integration, usually termed quadrature, are examined: Gaussian quadrature and sampling at regular intervals using the rectangular rule. This is illustrated in Fig. 2 where the 2 x

2 sampling points on the surface of an element are shown for both rules. For Gaussian quadrature, the sampling points and weighting factors can be found in the literature[9].

+ Rectangular Sampling Point

× Gaussian Sampling Point

FIG. 2--Gaussian and rectangular quadrature sampling points

Determination of Stresses

Volume integrals--In the volume integrals, the stresses have to be evaluated at the sampling points within the elements. In a finite element analysis the stresses are usually calculated at the Gaussian stress sampling points[10]. These are dependent on the type of element used. For example, for an 8-noded 2-D element there are 4 Gaussian stress sampling points. From these values, finite element programs usually extrapolate the stresses to the nodes and, typically, the stresses at nodes are averaged between elements.

Whilst it is possible to use the stresses calculated by the finite element program, this is restrictive as the number of points at which a finite element program evaluates the stresses is limited. Instead, in BRITPOST, the stresses are evaluated at the sampling points within the element using the nodal displacements calculated from the finite element analysis and the equation [11],

$$\sigma = \mathbf{D} \, \mathbf{B} \, \mathbf{a} \qquad (6)$$

where σ is the stress vector
           $\mathbf{D}$ is the elasticity matrix
           $\mathbf{B}$ is the strain shape function matrix
and $\mathbf{a}$ is the nodal displacement vector

Thus, it is possible to use any order of quadrature and either Gaussian or rectangular quadrature to evaluate the volume integral. Once the stresses at a point have been evaluated it is straightforward to determine the stress components required by the various theories.

Surface integrals--For the surface integrals it is necessary to determine the stresses in the plane of the surface. Whilst other programs such as CARES[3] use special membrane elements, BRITPOST evaluates the stresses directly.

However, the question arises as to how to do this as it is well known in finite element analysis that the worst stress predictions occur on the surfaces of a body. For most surfaces, there are no pressure or traction forces acting and yet an examination of the finite element output will show that the normal and shear stresses are not zero. Also the stresses will usually be given in the cartesian directions which will not, in general, be along the surface of the body.

Now, on the surface, the actual stress system can be described for the purpose of evaluating the surface stress integrals, by the principal stresses. Thus one possible option is to find the 3 principal stresses (for example by an eigenvalue routine or by solving the stress equation). By assuming that, if no pressure or surface tractions are acting, the principal stress nearest to zero is the out-of-plane

principal stress, then the other 2 principal stresses are the in-plane
principal stresses. These can then be used to evaluate the surface
integrals.

    Another method is to evaluate directly the in-plane stresses on
the surface. In finite element analyses, within a brick element there
are 3 local directions. Whilst these are orthogonal in the undistorted
element, in most elements in a finite element analysis the local
directions will not be orthogonal because of the distortion in the
element. However, 2 of the local directions will lie on the surface,
Fig. 3. Thus, it is possible by forming the cross product of the local
direction vectors on the surfacᴇ to find the direction vector of the
normal to the surface of the el ᴉent. Once this is known, 2 orthogonal
in-plane vectors can be found.

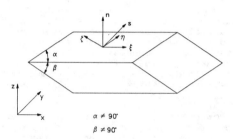

FIG. 3--A distorted brick element

    For example, consider the distorted brick shown in Fig. 3. Whilst
the 2 vectors ξ and η lie on the surface of the element, they are not
orthogonal. Also ζ is not normal to the surface. However, by forming
the cross product of the vectors ξ and η, the normal to the surface, n,
can be found. Once the vector n is known, by forming the cross product
with one of the in-plane vectors, for example ξ, a third orthogonal
vector s can be found. Thus, the orthogonal vector set (ξ, s, n) can be
calculated.

    Once the orthogonal vector set is found, the direction cosines of
the vectors relative to the cartesian coordinates can be determined and
hence the stresses in the new coordinate set (ξ, s, n) calculated. If
the transformation matrix between the global (x, y, z) coordinates and
the (ξ, s, n) coordinates is given as

$$\begin{Bmatrix} \xi \\ s \\ n \end{Bmatrix} = [T] \begin{Bmatrix} x \\ y \\ z \end{Bmatrix} \tag{7}$$

then

$$[\sigma_{\xi, s, n}] = [T] [\sigma_{x, y, z}] [T]^T \tag{8}$$

    This is readily programmed so that the in-plane surface stresses
$\sigma_{\xi\xi}$, σ_{ss} and $\sigma_{\xi s}$ can be calculated for a brick element. For a 2-D membrane
element the stresses can be calculated in a similar manner although the
calculations are far simpler.

NUMERICAL EXAMPLES

In the previous section, three failure theories, two methods of quadrature and two methods of determining the surface stresses were discussed. The results for the various options will now be compared on two models. The models chosen are from the WELFEP benchmarks. These are the notched beam and the notched bar, both subjected to 4-point bending. The results for the notched beam have already been reported[4]. These 2 examples have been chosen as they represent a 2-D and a 3-D analysis and both have a varying stress field caused by a stress concentration at the notch.

In the WELFEP results a mean nominal stress is used for comparing the analyses from different postprocessors. However, we prefer to use the 50% failure load factor. This is the factor by which the load given for the WELFEP tests must be multiplied so that there is a 50% probability of failure. This is chosen as it has a physical significance - usually loads are applied to specimens. Also, a similar measure can be used for the finite element analysis. In this case, as the number of elements in the models increase, the predicted maximum stress in the body changes. Thus, if a body fails at a single value of maximum stress then a load factor to give this maximum stress can be calculated. This situation, where a body fails at a given value of the maximum stress, corresponds to m=∞.

The finite element analyses have been performed with PAFEC although any finite element program could be used provided the elemental geometry and displacements can be obtained to use in the postprocessor.

<u>Notched Beam</u>

To allow a convergence study of the numerical procedures outlined above the WELFEP mesh has not been used although the same shape, loading and material properties have been used. The outline shape of the beam, the material properties and loading are shown in Fig. 4a and the base mesh shown in Fig. 4b. To check on convergence each element has then been divided into 2x2, 3x3, 4x4 and 6x6 elements as illustrated in Fig. 4c. The WELFEP mesh consists of 255 8-noded elements and 846 nodes whereas these meshes consist of 19, 76, 171, 304 and 684 8-noded elements and 78, 269, 574, 993 and 2771 nodes. (These meshes will be denoted as mesh 1, 1/2, 1/3, 1/4 and 1/6.)

It can be seen that there are no special surface elements and that the meshing around the stress concentration has not been highly refined in mesh 1.

Before considering the results from the various theories the values for σ_x and σ_y at the stress concentration point (point A in Fig 4b) as calculated by PAFEC are given in Table 1.

TABLE 1 -- <u>Stresses (in MPa) in notched beam at point A in Fig. 4b</u>

Mesh	σ_x	σ_y
1	692	71
1/2	723	48
1/3	716	28
1/4	713	18
1/6	711	9

As mentioned earlier, σ_y, which should be zero as it is normal to the free surface, is not zero. Thus, even before conducting any postprocessing, the basic stress analysis has errors associated with it. For mesh 1 σ_y = 71 MPa which is approximately 10% of the maximum stress. Even for the 1/6 mesh with 684 elements σ_y is still 9 MPa. Thus, it must be appreciated that although there may be errors associated with postprocessing, there will also be errors associated with the initial finite element analysis.

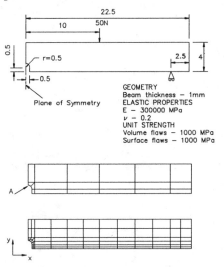

FIG. 4--a)   The notched beam geometry loading and material properties (dimensions in mm)   b)   the base mesh   c)   the 1/2 mesh

Postprocessing volume integrals--Three different failure laws have been analysed, the principle of independent actions (PIA), normal stress averaging (NSA) and the maximum non-coplanar strain energy release rate (MSE). It is found that all give similar results and so the principle of independent actions has been chosen to illustrate the results.

Figs. 5a and 5b show the convergence of the predictions for the 50% failure load factor as the finite element mesh is refined and as the postprocessing quadrature order is increased for m = 5 and m = 20 for PIA. Only the 1, 1/2 and 1/3 meshes have been plotted as beyond this level of mesh refinement the changes are small and the graph becomes congested. It can be seen that increasing the number of elements and increasing the number of sampling points all lead to increasing accuracy and that Gaussian quadrature is clearly superior to using rectangular quadrature. Apart from the base mesh for m = 20 there is little to be gained from using an quadrature order of more than 4.

The convergence of the predictions is shown in Fig. 6. In plotting Fig. 6 it has been assumed that the results for the 1/6 mesh with a quadrature order of 4 have converged. An examination of Fig. 6 shows that this is a reasonable assumption. It can be seen that, except for the case of m = 20 and mesh 1, the postprocessing predictions using Gaussian quadrature are better than the finite element results. This is reassuring as it indicates that if the errors associated with the finite element analysis can be tolerated, the postprocessing errors can be tolerated.

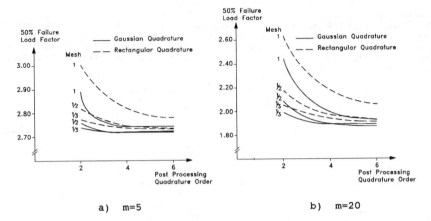

a)   m=5   b)   m=20

FIG. 5--Effect of quadrature order on 50% failure load factor for
PIA volume integral theory

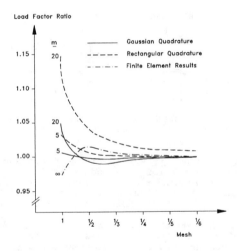

FIG. 6--Load factor ratio convergence as mesh is refined for PIA volume
integral theory

As in all cases using rectangular quadrature leads to greater
error, this will not be examined further for this model and volume
integrals. To compare the other two theories investigated, NSA and MSE,
the results for m = 20 are shown in Fig. 7. It can be seen by comparing
these results with those in Fig. 5b that the trends are the same. Apart
from the coarsest mesh, the results have effectively converged for an
integration order of 4. Again, assuming that the results for the 1/6
mesh and an integration order of 4 have converged, the convergence of
the various theories is shown in Fig. 8. It can be seen that the
convergence pattern is the same for all theories.

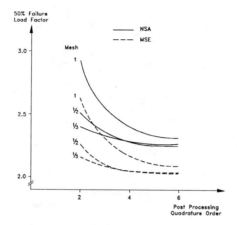

FIG. 7--Effect of quadrature order on 50% failure load factor for NSA
and MSE volume integral theories for m=20.

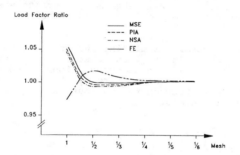

FIG. 8--Load factor ratio convergence as mesh is refined for volume
integral theories and m=20

Thus, it can be seen on the evidence of this model that to
accurately evaluate the volume integrals in the various failure theories
fourth order Gaussian quadrature should be used. Lower orders will
probably increase the errors.

Postprocessing surface integrals--When evaluating the surface
integrals for this model, because a thickness of 1mm has been specified,
nearly all the value of the surface integrals will come from the edge
rather than the side of the notched beam. When the results are examined
it is found that, except for mesh 1, the failure load factor predictions
vary very little for a given theory whether Gaussian or rectangular
quadrature is used and for quadrature orders from 1 to 4.

To examine this unexpected result, the postprocessor was modified
to print the stress in the direction parallel to the edge of the beam
near the notch. This is shown in Fig. 9 for the meshes 1, 1/2 and 1/3.
It can be seen that, although there is a stress concentration at the
corner of the notch the stress varies little near the notch.
Consequently, the need for more elements and more integration points to
capture the peak stress at the notch is reduced.

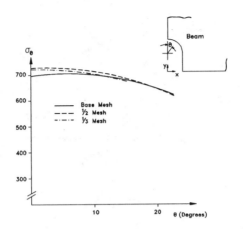

FIG. 9--Stress (in MPa) around notch of beam

To illustrate the results obtained, the 50% failure load factor is plotted in Fig. 10 for PIA and m = 20. For mesh 1, whilst the predictions vary with quadrature order, for a quadrature order of 2 or more, the variation is only 2%. When more elements are used, there is hardly any variation in the predicted load factors and a quadrature order of 1 will suffice.

Earlier, two ways of calculating the stresses parallel to the surface were discussed. It was found that for this example there was virtually no difference between the predictions given by the two methods. Because of the stress pattern at the notch this is not a good example to discriminate between the various techniques of determining the surface stress integrals.

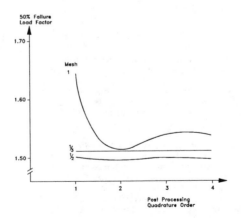

FIG. 10--Effect of quadrature order on 50% failure load factor for PIA surface integral theory for m=20 and Gaussian quadrature

Notched Bar

As for the 2-D notched beam, the mesh analysed is not that

specified by WELFEP. Again, a different mesh has been used to allow a convergence study to be performed. The geometry, loading and material data are given in Fig. 11a and the base mesh in Fig. 11b. Because of symmetry only a quarter of the bar is analysed. Each element in the base mesh has then been subdivided into 2 x 2 x 2, 3 x 3 x 3 and 4 x 4 x 4 elements. This gives meshes with 30, 240, 810 and 1920 20-noded elements and 217, 1309 , 3997 and 9004 nodes. These compare with the WELFEP mesh which has 264 elements and 1431 nodes. Again, these meshes are denoted as mesh 1, 1/2 ,1/3 and 1/4. The meshes consist entirely of standard 20-noded hexahedron elements with no special surface elements. To avoid any problems associated with the point load, the integrals are evaluated in the range $0 \leq z \leq 23.4$

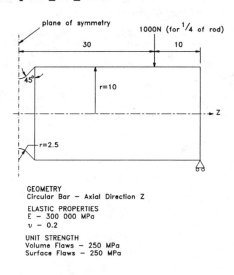

plane of symmetry   1000N (for $^{1}/_{4}$ of rod)

30   10

45°

r=10

r=2.5

GEOMETRY
Circular Bar — Axial Direction Z

ELASTIC PROPERTIES
E — 300 000 MPa
v — 0.2

UNIT STRENGTH
Volume Flaws — 250 MPa
Surface Flaws — 250 MPa

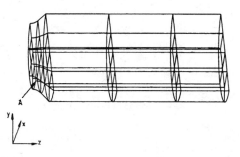

FIG. 11--a) The notched bar geometry, loading and material properties (dimensions in mm) b) The notched bar base mesh

At the stress concentration point, A in Fig. 11b, the values for σ_x, σ_y and σ_z as calculated by PAFEC are given in Table 2.

TABLE 2 -- <u>Stresses (in MPa) in notched bar at point A in Fig. 11b</u>

Mesh	σ_x	σ_y	σ_z
1	14.1	2.6	86.0
1/2	18.8	4.6	101
1/3	19.5	4.2	103
1/4	19.5	3.3	104

At point A σ_y, the stress normal to the surface, should be zero. As for the 2-D case, this is not so indicating the errors associated with the basic finite element analysis.

Postprocessing volume integrals--Again, as in the 2-D case, the three rules analysed, (PIA, NSA and MSE) all give similar results and so the results from the PIA analysis have been chosen to illustrate the trends. Figs. 12a and 12b show the convergence of the predictions for the 50% failure load factor for m = 5 and m = 20 as the finite element base mesh is refined and as the postprocessing quadrature order varies. It can be seen that as the mesh is refined and as the postprocessing quadrature order increases, the results are converging with Gaussian quadrature again being superior to rectangular quadrature

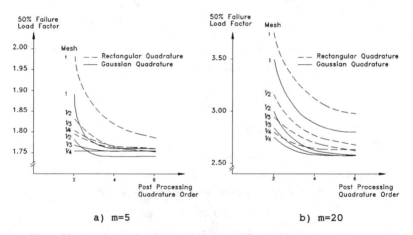

a) m=5   b) m=20

FIG. 12--Effect of quadrature order on 50% failure load factor for
PIA volume integral theory

For Gaussian quadrature and m = 5, the results have converged for fourth order quadrature. (As this is a volume integral this means that the numerical sampling has occurred at 64 internal points.) By refining the base mesh so that there are 64 times as many elements (1/4 mesh), the change in the 50% load factor is less than 2%. This compares with the change in the value of σ_z at point A Fig. 11 of 20% (see Table 2). When m = 20 it can be seen that even with a Gaussian quadrature order of 4, the results have not converged for the base mesh but have almost converged for the 3 refined meshes. Assuming that the converged 50% failure load factor is 2.58, the error from using 4 point Gaussian integration with mesh 1 is 11%. This should be compared to the change in the stress σ_z of 20%. For the other 2 failure rules evaluated, NSA and MSE, the results are given in Fig. 13 for Gaussian quadrature. It can be seen that the behaviour is similar to that for PIA.

If the 1/4 mesh is assumed to have converged with fourth order quadrature the load factor variation can be plotted and these are shown in Fig. 14. Again as in the first example, the errors from the post processed results are less than the error in the finite element analysis.

Postprocessing surface integrals--Before discussing the surface integral results, the stresses around the notch are examined. Again, the postprocessor was modified to calculate the stresses. The stresses were calculated on the central plane around the notch and the maximum principal in-plane stress is plotted in Fig. 15. It can be seen that as the mesh is refined the stresses are converging so that the maximum stress is on the centre line but the base mesh is not a very good representation. It can also be seen that near the centre of the notch the maximum in-plane principal stress is not changing rapidly. By comparing Figs. 9 and 15 it can be seen that the stress patterns for both test cases are very similar.

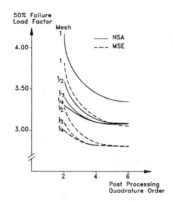

FIG. 13--Effect of quadrature order on 50% failure load factor for NSA and MSE for volume integral theories and m=20

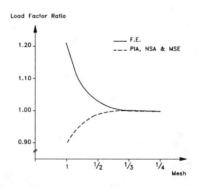

FIG. 14--Load factor ratio convergence as mesh is refined

Again, the results for each failure rule are very similar and so the results for PIA and m = 5 and m = 20 for Gaussian quadrature are plotted in Fig. 16. Results are given when the surface integral is evaluated by calculating the in-plane stresses, labelled σ_n in Fig. 16, and also when the surface stresses are assumed to be the 2 principal

stresses furthest away from zero, labelled σ_1. As the mesh is refined,
the 50% failure load factors calculated using the 2 methods of
determining the in-plane stress are converging although there are
differences (about 3% when m = 5) between the 2 methods for the base
mesh. (The 1/4 mesh results have not been plotted but are almost
identical to the 1/3 mesh.)

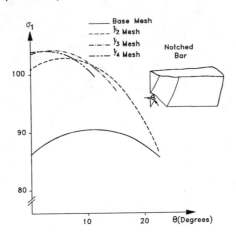

FIG. 15--Maximum principal in-plane stress (in MPa) on
central plane around notch

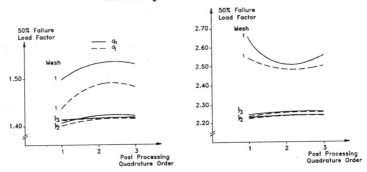

FIG. 16--Effect of quadrature order on 50% failure load factor for
PIA surface integral theory

As with the 2-D beam, a postprocessing integration order of 3 is
more than adequate and the results are very similar whether Gaussian or
rectangular quadrature is used because of the stress profile illustrated
in Fig. 15.

DISCUSSION

The purpose of this paper is to examine critically how various
failure laws for brittle materials can be evaluated efficiently and

accurately. Three different laws have been examined and it has been found that what is true for one rule is the same for the others.

When evaluating volume integrals, it is important that the postprocessing samples near to the point of maximum stress and as this will be on the surface, a low integration order will not capture this stress adequately. However, it has been found that even for coarse finite element meshes a Gaussian quadrature order of 4 suffices in both 2-D and 3-D. This is true for a small Weibull modulus (m = 5) or a higher value (m = 20). As a general rule, the higher the Weibull modulus, the greater the need to capture the maximum stress and hence the need for a higher quadrature order. It has also been found that Gaussian quadrature is more efficient than rectangular quadrature. Perhaps most importantly, if 4 point Gaussian quadrature is used, the accuracy of the failure predictions is generally better than the accuracy of the original finite element analysis. If an accurate failure prediction is to be made it is important that a sufficiently refined finite element mesh is used.

When evaluating surface integrals, the stresses are sampled on the surface and so the effect of the change in the quadrature order will be dependent on the stress variation on the surface. For the 2 cases considered, the surface stress varies slowly around the point of the stress concentration and so a low quadrature order suffices. Because of the slow variation in the stress, it has been found that both Gaussian and rectangular quadrature give very similar results.

To evaluate the surface stresses two methods have been considered for determining the in-plane stresses. Firstly, the direction cosines of the normal to the surface are determined and these are used to find a transformation matrix to convert the stresses from the global coordinate system to the normal and in-plane stresses. Secondly, the three principal stresses were determined and the one nearest to zero was assumed to be the normal stress with the other 2 principal stresses lying on the surface. For the 2-D example both methods gave almost the same result but for the 3-D bar there were differences of about 3% between the methods for the coarse mesh and when m = 5. When the finer meshes were analysed the results from the two methods were very similar. At present there is insufficient evidence to state categorically that one method is superior. This is an area which will be examined further. Certainly, it is far easier to write the computer code to determine the principal stresses particularly if they are found with an eigenvalue routine which already exists.

For evaluating surface stress integrals, on the evidence obtained from these and other examples, both Gaussian and rectangular quadrature give similar results. Generally, an order of 2 is sufficient but until further evidence is accumulated the authors would recommend a value of 3. As for volume integrals, the errors in the postprocessing of the finite element results with a surface integral is similar in magnitude to the errors of the finite element analysis provided the guidelines established above are followed. Whilst the results in this paper are limited to 2 geometries, the results obtained are similar to those the authors have found for other geometries.

On a first examination it is surprising that such low orders of Gaussian quadrature suffice when calculating the 50% failure load factor as the integral $\int \sigma^m \, dv$ has to be evaluated. The stress variation over an 8-noded 2-D element is often called quasi-linear. This means that in the undistorted element the strain variation in a direction is linear but when the stresses are evaluated, the stress in one direction is a function of strain in two directions e.g.

$$\sigma_x = \frac{E}{(1-v^2)} \, (e_x + v e_y) \qquad (9)$$

When the element is distorted, the situation becomes more complex. Thus σ^m can be a very high order polynomial. Using Gaussian quadrature, an integration order of n exactly integrates a polynomial of the order $(2n - 1)$ and it might be expected that to fully integrate Σ, a Gaussian quadrature of order, say, 20 would be required. However, it can be seen that this is not necessary.

This can be explained. For a 50% probability of failure then Eq 1 gives

$$0.5 = \exp\left\{ -\left(\frac{\sigma_{NOM}}{\sigma_o}\right)^m \Sigma_v^1 \right\} \qquad (10)$$

The only unknown is Σ_v^1. For simplicity, consider that Σ_v^1 is calculated for PIA and that only σ_1 is non-zero. Eq 5 now gives

$$\Sigma_v = \int_v \left(\frac{\sigma_1}{\sigma_{NOM}}\right)^m dv \qquad (11)$$

and if the loads are multiplied by a load factor L for a 50% probability of failure then

$$\Sigma_v^1 = \int_v \left(\frac{L\sigma_1}{\sigma_{NOM}}\right)^m dv = L^m \int_v \left(\frac{\sigma_1}{\sigma_{NOM}}\right)^m dv = L^m \Sigma_v \qquad (12)$$

However, as Σ_v^1 is given exactly by Eq 10 then any error that arises in evaluating Σ_v will affect L. For example, if Σ_v is only 90% of its true value then the error in L if m = 20 is $(1/0.90)^{1/20} = 1.005$, an error of 0.5%. So any error in evaluating Σ_v causes a much smaller error in load factor.

Whilst it can be argued that Σ needs to be evaluated accurately, for engineering purposes loads are usually required. This is reassuring as the errors involved can be seen to be substantially lower. Σ is only an intermediate stage in the calculation with no physical significance.

If more accurate results are required the first prerequisite is an accurate finite element analysis. Whilst this can be obtained by refining the mesh, the cost of this can be prohibitive. Consequently, as the majority of the contribution is usually associated with one element, the results in this single element may be refined using techniques such as those developed by Smart[12]. This will provide a more economical method.

A further restriction must be noted. For sharp corners, the stress will tend to infinity and as the mesh is refined the predictions will give a failure load tending to zero[13]. In this case, then a fracture mechanics approach must be used.

CONCLUSIONS

When evaluating material failure laws for brittle materials, provided that:

a)   for volume integrals the sampling is performed using 4-point Gaussian quadrature

b)   for surface integrals 3-point quadrature is used

then the error in calculating the load for a 50% probability of failure
is of a similar magnitude to that associated with the initial finite
element analysis. To obtain a more accurate prediction of failure loads
the finite element mesh must be refined.

REFERENCES

[1]   Smart, J., "The Determination of Failure Probability using
       Weibull Probability Statistics and the Finite Element Method," Res
       Mechanica, Vol. 31, 1990, pp 205-219.

[2]   Smart, J., and Fok S.L., "Calculation of Failure Probability for
       Brittle Materials," In: Mechanics of Creep Brittle Materials 2,
       Ed Cocks A.C.F. and Ponter A.R.S, Elsevier Applied Science, 1991,
       pp 268-281.

[3]   Powers L.M., Starlinger A. and Gyekenyesi J.P., "Ceramic
       Component Reliability with the restructured NASA/CARES Computer
       Program," Proceedings of the International Gas Turbine and
       Aeroengine Congress and Exposition, Cologne, 1992, ASME paper 92-
       GT-383.

[4]   Dortmans L., Thiemeier Th., Brückner-Foit A. and Smart J.,
       "WELFEP: A Round Robin for Weakest Link Finite Element
       Postprocessors," Journal of the European Ceramic Society, Vol 9,
       1993, pp 17-21.

[5]   Thiemeier Th., Brückner-Foit A. and Kohler H., "Influence of the
       Fracture Criterion on the Failure Prediction of Ceramics Loaded in
       Biaxial Flexure," Journal of the American Ceramic Society, Vol.11,
       (1) 1991, pp 48-52.

[6]   Weibull, W., "A Statistical Theory of the Strength of Material,"
       Ingeniurs Vetenskaps Akadamien Handlinger, Vol. 151, 1939, pp 1-
       45.

[7]   Hellen T.K. and Blackburn W.S., "The Calculation of Stress
       Intensity Factors for Combined Tensile and Shear Loading,"
       International Journal of Fracture, Vol. 11, 1975, pp 605-617.

[8]   Stanley P., Fessler H. and Sivill A.D., "An Engineer's Approach
       to the Prediction of Failure Probability of Brittle Materials,"
       Proceedings of the British Ceramic Society., Vol. 27, 1973, pp
       453-487.

[9]   Stroud A.H. and Secrest D., Gaussian Quadrature Formulas,
       Prentice Hall, 1966.

[10]   Barlow J., "Optimal Stress Locations in Finite Element Models,"
       International Journal of Numerical Methods in Engineering,
       Vol. 10, 1976, pp 243-51.

[11]   Zienkiewicz O.C. and Taylor R.L., The Finite Element Method,
       McGraw Hill, 1989.

[12]   Smart J., "On the Determination of Boundary Stresses in Finite
       Elements," Journal of Strain Analysis, Vol. 22, 1987, pp 87-96.

[13]   Smart J., "The Failure of Graphite Channel Sections,"
       Proceedings of the International Conference on Computer Aided

Assessment and Control of Localized Damage, Vol. 3 Computational
Mechanics Publications, 1990, pp 185-194.

Theo Fett[1] , Dietrich Munz[2]

LIFETIME PREDICTION FOR CERAMIC MATERIALS UNDER CONSTANT AND CYCLIC LOAD

REFERENCE: Fett, T. and Munz, D., "Lifetime Prediction for Ceramic Materials Under Constant and Cyclic Load," Life Prediction Methodologies and Data for Ceramic Materials, ASTM STP 1201, C. R. Brinkman and S. F. Duffy, Eds., American Society for Testing and Materials, Philadelphia, 1994.

ABSTRACT: Subcritical crack growth under static and cyclic loads is an important failure mode and often responsible for delayed failure. At least for subcritical crack extension under constant load it is a well-known fact that the crack growth behaviour of natural cracks differs significantly from that of macroscopic cracks, especially for such materials exhibiting a pronounced R-curve behaviour. The influence of bridging surface interactions - responsible for R-curves in alumina (Al_2O_3) - will be discussed with respect to strength and static lifetimes. Experimental results obtained for both loading cases are presented for alumina. Measured lifetimes are significantly shorter than the predictions and a reduction of lifetimes is found for increasing frequencies. From these results an effect of cycles becomes obvious and it can be concluded that a real cyclic fatigue effect occurs.

KEYWORDS: Static fatigue, cyclic fatigue, lifetime prediction, R-curve effect, bridging interactions, coarse-grained alumina (Al_2O_3)

Crack growth under constant and cyclic loads is an important failure mode and responsible for delayed failure of ceramic materials. Therefore, the determination of subcritical crack growth data in natural flaws is necessary to allow lifetime predictions of real ceramic components. At least for subcritical crack extension under constant load it is a well-known fact that the crack growth behaviour of natural cracks differs significantly from that of macroscopic cracks, especially for such materials exhibiting a pronounced R-curve behaviour. Since the same effects may also be present in the case of cyclically loaded ceramics, it is necessary to get information on growth behaviour of natural cracks (introduced during fabrication and surface machining) for both loading cases.

In the first part, lifetime predictions for components under static (or moderately changing) loads will be briefly discussed taking into consideration R-curve effects due to crack bridging interactions. The second part will deal with the problem how a real cycle effect can be detected. Experimental results obtained for coarse-grained alumina (Al_2O_3) will be given in this context. Finally, a model will be discussed that explains cyclic fatigue effect in ceramics with R-curve behaviour by interaction of well-known subcritical crack growth and a reduction of bridging stresses due to cycles.

[1] Research Engineer, Kernforschungszentrum Karlsruhe, Institut für Materialforschung II.

[2] Professor, Universität Karlsruhe, Institut für Zuverlässigkeit und Schadenskunde im Maschinenbau, and Kernforschungszentrum Karlsruhe, Institut für Materialforschung II, Postfach 3640, W 7500 Karlsruhe 1, Germany

LIFETIMES OF MATERIALS WITH SUBCRITICAL CRACK GROWTH

Ceramics without R-curve behaviour

In this investigation the crack growth relation is described by the usual power law between the crack growth rate v and the stress intensity factor K_I

$$v = AK_I^n \tag{1}$$

The parameters A, n are assumed to be known. The determination of these parameters will not be the topic of the paper. Combined with the definition of the stress intensity factor

$$K_I = \sigma \sqrt{a}\, Y \tag{2}$$

($Y \simeq 1.3$, σ = applied stress, a = crack size) eq.(1) yields the lifetime formula

$$t_f \simeq \int_{a_i}^{a_c} \frac{da}{v} = \frac{1}{A} \int_{a_i}^{a_c} \frac{da}{[\sigma Y \sqrt{a}\,]^n} \tag{3}$$

with the initial crack size a_i and the critical crack size a_c or in the integrated form

$$t_f = B\sigma_c^{n-2}\sigma^{-n}\left(1 - (\sigma/\sigma_c)^{n-2}\right) \simeq B\sigma_c^{n-2}\sigma^{-n} \tag{4}$$

where

$$B = \frac{2}{AY^2(n-2)K_{I0}^{n-2}} \tag{5}$$

with the inert strength σ_c (the strength in absence of subcritical crack growth) and stress intensity factor K_{I0} at the onset of unstable crack extension. These interrelations are very clear and therefore often used in the literature including Weibull statistics for lifetime predictions. If the inert strength σ_c is given by a Weibull distribution with the cumulative frequency $F(\sigma_c)$ it holds

$$F(\sigma_c) = 1 - \exp\left[-(\sigma_c/\sigma_{c0})^m\right] \tag{6}$$

where σ_{c0} and m are the Weibull parameters. From eq.(4) one can conclude that also the lifetime will be Weibull-distributed according to

$$F(t_f) = 1 - \exp\left[-(t_f/t_{f0})^{m^*}\right] \tag{7}$$

with the Weibull parameters t_{f0}, m^* defined by

$$t_{f0} = B\sigma_{c0}^{n-2}\sigma^{-n} \quad , \quad m^* = m/(n-2) \tag{8}$$

The procedure of lifetime prediction under static (i.e. under constant) load calls for knowledge of the parameters A (or B) and n. The mostly applied method of determining these parameters is the dynamic bending test where bending strength measurements are performed at several loading rates. If σ_f is the strength obtained, it holds for $\sigma_f \lesssim 0.9\sigma_c$

$$(\sigma_f)^{n+1} = B\sigma_{c0}^{n-2}(n+1)\dot\sigma \tag{9}$$

As an example lifetime predictions were performed for Al_2O_3 (99.6% Al_2O_3, density = 3.92 g/cm^3, mean grain size = 3 μm, K_{Ic} = 4.5 MPa$\sqrt{m}$, σ_c = 280 MPa, m = 14) in water at room temperature. In order to determine the crack growth parameters dynamic bending tests were carried out in the same environment at four loading rates. The results of strength measurements are plotted in the insert of fig.1 as the median value of σ_f (in MPa) versus the stress rate $\dot\sigma$ (in MPa/s). For the highest and for the lowest loading rates each test series includes 20 specimens and for the intermediate test series 10 specimens each. The Weibull moduli for the extremal stress rates were found to be m = 17.7 for $\dot\sigma$ = 0.00345 MPa/s and m = 19 for $\dot\sigma$ = 316 MPa/s.

From the slope of the regression line we find with eq.(9): $n = 52$ and $lg(B\sigma_c^{n-2}) = 121.2$. The predicted lifetimes resulting from eq.(4) are entered in fig.1 as a solid line. The results of lifetime measurements under constant load made in the same loading arrangement are entered in fig.1 as circles. The quantity F in fig.1 is the cumulative failure probability and can be calculated from $F = (i - 1)/N$ where for a given stress i is the number of specimens which failed at times $\leq t_f$. N is the total number of tests. The agreement between prediction and measurement is very good. The authors have experienced that in case of Al_2O_3 such encouraging results can be achieved only with fine-grained materials.

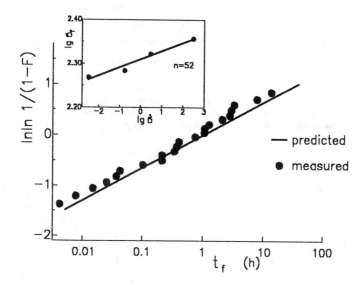

Fig.1 Lifetime prediction for a fine-grained Al_2O_3 in static bending tests, based on the results of dynamic bending tests (insert); both tests performed in water (20°C).

Ceramics with R-curve behaviour

In materials without R-curve behaviour the crack-tip stress intensity factor $K_{I\,tip}$ - responsible for crack growth - is identical with the externally applied stress intensity factor $K_{I\,appl}$. In ceramics with R-curve behaviour the crack tip is shielded by the shielding stress intensity factor K_{Is}. Especially in case of coarse-grained Al_2O_3, the shielding effect is caused by crack-surface interactions in the wake of a propagating crack. In this case one can replace the shielding stress intensity factor by the so-called bridging stress intensity factor K_{Ibr}. Therefore, the crack tip stress intensity factor results as [1]

$$K_{I\,tip} = K_{I\,appl} + K_{Is} = K_{I\,appl} + K_{Ibr} \quad , \quad K_{Is}, K_{Ibr} < 0 \tag{10}$$

This relation is the basis for understanding R-curve influences on the lifetime behaviour. Whilst in a controlled fracture test stable crack propagation occurs with a constant crack tip stress intensity factor, namely $K_{I\,tip} = K_{I0}$, during subcritical crack growth in a lifetime test under constant load, the value of $K_{I\,tip}$ must change with $K_{I\,appl}$ and K_{Ibr} according to eq.(10). This behaviour is explained by fig.2. At the beginning of crack extension the crack-tip stress intensity factor decreases significantly with increasing crack length, passes a minimum value, and increases again. The applied stress intensity factor $K_{I\,appl}$ (dashed curve) increases monotonically and the difference between these two curves is the bridging stress intensity factor K_{Ibr}.

In [2] the authors have derived an approximative formula to calculate the bridging stress intensity factor for small circular cracks which can be expressed in case of stable crack propagation

$$K_{lbr} \simeq K_{lbr,max}\left[1 - \exp\left(-\frac{\sigma_0 Y}{K_{lbr,max}}\sqrt{a - a_0^2/a}\right)\right] \tag{11}$$

where $K_{lbr,max}$ is the maximum bridging stress intensity factor, a_0 is the initial crack size free of crack surface interactions, and σ_0 is the maximum stress in the bridging stress vs. displacement relation $\sigma_{br} = f(\delta)$ for which the authors proposed the relation [3]

$$\sigma_{br} = \sigma_0 \exp(-\delta/\delta_0) \tag{12}$$

The maximum bridging stress intensity factor reads in terms of eq.(12)

$$K_{lbr,max} = \frac{\sigma_0 Y^2 E\pi\delta_0}{4.56(1 - v^2)K_{l0}} \tag{13}$$

where K_{l0} is the crack-tip stress intensity factor necessary for stable crack propagation, E is the Young's modulus, and v is the Poisson ratio. If crack propagation with a stress intensity factor different from K_{l0} is considered (e.g. subcritical crack growth), this value must replace K_{l0} in (13).

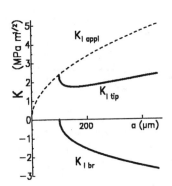

Fig.2 Development of the stress intensity factors of eq.(10) in a stable crack growth test.

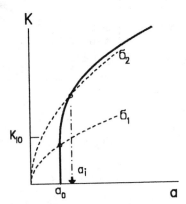

Fig.3 Definition of initial crack size for eq.(14).

The lifetime t_f results for a given stress σ as

$$t_f = \int_{a_i}^{a_c} \frac{da}{v} = \frac{1}{A}\int_{a_i}^{a_c} \frac{da}{\left[\sigma Y\sqrt{a} - K_{lbr}\right]^n} \tag{14}$$

In (14) a_i is that crack size at which slow crack growth starts and a_c is the critical crack size. For ceramics without an R-curve behaviour a_i is identical with the crack size a_0 of the existing crack. In case of a strong R-curve the specimen can be exposed to such high stresses that at the beginning of the test there will first be stable crack extension with $K_{l\,tip} = K_{l0}$ before at a_i subcritical crack growth with $K_l < K_{l0}$ occurs. The meaning of the crack sizes a_i and a_c is illustrated in fig.3. It should be mentioned that this effect will influence the lifetime only at stresses close to the strength of the specimen.

At the point of failure ($a = a_c$) the crack tip stress intensity factor $K_{I\,tip}$ reaches K_{I0} and, consequently, the subcritical crack growth rate becomes extremely high. Therefore, the integral in eq.(14) may be extended to infinity without noticeable loss of accuracy. Only in case of negligible crack extensions, i.e. $a_c/a_0 \simeq 1$, this approximation will fail. With this assumption one obtains

$$t_f \simeq \frac{1}{A} \int_{a_i}^{\infty} \frac{da}{[\sigma Y \sqrt{a} - K_{Ibr}]^n} \tag{15}$$

Influence of bridging stresses on the n-value

As shown in fig.4, the lifetime curve including bridging stresses is flatter than the curve obtained in the absence of bridging effects, i.e., the apparent power law exponent n', defined by lifetimes obtained at different stress levels

$$n' = -\frac{d \log t_f}{d \log \sigma} \tag{16}$$

The difference between the two n-values (n, n') can easily be shown. Therefore, we use the approximation given by eq.(15).

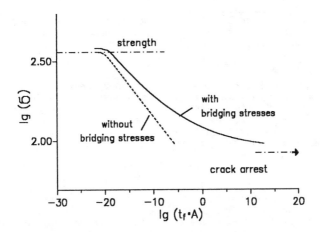

Fig.4 Lifetimes in static tests for $n = 19$ [4] including R-curve behaviour.

Differentiation of eq.(15) with respect to the stress σ yields

$$\frac{d t_f}{d \sigma} = -n \int_{a_i}^{\infty} \frac{Y \sqrt{a}\ da}{[\sigma Y \sqrt{a} - K_{Ibr}]^{n+1}} \tag{17}$$

This can be written by use of the mean value theorem for integrals as

$$\frac{d\,t_f}{d\,\sigma} = -\,n\,\frac{Y\sqrt{a'}}{\sigma Y\sqrt{a'} - K_{Ibr}'}\,t_f \tag{18}$$

where the primes indicate a certain crack size in the range $a_i < a' < a_c(=\infty)$. Introducing logarithmic derivatives yields

$$\frac{d\,\log(t_f)}{d\,\log(\sigma)} = -\,n\,\frac{\sigma Y\sqrt{a'}}{\sigma Y\sqrt{a'} - K_{Ibr}'} \tag{19}$$

From eqs.(16) and (19) it results for the static tests

$$n' = n\,\frac{\sigma Y\sqrt{a'}}{\sigma Y\sqrt{a'} - K_{Ibr}'} \;>\; n \tag{20}$$

Finally, we can conclude

$$n' \ge n \tag{21}$$

PROCEDURES APPLIED IN EVIDENCING CYCLIC FATIGUE EFFECTS

In order to be able to decide whether or not cyclic fatigue produces an effect in a ceramic material, two types of test are recommended:

1. The comparison of cyclic lifetimes determined in experiments with predictions from static tests made on the basis of K_I-governed subcritical crack growth allows an effect of the cycles to be detected [5].
2. A second method is the measurement of cyclic lifetimes with very different frequencies, but identical upper and lower stresses and identical shape of the amplitudes. In case of a real fatigue effect the lifetime must decrease with increasing frequency.

Predictions of lifetimes in cyclic tests

Since delayed failure of mechanically loaded ceramic components at moderate temperatures can be caused by subcritical crack growth as well as by a real fatigue damage, a first possibility to prove a cyclic effect is to compare experimental lifetime data with predictions on the basis of pure subcritical crack growth. These predictions may be based on a power law representation of subcritical crack growth as given by (1), where crack growth of statically loaded cracks is only possible under tension

$$v = \begin{array}{ll} A K_I^n & \text{for } K_I > 0 \\ 0 & \text{for } K_I \le 0 \end{array} \tag{1a}$$

In this context it should be emphasized that crack-growth phenomena under compressive loads as found by Ewart and Suresh [6] are cyclic effects and have to be neglected in predictions based on static lifetime measurements. Evidently, such effects may cause deviations between measured and predicted cyclic lifetimes and will therefore be detected.
For a sinusoidal stress with the mean value $\bar{\sigma}$ and the amplitude σ_a

$$\sigma(t) = \bar{\sigma} + \sigma_a \sin(\frac{2\pi}{T}\,t) \tag{22}$$

the lifetime in a cyclic test t_{fc} results from the lifetime t_{fs} obtained from a static test by [7]

$$t_{fc} = t_{fs}\left(\frac{\sigma_s}{\sigma_a}\right)^n \frac{1}{h(\bar{\sigma}/\sigma_a, n)} \tag{23}$$

$$h(\frac{\bar{\sigma}}{\sigma_a}, n) = \frac{1}{\pi}\int_{-\alpha}^{\pi/2}[\bar{\sigma}/\sigma_a + \sin\varphi]^n d\varphi \quad \text{with} \quad \alpha = \arcsin(\bar{\sigma}/\sigma_a) \tag{24}$$

If the lifetime tests are performed in bending, eq.(23) has to be modified by

$$t_{fc} = t_{fs}\left(\frac{\sigma_s}{\sigma_a}\right)^n \frac{1}{h(\overline{\sigma}/\sigma_a, n)} \lambda(R) \qquad (25)$$

with $R = \sigma_{min}/\sigma_{max}$. The function $\lambda(R)$ takes into account different effective surfaces (or volumes) in static and cyclic tests.

Whereas in the static bending tests only one side is exposed to tensile loading, failure can occur on both sides in the alternating bending test, i.e., the effective surface of the specimens in cyclic tests is twice the surface in static tests. This has to be taken into consideration for the failure prediction of the specimens with natural flaws. Consequently, special values of $\lambda(R)$ become

$$\lambda(R > 0) = 1 \qquad \lambda(R = -1) = (1/2)^{(n-2)/m}$$

The quantity m is the Weibull modulus for inert strength data. In case of a test with $R = -1$ the function h is

$$h(0,n) = \frac{1}{2\sqrt{\pi}} \frac{\Gamma(\frac{n}{2} + \frac{1}{2})}{\Gamma(\frac{n}{2} + 1)} \qquad (26)$$

which can be approximated for n>15 by [7]

$$h(0,n) \simeq \frac{0.395}{\sqrt{n}} \qquad (27)$$

EXPERIMENTAL RESULTS

Experimental results helping to decide whether or not a cyclic fatigue effect exists in Al_2O_3 will be reported in the following section. Measurements were carried out on 3.5x4.5x50mm specimens from two batches of 99.6% Al_2O_3 (Frialit/Degussit, Friedrichsfeld AG, Mannheim, FRG) with a mean grain size of $\simeq 20\mu$m.

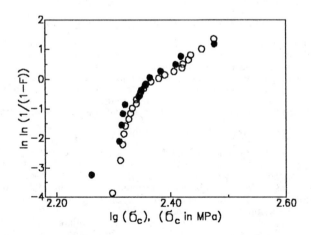

Fig.5 Inert strength σ_c of coarse-grained Al_2O_3 in Weibull-representation.

The specimens were roughly ground which resulted in a relatively low strength. Such a surface state ensures that all specimens will fail due to surface cracks. After manufacture the specimens were annealed in vacuum for 5 hours at 1200°C. Strength tests were performed at a loading rate of ≃200MPa/s. The results are plotted in fig.5, where the open circles correspond to batch 1 and the solid circles to heat 2. There are distinct deviations from a linear Weibull distribution. This effect is typical of materials with a pronounced R-curve behaviour [8] which has been proved for the coarse-grained Al_2O_3 in [3]. Cyclic fatigue tests were carried out in alternating bending tests loaded with loudspeakers in a cantilever arrangement. The testing device applied in this investigation is described in detail in [9], [10].

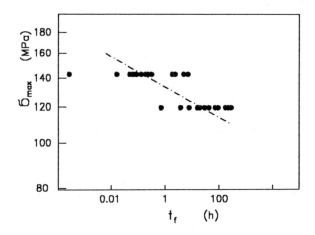

Fig.6 Lifetimes measured in cyclic tests with $R = -1$, $f = 50Hz$.

Cyclic lifetime results and predictions from static tests (batch 1)

In fig.6 the results of the cyclic tests are shown for a frequency of 50Hz and an R-ratio of $R = -1$. The slope of the dash-dotted straight line gives an exponent of $n = 28.6$ for a power-law description of cyclic crack growth similar to eqs.(1) and (4).
Results of tests with constant stress are shown in fig.7. From these data $n_s = 39$ was obtained, which is significantly higher than for the cyclic tests. Figure 8 shows the prediction from the static tests using eq.(25). The predicted and measured lifetimes are significantly different. It has to be concluded from these results that the cyclic fatigue effect is very strong.

Cyclic lifetime results and predictions from static tests (batch 2)

Figure 9 shows the lifetimes obtained for a maximum stress of $\sigma_{max} = 175MPa$ and frequencies of 0.2, 2, and 20Hz. As can be seen, the lifetimes decrease with increasing frequency. This is an indication that not only subcritical crack growth may be the reason for failure in cyclic tests. In the static tests with $\sigma = 175$ MPa all specimens survived until 100h (fig.10). In addition, results of a second series of static tests performed at an increased stress level of 194 MPa are shown. In this case a number of specimens survived, too.
From the static lifetimes shown in fig.10 the cyclic lifetimes have been calculated for $\sigma_{max} = 175MPa$. These predictions are represented in fig.11 together with the experimentally determined cyclic lifetimes. It becomes obvious that the experiments yield distinctly lower lifetimes than predicted. Also from this result it must be concluded that the influence of cycles is strong. Figure 12 gives the same results as fig.9, dependent on the number of cycles to failure. In this representation an influence of the frequency becomes evident, too.

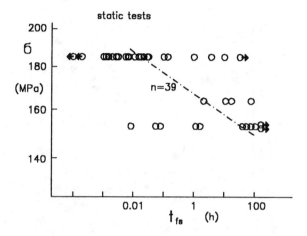

Fig.7 Lifetimes measured in static tests.

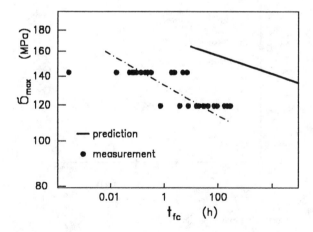

Fig.8 Comparison of measured and predicted cyclic lifetimes.

Finally, it can be concluded that the coarse-grained 99.6%-Al_2O_3 exhibits a cyclic effect that exceeds the effect expected from subcritical crack growth in static tests. The conspicuous disagreement between predictions based on static lifetimes and the experiments as well as the influence of frequency on the number of cycles to failure are significant indications of this fact.

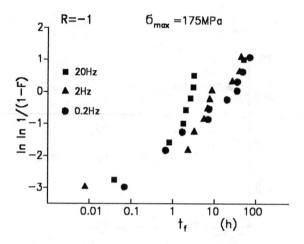

Fig.9 Lifetimes under alternating bending load (R = -1) with maximum stress σ_{max} = 175MPa.

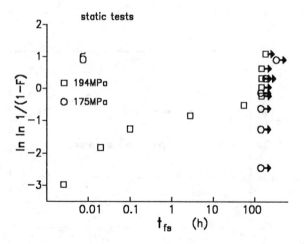

Fig.10 Lifetimes in static bending tests.

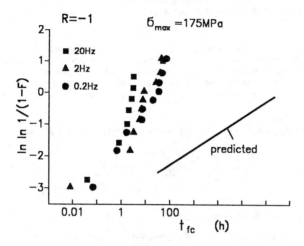

Fig.11 Lifetimes in cyclic tests compared with predictions on the basis of static lifetime tests; (straight line: prediction from fig.4).

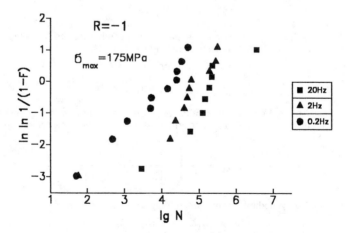

Fig.12 Results of fig.9 represented as a function of the number of cycles to failure.

INFLUENCE OF R-CURVE EFFECTS ON LIFETIMES IN CYCLIC TESTS

A lifetime prediction for cyclically loaded ceramics becomes complicate in the presence of R-curve effects due to crack bridging interactions. In cyclic tests these crack interactions are assumed to diminish gradually with increasing number of cycles [11] as has been shown by in-situ microscopic examinations [12], [13]. In this case the crack tip is exposed to higher loading in cyclic tests than in static tests.

Since for small cracks in coarse-grained Al_2O_3 with $a < 100\mu m$ the maximum crack opening is small compared with δ_0, one can approximate [14]

$$\sigma_{br} \simeq \sigma_0 \tag{28}$$

In case of a cyclic loading the surface interactions may be reduced by the cycles. The number of bridging events will be reduced and in terms of bridging stresses the stress parameter σ_0 decreases. In order to model the general behaviour we will assume that the decrease of the maximum value of bridging stresses is proportional to the number of cycles (N) and to the actual value of σ_0 which reads in the integrated form

$$\sigma_0 = \sigma_{00} \exp(-\alpha N) \tag{29}$$

With (28) the cycle-dependent bridging stress intensity factor K_{Ibr} results as

$$K_{Ibr} = \frac{2\sigma_{00}}{\sqrt{\pi a}} \int_{a_0}^{a} \frac{\sigma_{br} \, r \, dr}{\sqrt{a^2 - r^2}} \tag{30}$$

Now we will consider two limit cases.

- **Case 1:** The bridging interactions are unaffected by the cycles, i.e., $\alpha \to 0$. This case also describes the bridging stress intensity factor for static load.

$$K_{Ibr,1} = \frac{2\sigma_{00}}{\sqrt{\pi a}} \sqrt{a^2 - a_0^2} \tag{31}$$

- **Case 2:** Only a few cycles are necessary to dissolve the crack surface interactions newly created during crack propagation, i.e., $\alpha \to \infty$.

$$K_{Ibr,2} = 0 \tag{32}$$

In order to make the calculations as transparent as possible for the following considerations a step-shaped load-time history is chosen for the cyclic tests

$$\sigma = \begin{cases} \sigma_0 & for \quad 0 < t < T/2 \\ -\sigma_0 & for \quad T/2 < t < T \end{cases} \tag{33}$$

resulting in step-shaped stress intensity factors. Considering the statistical effect - as used in eq.(25) - the lifetime t_f in a static test performed with the stress σ results as

$$t_{f \, static} = \frac{1}{A} \int_{a_0}^{\infty} \frac{da}{\left[\sigma Y \sqrt{a} - K_{Ibr,1} \right]^n} \tag{34}$$

The lifetime for the cyclic limit case 1 (unaffected bridging stresses) is

$$t_{f \, cycl,1} = 2\lambda(R) \, t_{f \, static} \tag{35}$$

and in case 2 (completely vanished crack surface interactions) we obtain

$$t_{f \, cycl,2} = \frac{2}{A} \lambda(R) \int_{a_0}^{\infty} \frac{da}{\left[\sigma Y \sqrt{a} \right]^n} \tag{36}$$

The factor 2 enters eqs.(35) and (36) due to the special step-shaped loading history.

Figure 13 shows the median values of static lifetimes (dashed line), predicted with eq.(34) using the data-set

$$n = 20 \,, \quad m = 10 \,, \quad \sigma_{00} = 100 MPa \,, \quad a_0 = 100 \mu m$$

and the two limit cases of cyclic fatigue (solid lines) computed with eqs.(35) and (36). The range where the real cyclic lifetimes have to be expected is hatched. As can be seen from fig.13, the n-value (i.e. the negative reciprocal slope of the curves) is lower for the limit case 2 than for the limit case 1. This can be directly concluded from eqs.(21),(35) and (36) and

$$n_{static} > n_{cyclic} \tag{37}$$

A direct consequence of eq.(37) is a reduction of scatter of lifetimes in cyclic tests. Let m_{σ_c} be the Weibull-modulus of the inert strength σ_c and m^* the Weibull-modulus obtained in lifetime tests; then it holds

$$m^* = \frac{m_{\sigma_c}}{n - 2} \tag{38}$$

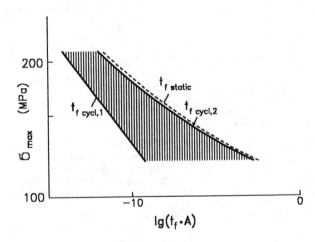

Fig.13 Lifetime predictions for static (dashed line) and cyclic (hatched area) tests calculated with $n = 20$, $\sigma_0 = 100 MPa$ and a median value of the crack distribution of $a_0 = 100 \mu m$;

For the static and cyclic lifetime tests one obtains

$$\frac{m^*_{cyclic}}{m^*_{static}} = \frac{n_{static} - 2}{n_{cyclic} - 2} \tag{39}$$

and from eq.(37)

$$m^*_{cyclic} > m^*_{static} \tag{40}$$

i.e., the scatter of cyclic lifetimes is less than the scatter of static lifetimes.

SUMMARY

A method is described to extend the procedure of lifetime calculations under static load which is well-known for materials without increasing crack growth resistance, to materials with a rising R-curve.

The procedure of lifetime predictions for cyclically loaded ceramics has been used to detect the influence of cycles on crack propagation. From the comparison of measured cyclic lifetimes and lifetimes predicted on the basis of subcritical crack growth it can be concluded that the coarse-grained 99.6%-Al_2O_3 exhibits a cyclic effect that exceeds the effect expected from subcritical crack growth in static tests. The same conclusion can be drawn from the influence of frequency on the number of cycles to failure which should not occur in the absence of cycle effects. Finally, a region of lifetimes has been predicted, where cyclic lifetimes have to be expected for materials with R-curve effects caused by crack-surface interactions.

REFERENCES

[1] Mai, Y., Lawn, B.R., Crack-interface grain bridging as a fracture resistance mechanism in ceramics: II. Theoretical fracture mechanics model, Journal of the American Ceramic Society 70(1987),289.

[2] Fett, T., Munz, D., Stress intensity factors for circular cracks with crack surface interactions due to bridging stresses, International Journal of Fracture 55(1992)R73-R79.

[3] Fett, T., Munz, D., Evaluation of R-curves in ceramic materials based on bridging interactions, KfK-Report 4940, Kernforschungszentrum Karlsruhe, 1991.

[4] Fett, T., Munz, D., Subcritical crack growth of macro- and microcracks in ceramics, "Fracture Mechanics of Ceramics" (Eds. R.C. Bradt, D.P.H. Hasselman, D. Munz, M. Sakai, V.Y. Chevchenko), Vol.9, 219-233, Plenum Press, New York, 1992.

[5] Evans, A.G., Fuller, E.R., Crack propagation in ceramic materials under cyclic loading conditions, Metallurgical Transactions 5 (1974)27-33.

[6] L. Ewart, S. Suresh, Crack propagation in ceramics under cyclic loads, Journal of Materials Science 22(1987)1173-92.

[7] Fett, T., Munz, D., Subcritical crack extension in ceramics, MRS Int'l. Mtg. on Adv. Mats. 5, Mat. Res. Soc.,1989,505-523.

[8] Fett, T., Munz, D., Influence of R-curve effects on lifetimes for specimens with natural cracks, Proceedings of the Conference on FRACTURE PROCESSES IN BRITTLE DISORDERED MATERIALS, Noordwijk, June 19-21, 1991, pp.365-374.

[9] Fett, T., Martin, G., Munz, D., Thun, G., Determination of da/dN-ΔK-curves for small cracks in alumina in alternating bending tests, Journal of Materials Science 26(1991)3320-28.

[10] Fett, T., Munz, D., An indirect method for the determination of the da/dN-ΔK-curves for ceramic materials, "Fracture Mechanics of Ceramics" (Eds. R.C. Bradt, D.P.H. Hasselman, D. Munz, M. Sakai, V.Y. Chevchenko), Vol.9, 559-568, Plenum Press, New York, 1992.

[11] Lathabai, S., Rödel, J., Lawn, B.R., Cyclic fatigue from frictional degradation at bridging grains in alumina, Journal of the American Ceramic Society 74(1991)1340-48.

[12] Frei, H., Grathwohl, G., New test methods for engineering ceramics - in-situ microscopy investigation, Ceramic Forum International, cfi, 67(1991), 27-35.

[13] Rödel, J., Kelly, J., Lawn, B.R., In situ measurements of bridged crack interfaces in the scanning electron microscope, Journal of the American Ceramic Society 73(1990)3313-18.

[14] Fett, T., Munz, D., Differences between static and cyclic fatigue effects in alumina, Journal of Materials Science Letters 12(1993)220-222.

Jacques L. Lamon [1]

PROBABILISTIC FAILURE PREDICTIONS IN CERAMICS AND CERAMIC MATRIX FIBER REINFORCED COMPOSITES

REFERENCE: Lamon, J. L., "Probabilistic Failure Predictions in Ceramics and Ceramic Matrix Fiber Reinforced Composites," Life Prediction Methodologies and Data for Ceramic Materials, ASTM STP 1201, C. R. Brinkman and S. F. Duffy, Eds., American Society for Testing and Materials, Philadelphia, 1994.

ABSTRACT : An approach to predictions of failure in ceramics and of failure-damage in Ceramic Matrix Composites (CMC) is discussed. For ceramics, particular emphasis was placed upon the populations of fracture inducing flaws. The flaw strength parameters, that are used to infer the failure of various loading geometries were estimated in view of fracture origins identified by scanning electron microscopy, using analytical as well as computerized methods. The proposed approach for CMC considers the microcomposite geometry. The Weibull model as well as a more fundamental model based upon the elemental strength concept were used.

KEY WORDS : failure probability, ceramics, ceramic matrix composites, Weibull distribution, multiaxial model, scale factors

----

Ceramic materials contain microstructural flaws of various types, shapes and sizes that appear throughout the lifetime of components : first during processing (preexisting flaws : pores, voids, cracks, inclusions, etc ...), then during surface finish operations (microcracks and cracks) and finally during service (in hostile environments or at high temperature).

Failure and reliability are controlled by flaws. Under high stresses, catastrophic flaw extension is caused by the most severe flaw in the structure. Under low stresses, delayed failure may occur, even if the preexisting flaws are

---

[1] Research Director, Laboratoire des Composites Thermostructuraux,
UMR 47 CNRS-SEP-UB1, Domaine Universitaire, 3 Allée de La Boétie, F-33600 Pessac

not severe under the current applied stress state. Slow growth of the flaws results from chemical or physical phenomena. Failure occurs when one of the flaws becomes critical, leading to a crack that will propagate at a more or less rapid rate.

The presence of flaws is unavoidable. They are still present in toughened ceramics such as fiber reinforced ceramic matrix composites (CMC), although an enhanced damage tolerance is obtained. CMC exhibit a typical non-linear stress-strain behavior resulting from damage induced by matrix cracks initiated in the matrix and arrested by the fibers at fiber/matrix interfaces. The three important basic features of the stress-strain behavior of CMC are the stresses at matrix cracking and at matrix cracking saturation, and the ultimate strength determined by fiber bundle failure. Matrix-damage and bundle failure are induced by microstructural defects, and they involve non- interactive brittle events occurring sequentially [1].

Probabilistic-statistical approaches provide the essential relationships between the three basic criteria for fracture in ceramics : fracture inducing flaw populations, stresses, and failure probability. We can go around this triangle in any way for failure predictions.

In monolithic ceramics, failure probability is usually predicted from strength data measured on specimens having well defined geometry and stress-state. Three - or four - point bending of bars is the most frequently employed test. Analysis of strength data scatter provides statistical parameters that describe or reflect the fracture inducing flaw populations. In some cases, the nature and location of flaw populations are identified by fractographic examination of the broken test specimens. In many cases, multiple populations are activated. Subsequent failure predictions as well as determination of flaw strength parameters are carried out using various models. Post-processors have also been developed by researchers for failure prediction purposes [2, 3].

By contrast, probabilistic analysis of failure in CMC has not been the subject of many extensive studies.

The present paper examines failure predictions in ceramics and in CMC with a dual intent : first determine the influence of the flaw strength parameters upon failure probability for monolithic ceramics and, second, propose a probabilistic approach to description of matrix damage and failure in CMC. The flaw strength parameters of various ceramics were extracted from various sets of strength data measured in-house or available in the literature, considering various loading geometries. Accuracy in failure predictions was thus evaluated from the scatter in flaw strength parameters induced by loading geometry. Primary emphasis was placed on the scale factors due to the loading geometry independence exhibited by shape parameters. Multiple populations of fracture-inducing flaws were identified by scanning electron microscopy. Failures of ceramics were analyzed and predicted using the Weibull model and the model based upon the elemental strength concept [4]. Probabilities were computed using the CERAM post processor developed for design purposes [2]. CERAM performs 2- or 3- dimensional analyses. It evaluates failure from surface -and/or volume- located flaws (7 different flaw populations can be specified in each case). Flaw strength parameters may be time-dependent to account for environmental effects.

Failure-damage predictions for CMC used the microcomposite geometry, which represents the elementary cell of the composite. A microcomposite consists of a single fiber coated with an interphase material and then with the ceramic matrix.

FAILURE PROBABILITY ANALYSIS - GENERAL ASPECTS

Failure probability - strength equations

Failure probability relations in the model based upon the elemental strength concept are obtained from the following fundamental equation [5, 6] :

$$P = 1 - \exp - \int_V dV \int_O^S g(S)\, dS \tag{1}$$

where $g(S)$ represents the number of flaws per unit volume with an elemental strength between S and $S + dS$. The equations of failure probability are derived from the distribution of elemental strengths that provide a description of fracture-inducing flaw populations. Elemental strengths represent the strengths of volume elements, each containing a single flaw. A multiaxial elemental strength was defined by a combination of local stresses acting upon the flaws considering concepts of non-coplanar crack extension [7]. A simple Weibull-type power function was assumed for the distribution of elemental strengths. Finally, the following failure probability equations were obtained respectively for surface - and volume - located failure origins :

$$P_{MS} = 1 - \exp\left[- \int_A \left(\frac{\sigma_1}{\sigma_{OMS}}\right)^{m_s} I_s\left(m_s, \frac{\sigma_2}{\sigma_1}\right) dA\right] \tag{2}$$

$$P_{MV} = 1 - \exp\left[- \int_V \left(\frac{\sigma_1}{\sigma_{OMV}}\right)^{m_v} I_v\left(m_v, \frac{\sigma_2}{\sigma_1}, \frac{\sigma_3}{\sigma_1}\right) dV\right] \tag{3}$$

where the subscripts S and V refer to surface and volume respectively.

m_s and m_v are the shape parameters and σ_{OMS} and σ_{OMV} are the scale factors. The functions I_v and I_s are integrals accounting for flaw shear sensitivity and orientation respective to principal stresses σ_1, σ_2 and σ_3 ($\sigma_1 > \sigma_2 > \sigma_3$). σ_2 and σ_3 may be compressive. σ_1 as well as the multiaxial elemental strength S_E are ≥ 0.

Failure probabilities were also evaluated using the following well-known Weibull equations for failure under uniaxial stress states and under polyaxial stress states respectively (Principle of Independent Action)[1] :

$$P_{WV} = 1 - \exp - \int_V (\frac{\sigma}{\sigma_{OWV}})^{m_v} dV \tag{4}$$

$$P_{WV} = 1 - \exp - \int_V [\frac{\sigma_1^{m_v} + \sigma_2^{m_v} + \sigma_3^{m_v}}{\sigma_{OWV}^{m_v}}] dV \tag{5}$$

A different meaning is attributed to the statistical parameters (particularly the scale factors σ_O) appearing in both models. σ_{OMS} and σ_{OMV} describe the distribution of multiaxial elemental strengths at a microstructural scale. They are thus pertinent to the fracture inducing flaws. The Weibull scale factors σ_{OWS} and σ_{OWV} measure the scatter in strength data considered at a macroscopic scale.

Determination of statistical parameters in ceramics

Statistical parameters are estimated from experimental distributions of strength data which are established using the ranking statistics method. The measured strengths are ordered from lowest to highest. The ith result in the set of samples is assigned a cumulative failure probability P_i, calculated using an estimator. In the present paper the following estimator was used for monolithic ceramics : $P_i = i/N + 1$ (N is the sample size). $P_i = \frac{i - 0.5}{N}$ was preferred for CMC due to smaller numbers of samples (15 specimens).

When concurrent multiple flaw populations were identified by Scanning Electron Microscopy, strength distributions were separated using the censored data method proposed by Johnson [8]. This method determines a new rank i' for the strengths of each population by calculating a new increment Δ as soon as one or more censored strengths are encountered in the sequence of test data.

In a first step, the statistical parameters were estimated by fitting theoretical probability-strength equations to stress data distributions. Various available methods were used including :
- linear regression analysis (LRA) of loglog plots of strength data (for estimation of the shape and scale factors)
- maximum likelihood estimation (MLE) (for scale factor determination)
- and average strength estimation (ASE) (for scale factor determination).

---

[1] similar equations can be used for surface-located failure origins

The relevant equations are given in Appendix. The use of another method (such as MLE) for determination of shape parameters was not necessary for the analysis, since a limited scatter was obtained with LRA, and since the important factor for evaluation of failure predictions is thus the scale factor scatter.

Then, since the application of these methods is restricted to simple stress-states, an alternative method based upon failure probability computations using the CERAM post-processor was also employed. This method can be easily applied to derive flaw strength parameters from failures observed with complex loading geometries. The scale factors are obtained using the following equation for volume as well as for surface failure origin :

$$\sigma_{OM} = \sigma_{OA} \left[\frac{\text{Ln} (1 - P_{CERAM})}{\text{Ln} (1 - P_{exp})} \right]^{1/m} \qquad (7)$$

where P_{CERAM} is the failure probability computed using CERAM for a dummy scale factor σ_{OA}. P_{exp} is the corresponding failure probability obtained experimentally for the same maximum stress in the testspecimen. Failure load was also used for comparison.

Prediction of failure - damage in CMC

Equations of failure probability for the matrix and the fiber were derived from the simple Weibull equation for volume failure origins (equation 4). Matrix damage and interface debonding affect the applied uniform uniaxial stress state by inducing local peak stresses in the fiber and stress drops in the matrix [1].

Introducing the stress states induced by matrix damage into equation (4) led to the following equations respectively for the probability of occurrence of the nth matrix crack and for the probability of fiber failure in the presence of (n-1) matrix cracks [1] :

$$P_M(n) = 1 - \exp\left[- S_M \left(\frac{\sigma}{\sigma_{0M}} \right)^{m_M} L \left(1 - \frac{2m_M}{1+m_M} \sum_1^{n-1} \frac{\ell_i}{L} \right) \right] \qquad (8)$$

$$P_F(n-1) = 1 - \exp\left[- \left[\frac{(a+1)V_M}{aV_F} \cdot \frac{\sigma_M}{\sigma_{0F}} \right]^{m_F} S_F L \left[\left(\sum_1^n \ell_i + \sum_2^{n-1} \ell_i \right) \frac{1}{L} B + \left(\frac{1}{1+a} \right)^{m_F} \right] \right]$$

$$\text{with } B = \frac{1}{1+m_F} \frac{a+1}{a} \left[1 - \left(\frac{1}{1+a} \right)^{m_F+1} \right] - \left[\frac{1}{1+a} \right]^{m_F} \qquad (9)$$

where the subscripts M and F refer to matrix and fiber respectively. S_M and S_F are the cross section areas, L is the gauge length, l_i the debond length associated with the ith matrix crack, σ_{OM}, m_M, σ_{OF}, m_F are the statistical parameters, a = $\dfrac{EmVm}{EfVf}$, and σ_M is the stress applied to the matrix. E_m and E_f are Young's moduli of the matrix and of the fiber. V_m and V_f are volume fractions of matrix and fiber.

The statistical parameters pertinent to the matrix and to the fiber are derived from distributions of matrix cracking stresses and of ultimate strengths measured on microcomposite samples tested in tension. Matrix cracking is evidenced by features of the load - displacement curves and by acoustic emission [9]. In the present paper, statistical parameters were derived by linear regression analysis of loglog plot of strength data.

FAILURE-DAMAGE PREDICTIONS IN MONOLITHIC CERAMICS AND IN CMC

Monolithic ceramics : influence of the scale factor upon failure predictions

Table 1 summarizes the sets of strength data which were treated, including data measured on silicon nitride ceramics (referred to as SN1 and SN2) and data available in the literature on alumina and SiC ceramics [10, 11].

The silicon nitride test bars were machined out of a single piece (SN1 ceramic) or out of billets and piston pins (SN2 ceramic). These latter specimens had been taken from the interior (heart samples) and from the surface (skin samples) of the pins. They were tested in 4 point-bending. The specimens were polished and chamferred, except the skin samples for which the initial machining finish was kept in the tensile surface.

Examination of the SN1 and SN2 broken testspecimens revealed the presence of various populations of fracture inducing flaws (Table 2):

a) machining cracks and pores created during processing were detected in the SN1 samples. The pores were predominantly located in the surface of the 3-point bending specimens, and both in the surface and in the volume of the 4-point bending bars.
b) inclusions rich in silicon and voids predominantly located within the interior (except in the disks) were observed in the SN2 samples. Failure of the heart samples was dictated by inclusions rich in silicon (50%) or iron (50% of failures). Failures initiated from surface flaws induced by machining in the skin samples.

The separation of strength distributions was conducted to conform with the presence of bimodal surface - and volume-located flaw populations in the silicon nitride samples. In the SN1 samples, the initial strength distributions were first separated according to the presence of machining cracks and pores considering the machining cracks as a censored group. Then the distribution of strength data relative to the pores which was determined in 4 point bending

Table 1--Loading geometries and ceramics examined in the present paper. The sample sizes[5] are given within brackets ().

Material Geometry	SN1	SN2	Al$_2$O$_3$ [11]	SiC [10]
3 pt bending [1] 1 long span	40x4x4 (76)	60x3x4 (50)	38x2.5x5	
3 pt bending [1] 2 intermediate span	10x4x4 (76)			
3 pt bending [1] 3 short span	8x4x4 (76)			
4 pt bending [2]	20x40x4x4 (79)	12.5x24x3x4 (124)	19x32x2.5x5	6.35x19.05x2.6x3.1
Biaxial flexure (3) of Disks		30x3 (50)	31.75x2.5	
Compression of C-Rings (4)				9.3x12.5x5.2

[1] span length x height x width (mm)
[2] upper span x lower span x height x width (mm)
[3] diameter x height (mm)
[4] external radius x internal radius x width (mm)
[5] sample sizes were available only for those specimens tested in-house.

was subsequently separated into two distributions pertinent to the surface - and volume-located failure origins. The undetermined fracture origins that were obtained only in a very limited fraction of samples, were treated as censored data. Failure data for the alumina and SiC specimens were assumed to be induced by a single population of flaws as suggested by linearity of loglog plots of fracture distributions [10, 11]. Volume flaws were observed by Ferber and coworkers on several samples of SiC [10]. Surface fracture origins were assumed for the alumina samples [11].

Estimation of statistical parameters used the analytical methods for the SN1 ceramic only. The computerized method using CERAM was then applied to all the loading geometries of table 1. Stress analyses were performed using the NIKE finite element code (developed by Lawrence Livermore Laboratory (1983)).

Table 2--<u>Fracture origins identified by scanning electron microscopy in the silicon nitride samples</u>

	Machining cracks	voids	inclusions	surface	volume
SN1					
3 pt Bending 1	42	29		26/29	3/29
3 pt Bending 2	56	17		14/17	3/17
3 pt Bending 3	52	18		18/18	
4 pt Bending	41	27		19/27	8/27
SN2					
3 pt Bending		53%	47%	3%	97%
4 pt Bending					
Billets		36%	64%	12%	88%
Heart spec.			100%		
Skin spec.	95%				
Disks				54%	46%

<u>Influence of the statistical parameters</u> Table 3 shows that the shape parameters pertinent to the population of pores in SN1 samples exhibit a limited scatter as a function of loading geometry. Elimination of the failures caused by machining flaws tremendously reduced the scatter observed on the initial distribution. However, it is worth noting that, despite the presence of bimodal flaw populations in the SN1 samples, initial strength distributions measured in three point bending exhibited relatively high and comparable shape parameters, which could give the erroneous impression that failure was dictated by a single population of flaws.

The scale factors estimated for the population of surface - located pores are sensitive to the **probabilistic model and to the method of estimation**. The scale factors relevant to the Weibull model are generally larger than those pertinent to the Multiaxial Elemental Strength Model. Ratios $\frac{\sigma_{OM}}{\sigma_{OW}}$ were about 0.9. This trend was confirmed with the data obtained for the other ceramics under various loading geometries (Table 4).

The scatter in scale factors induced by the **method of estimation** was about 10% for a given loading geometry. The Maximum Likelihood Estimation and the Average Strength Estimation methods provided the lowest scale factors, whereas the Linear Regression Analysis method led to the highest. The CERAM-based scale factors depend upon mesh size. Scale factors determined with a coarse mesh may be higher than those obtained by Linear Regression Analysis. Optimized meshes with four mode elements were constructed for computations. Table 3 shows that identical results were obtained when considering the maximum stress in the testspecimen or the failure load.

Table 3--<u>Shape parameters and scale factors (pertinent to pores) estimated for the SN1 ceramic (MPa m$^{3/m_v}$)</u>

σ_{OWV} and σ_{OMV} are given within brackets

Scale factors			3 pt B 1	3 pt B 2	3 pt B 3	4 pt B
MLE		σ_{OWS}	291	249	235	283 (207)
		σ_{OMS}	268	227	215	257 (182)
ASE		σ_{OWS}	292	252	238	290 (209)
		σ_{OMS}	269	230	217	263 (184)
LRA		σ_{OWS}	359	332	319	341 (244)
		σ_{OMS}	330	303	291	310 (215)
CERAM computations (1)		σ_{OWS}	332	308	298	326 (235)
		σ_{OMS}	305	282	275	297 (208)
CERAM computations (2)		σ_{OWS}	332			
		σ_{OMS}	305			
shape parameters	initial distribution		11.2	12.9	11.8	6.4
	pores m_s m_v		8.6	8.0	8.0	7.5 11.3

(1) P exp given by the maximum stress in the specimen
(2) P exp given by the failure load.

Comparison of the scale factors given by table 3 clearly shows that failure predictions require use of relevant scale factors. Discrepancy in scale factors as a function of loading geometry reflects the accuracy of predictions from strength data measured for a given loading geometry. Therefore computations from scale factors estimated by one of the analytical methods will lead either to overestimations (based upon the MLE and the ASE methods) or to underestimations (based upon the LRA method) of failure probability. Analytical predictions from scale factors estimated either with the MLE or the ASE methods or with the CERAM based method will be overestimated. Incorporation of the Weibull scale factors into the Multiaxial Elemental Strength Model will lead to underestimations of failure and vice-versa.

The scale factor discrepancy induced by the **loading geometry** depends upon the method of estimation. The larger discrepancy was obtained with the MLE and ASE methods ($\approx$ 18.5%). The LRA method led to an 11% scatter, whereas CERAM computations led to a scatter either > 10% (with the Weibull model) or < 10% (with the Multiaxial Elemental Strength Model). These results indicate that the most satisfactory failure predictions will be provided by the computerized approach. Similar sensitivity of statistical parameters to analytical methods of estimation was pointed out by Leon and Kittl for glass rods [12].

The above trends in failure predictions were confirmed on the other ceramics considered here. Table 4 shows that the scale factors determined using the CERAM based method depend upon the loading geometry and the probabilistic model. Results exhibited a certain discrepancy with those determined analytically by other authors [10, 11]. The accuracy in predictions which is suggested by the narrow scale factor scatter was checked by computing strength distributions using CERAM from the statistical parameters estimated on a given loading geometry : 3-point bending for SN_2, 4-point bending for Alumina and SiC. Agreement with experimental data was satisfactory as illustrated in [13] and in figure 1 which shows probabilities computed for SiC C-Rings subjected to a diametral compression [2]. Underestimation of failure probability by Ferber and coworkers may be attributed to overestimation of the scale factors. The Weibull model generally underestimated failure probability.

Evaluation of dependence of failure predictions upon statistical parameters--The incidence of scale factor scatter upon failure predictions was evaluated using the following equation derived from the above failure probability equations :

$$\frac{1}{m}(\frac{dP}{P}) = \frac{(1-P)}{P} \text{Ln}(1-P)(\frac{d\sigma_0}{\sigma_0}) \qquad (10)$$

Figure 2 shows that the dependence of failure probability upon the scale factor is significantly sensitive to the level of failure probability. This dependence is particularly important at low failure probabilities. It is enhanced by a large value of the shape parameter.

For shape parameters smaller than 10, as observed with most of the ceramics considered here, a 10% uncertainty in the scale factors will lead to a discrepancy in the predicted failure probabilities comprised between - P and - 0.15 P. For shape parameters larger than 20, a 10% uncertainty will now lead to an uncertainty in failure predictions between - 2.2 P and - 0.4 P.

Low failure probabilities are thus very sensitive to the scale factor. Therefore, this implies that an important effort in determination of scale factors is required for reliable failure predictions in the range of low probabilities.

Ceramic matrix composites Probabilities of occurrence of cracks in the matrix and in the fiber were determined for SiC/SiC microcomposite test specimens under tensile loading conditions (gauge length 25 mm, diameter 25 μm, fiber diameter 15 μm). Identical debond lengths were assumed at each matrix cracking event. The following statistical parameters were estimated using a batch of 15 microcomposite specimens [14]:

Table 4--<u>Scale factors obtained using the CERAM-based method. Also given are the data estimated in the literature for Alumina and SiC, using analytical</u> methods

| | Shape parameters | | | Scale factors (MPa m$^{3/m}$) | |
	all data	surface	volume	surface	volume
SN2 3pt bending	9.8		9.8		$\sigma_{OW} = 97$ $\sigma_{OM} = 84$
4 pt bending : billet samples	8.8	9.9	8.7		$\sigma_{OW} = 86.9$ $\sigma_{OM} = 74.5$
heart samples	9.8		9.8		$\sigma_{OW} = 97$
skin samples	18.5	18.5		$\sigma_{OW} = 310$ $\sigma_{OM} = 298$	$\sigma_{OM} = 84.1$
Disks	8.4	7.9	8.7		$\sigma_{OW} = 80$ $\sigma_{OM} = 72.2$
Al$_2$O$_3$ [11] 4 pt Bending	23.8	23.8		$\sigma_{OW} = 242$ [11] $\sigma_{OW} = 211$ $\sigma_{OM} = 205$	
3 pt Bending	23.4 [11]	23.4		$\sigma_{OW} = 212.5$ $\sigma_{OM} = 207$	
Disks	22 [11]	22		$\sigma_{OW} = 218$ $\sigma_{OM} = 208$	
SiC [9] 4 pt bending	7.87 [10]		7.87		$\sigma_{OW} = 122$ [10] $\sigma_{OW} = 29$ $\sigma_{OM} = 25$
C-Ring	8.04 [10]		8.04		$\sigma_{OW} = 101$ [10] $\sigma_{OM} = 24$

$m_m = 3.8$ $\sigma_{OM} = 0.5 \text{ MPa m}^{0.8}$
$m_f = 7.3$ $\sigma_{Of} = 38 \text{ MPa m}^{0.4}$

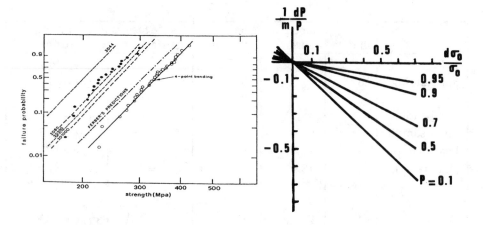

Figure 1 : Failure probability predicted from 4 point bending data using CERAM (2D and 3D analyses) and various meshes, for the C-Ring geometry (Multiaxial Elemental Strength Model) [2]

Figure 2 : Dependence of failure probability predictions upon scatter in scale factors

Incorporating these parameters into equations (8) and (9) allowed one to calculate P_M and P_F as a function of applied force (figure 3). It was assumed that matrix cracks are created at a 50% probability to illustrate the average behavior of microcomposites. Figure 3 shows that matrix damage exhibits first the higher probability of occurrence ($P_M > P_F$). The risk of matrix cracking increases with the applied stress. It drops at crack formation whereas failure probability of the fiber increases concurrently. This behavior is in agreement with logical expectation since further matrix cracking concerns smaller volume elements and it raises up the stresses in the fiber.

Transition from matrix cracking ($P_M > P_F$) to fiber failure preponderance ($P_F > P_M$) marks saturation in matrix cracks. Beyond this point, the mechanical behavior is dictated by the fracture resistance of the fiber.

Comparison of P_M (n) and P_F (n - 1) determines the characteristics of matrix damage at saturation including stress, crack density and crack spacing as a function of interfacial failure [1]. The trends anticipated by figure 4 are supported by current observations of the mechanical behavior of CMC as a function of interfacial properties. It is thus predicted on figure 4 that in the presence of a weak interface (important debond length) the saturation stress is decreased and non-linearity of the stress-strain behavior is caused by a little number of cracks. On the contrast strong interfaces enhance matrix cracking and saturation occurs at a much higher stress.

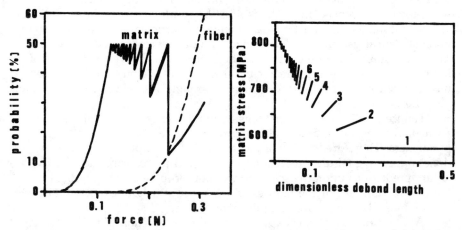

Figure 3 : Typical stress-probability relations for a SiC/SiC microcomposite exhibiting 12 matrix cracks at saturation

Figure 4 : Influence of debond length upon the stress at saturation as a function of the number of matrix cracks for a SiC/SiC microcomposite

CONCLUSION

The statistical parameters and more particularly scale factors depend upon several factors which can be grouped into 2 main families :

- material independent factors which can be easily controlled, including the method used for the scale factor determination, the probabilistic model, and the stress-analysis, and

- a material dependent factor, namely material reproductibility, which has been controlled rather inefficiently up to now with most ceramics.

The scatter in the scale factors induced by factors of the former group can be reduced to reasonable bounds by optimizing the stress analysis and using a computerized methodology for estimation of scale factors and for failure predictions. Optimization of stress analysis requires selection of an appropriate mesh with small elements in the regions subject to high stresses.

In the present paper, particular emphasis was placed upon the influence of flaw populations and the relevant scale factors. Origin of fracture was systematically identified using SEM fractographic analysis, and the distributions of strength data were separated accordingly to determine the statistical parameters pertinent to the pre-existent populations of fracture inducing flaws. A certain scatter in scale factors was observed, depending upon the method of estimation, the probabilistic model and the loading geometry. It was shown that consistent methods must be employed for failure predictions and for estimation of statistical parameters. In particular, predictions relying upon statistical parameters determined using analytical methods may not be safe. The computerized method provided more satisfactory predictions of failure. Moreover, the scale factor pertinent to the Weibull model cannot be incorporated in the Multiaxial Elemental Strength Model for failure prediction purposes, and vice versa.

Failure predictions are very sensitive to the scale factor. This effect is particularly significant at low failure probabilities and for high shape parameters. Therefore, one must recognize that an important effort should be done to improve failure prediction methods and to decrease the sensitivity of failure predictions upon the scale factor, in order to be able to predict low failure probabilities, or the failures which are routinely unexpected, but which are the most dangerous for ceramic components.

Application of statistical approaches to the prediction of failure damage of CMC was presented. These materials are expected to exhibit larger damage tolerance than ceramics. The trends which were anticipated are supported by experimental observations on CMC.

APPENDIX

Equations used for the determination of statistical parameters

Equations are given for volume-located fracture origins only. Similar equations for surface-located fracture origins can be easily established.

a) Linear regression analysis :

$$\sigma_{OWV} = [\frac{K_W V}{e^A}]^{1/mv} ; \qquad \sigma_{OMV} = [\frac{K_M I_V (mv, \frac{\sigma_2}{\sigma_1}, \frac{\sigma_3}{\sigma_1}) V}{e^A}]^{1/mv}$$

where K_W and K_M are factors accounting for the stress state and geometry. A is the intercept in a loglog plot of failure probability versus log of stress.

b) Maximum likelihood estimation

$$\sigma_{OWV} = V_{EW}^{1/mv} [\frac{1}{N} \sum_1^N S_{max}^{mv}]^{1/mv}$$

$$\sigma_{OMV} = V_{EM}^{1/mv}\, I_V^{1/mv}\,(m_v, 0, 0)\,[\frac{1}{N}\sum_{1}^{N} S_{max}^{mv}]^{1/mv}$$

where V_{EW} and V_{EM} are the effective volumes.

c) Mean Strength Estimation

$$\sigma_{OWV} = \bar{S}_{max}\, V_{EW}^{1/mv}\, /\Gamma\,(1 + \frac{1}{m_v})$$

$$\sigma_{OMV} = \bar{S}_{max}\, V_{EM}^{1/mv}\, I_V^{1/mv}(m_v, 0, 0)\, /\Gamma\,(1 + \frac{1}{m_v})$$

where Γ is the gamma function.

d) Effective volumes

$$V_{EW} = \int_V \frac{\sigma_1^{mv} + \sigma_2^{mv} + \sigma_3^{mv}}{S_{max}^{mv}}\, dV$$

$$V_{EM} = \int_V (\frac{\sigma_1}{S_{max}})^{mv}\frac{I_V(mv, \sigma2/\sigma1, \sigma3/\sigma1)}{I_V(m_v, 0, 0)}\, dV$$

REFERENCES

[1] Guillaumat, L. and Lamon, J., "A Probabilistic Approach to the Failure of Ceramic Matrix Composites (CMC) : Analysis of the Influence of Fiber/Matrix Interfaces" Proceedings of the Fifth European Conference on Composite Materials ECCM V, Ed. Bunsell A.R. et al., EACM, Bordeaux (France), 1992, pp. 585-590.

[2] Lamon, J., Pherson, D. and Dotta, P., "2 D and 3 D Ceramic Reliability Analysis using CERAM. Statistical Post Processor Software" Technical Documentation, Battelle Geneva Laboratories, 1989.

[3]   Gyekenyesi, J.P., "CARES - A Post-Processor Program to MSC/NASTRAN for the Reliability Analysis of Structural Ceramic Components", NASA Technical Memorandum 87188, 31st **International Gas Turbine Conference and Exhibit**, American Society of Mechanical Engineers, Dusseldorf (West Germany), 1986.

[4]   Lamon, J. and Evans, A.G., "Statistical Analysis of Bending Strengths for Brittle Solids : A Multiaxial Fracture Problem", Journal of the American Ceramic Society, vol. 66, N°3, 1983, pp. 177-182.

[5]   Freudenthal, A., Fracture, Vol. II, Ed. Liebowitz H., Academic Press, New York, 1969, pp. 592-621.

[6]   Matthews, J.R., McClintock, F.A. and Shack, W.J., "Statistical Determination of Surface Flaw Density in Brittle Materials", Journal of the American Ceramic Society, Vol. 59, N° 7-8, 1976, pp. 304-308.

[7]   Hellen, T.K. and Blackburn, W.S., "The Calculation of Stress Intensity Factors for Combined Tensile and Shear Loading", International Journal of Fracture, Vol. 11, 1975, pp. 605-617.

[8]   Johnson, L.G., "The Statistical Treatment of Fatigue Experiments", Elsevier, New York, 1964.

[9]   Lamon, J., Rechiniac, C. and Corne, P.,"Determination of Interfacial Properties in Ceramic Matrix Composites using Microcomposite Specimens", Proceedings of the Fifth European Conference on Composite Materials ECCM V, Ed. Bunsell A.R. et al., EACM, Bordeaux, 1992, pp. 895-900.

[10]   Ferber, M.K., Tennery, V.J., Waters, S.B. and Ogle, J., "Fracture Strength Characteriztion of Tubular Ceramic Materials Using a Simple C-Ring Geometry", Journal of Materials Science, Vol. 21, 1986, pp. 2628-2632.

[11]   Shetty, D.K, Rosenfield, A.R., Duckworth, W.H. and Held, P.R., "A Biaxial-Flexure Test for Evaluating Ceramic Strengths" , Journal of the American Ceramic Society, Vol. 66, N°1, 1983, pp. 36-42.

[12]   Leon, M. and Kittl, P., "On the Estimation of Weibull's Parameters in Brittle Materials", Journal of Materials Science, Vol. 20, 1985, pp. 3778-3782.

[13]   Lamon, J., "Statistical Approaches to Failure for Ceramic Reliability Assessment", Journal of the American Ceramic Society, Vol. 71, N°2, 1988, pp. 106-112.

[14]   Lamon, J. Lissart, N., Rechiniac, C., Roach, D.H., Jouin, J.M., "Mechanical and Statistical Approach to the Behavior of CMCs", <u>Proceedings of the 17th Annual Conference on Composites and Advanced Ceramics</u>, The American Ceramic Society, Cocoa Beach (Florida), 1993 (in press).

Huibert F. Scholten,[1] Leonardus J. Dortmans,[1] and Gijsbertus de With[1,2]

APPLICATION OF MIXED-MODE FRACTURE CRITERIA FOR WEAKEST-LINK FAILURE PREDICTION FOR CERAMIC MATERIALS

REFERENCE: Scholten H. F., Dortmans, L. J., and de With, G., "Application of Mixed-Mode Fracture Criteria for Weakest-Link Failure Prediction for Ceramic Materials," Life Prediction Methodologies and Data for Ceramic Materials, ASTM STP 1201, C. R. Brinkman and S. F. Duffy, Eds., American Society for Testing and Materials, Philadelphia, 1994.

Abstract: A set of combined experimental and numerical data is presented for the prediction of multiaxial strength for ceramics. Uniaxial and biaxial bend tests were performed on ten different materials. The strength predicted with various mixed-mode fracture criteria was compared with the measured values. A main conclusion was that with the introduction of an additional parameter, a "size-independent strength", all tests were predicted within 3 % accuracy. However, different criteria had to be used, which could not be interchanged between the various materials. The "size-independent strength" parameter is interpreted as a measure for the applicability of the weakest-link concept. Its physical meaning is yet uncertain, but predictions on materials for which the porosity was less than one percent showed that for these materials the deviations are largest. This could indicate that basic assumptions in the weakest-link models applied with respect to defect density are violated for these materials.

Keywords: Testing, Microstructure, Fractography, Mixed-Mode Fracture Criterion, Ceramics

---

[1]) Centre for Technical Ceramics, P.O. Box 595, 5600 AN Eindhoven, The Netherlands.

[2]) Also affiliated with Philips Research Laboratories, P.O. Box 80000, 5600 JA Eindhoven, the Netherlands.

Reliability concepts for strength prediction of brittle materials such as ceramics are generally based on weakest-link models. The basis of this formulation is that brittle fracture nucleates at defects, which act as a weakest link [1]. Many attempts have been made to use this concept for the prediction of strength of ceramics, in which the authors have extended or reformulated weakest-link theories by means of micro-mechanical fracture models e.g.[2,3,4,5].

The aim of these models is to obtain a method of predicting strength data from test results. Early research generally focused on a single failure criterion [6] like for ductile materials. From recent work, however, the question arises whether there exists a single failure criterion valid for all brittle materials e.g. [7,8].

In order to answer this question, it is necessary to pay attention to the defects from which brittle fracture originates. In ceramic materials there is a large variation in micro- and defect structures concerning grain size, grain shape, microfractures and porosity. Hence, it is the authors belief that this variation reflects itself upon the failure criterion which should be employed for the prediction of multiaxial strength, and that it is unlikely that a single failure criterion is valid for all ceramics.

In the present work, an attempt is made to establish relationships between microstructure and the appropiate mixed-mode fracture criteria. The strategy chosen is schematically represented in figure 1.

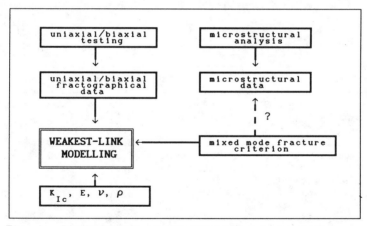

Fig. 1:--Block diagram of experimental and numerical methods to measure and predict strength data.

The strength data, obtained with the three (3PB)- and four-point (4PB) bend test, ball-on-ring (BOR) and ring-on-ring (ROR) tests are combined with results from fractography and microstructural analysis. These data are used in a weakest-link model to see which fracture criterion best fits experimental data.

MATERIALS AND MECHANICAL TESTING

Materials

With respect to the materials used for this study a number of selection criteria were pursued:

- the whole group of materials had to be diverse in microstructure,
- all materials should reveal brittle fracture at room temperature and not contain strengthening properties which violate the weakest-link theory,
- all specimens of a particular material should be processed from a single processing batch and machined equally to minimize variation in composition and defect structure,

The materials used in this work were, with one exception, commercially available or already in use in certain applications. The final selection consisted of: three aluminas, a hot-isostatically pressed silicon carbide, two NiZn-ferrites, a modified barium-titanate, a glass-ceramic and two refractory ceramics.

The materials were characterized with respect to both physical and microstructural parameters (Table 1). The microstructures have been quantified by the porosity [9], P, and and the mean linear intercept length [10], G_{mli}, and are further described qualitatively in the discussion of the results of this study.

TABLE 1--Physical and microstructural properties of the materials.

Material	K_{Ic}[#] (MPa√m)	E[$] (GPa)	ν[$]	ρ(g/cm^3)	G_{mli} [μm]	P[%]
Alumina I	3.9	369	0.24	3.85	7.9	5
Alumina II	4.5	377	0.24	3.89	3.1	3
Alumina III	3.5	313	0.23	3.69	10.6	10
Ferrite I	1.7	176	0.33	5.10	5.9	6
Ferrite II	1.3	128	0.29	4.37	2.9	25
HIPSIC	3.0	442	0.17	3.17	2.6	< 1*
Glass-ceramic	1.5	64	0.25	2.52	10*	< 1*
Barium-titanate	2.1	208	0.28	4.41	2*	< 1*
Refractory I	< 1*	23	0.25*	2.65	50*	15
Refractory II	< 1*	15	0.25*	1.89	100*	37

#: Determined with the chevron-notched beam method [11]
$: Determined ultrasonically with the pulse-echo method using longitudinal waves at 5 MHz and transverse waves at 20 MHz [12].
*: estimated value

The machining into "test-ready" samples was fairly equal for all materials, irrespective to the fact that some of them have been machined in different workshops. The surface finish as well as the final dimensions of the specimens was nominally the same. They are presented in Table 2.

TABLE 2--Nominal dimensions and surface finish of the specimens.

Bars[#]			Disks[&]		
length (1)	50	mm	diameter (ø)	30	mm (ROR)
width (w)	3.5	mm	(ø)	20	mm (BOR)
heigth (h)	4.5	mm	thickness (t)	1.5	mm
chamfer (ch)	0.1	mm			
Roughness (R_a)			$\simeq 0.3$ μm		
Flatness (f)			≤ 5.0 μm		
Parallel-faced			≤ 5.0 μm		

[#]: 80(1) × 10(w) × 10(h) mm for the refractory ceramics
[&]: 77.5(ø) × 10(t) mm for both BOR and ROR disks of the
refractory ceramics

Mechanical testing

The 3PB tests were generally performed on a span length of 20 mm.
For ALUMINA I, this test was also carried out at a span length of 40
mm. The 4PB test was performed with inner span length 20 mm and outer
span length 40 mm.

Ball-bearings were used for the biaxial jigs, which were 12 mm in
diameter for the BOR test and 12 mm and 20 mm for the ROR test.

Because of their large grain sizes, scaled-up versions of the tests
jigs were designed for the refactory ceramics. These materials were
tested at span lengths of 60 mm (3PB test), 60 and 30 mm (4PB test) and
diameters of 50 mm (BOR test) and 50 mm and 30 mm (ROR test).

In reference [13] the performance of the tests jigs is described in
detail. The main conclusions of this study were that all tests yielded
reproducible results within one or two percent accuracy. The prime
experimental condition for accurate testing is the application of free
rollers for the uniaxial test and ball-bearings for the biaxial test.
For three-point bending, the Seewald-von Karman correction for wedging
stresses has to be applied. The analytical solutions which were used to
calculate the nominal outer fiber stresses are given in [8].

All strength tests were carried out under equal conditions. The
testing speeds were selected such that these resulted in an outer fiber
strain rate of 5×10^{-4} s^{-1} for all tests on each material. The
humidity was kept low by means of dry N_2 (dew point $\leq - 35$ C^o). All

tests were carried out at room temperature.

THEORY

Weibull statistics

The results of the strength tests were statistically interpreted using the two-parameter Weibull equation for failure probability:

$$P_i(S_i) = 1 - \exp\left[-\left(\frac{S_i}{S_o}\right)^m\right] = 1 - \exp\left[-\left(\frac{1}{m}!\right)^m \left(\frac{S_i}{\bar{S}}\right)^m\right] \tag{1}$$

Here, P_i represents the cumulative failure probability at the nominal stress S_i (the outer fiber stress in the bend tests), m the Weibull modulus, S_o the characteristic strength, $\bar{S}$ the mean nominal fracture stress of a test batch and $m! = \Gamma(1 + \frac{1}{m})$, with Γ denoting the gamma function.

For the determination of the parameters m and S_o, a least-squares regression analysis was used in combination with a weight factor [8,14]. The strength data were ranked in ascending order and assigned a failure probability according to:

$$P_i = \frac{i - 0.5}{N} \tag{2}$$

where i is the rank number and N the total number of specimens. Although the choice of P_i and the weight factor are not standardized, other fit procedures (i.e. maximum likelihood method [14]) did not result in significant differences.

Strength predictions

Following the work of Thiemeyer et al. [7], the condition for fracture at any defect within the material was formulated as:

$$K_{eq} = \sigma_{eq} (Y/Z) \sqrt{a} \geq K_{Ic} \tag{3}$$

where K_{eq} represents an equivalent value of the stress intensity at a flat defect with size a, under simultaneous K_I, K_{II} and K_{III} loading, (Y/Z) is a geometry constant and K_{Ic} the fracture toughness of the material. The value of σ_{eq} represents an equivalent stress at the defect, which can be calculated from both normal and shear stresses on the crack plane using a mixed-mode fracture criterion. The following criteria have been used in this work:

NSA : normal stress averaging or mode I failure [1]
COP : coplanar energy release rate [15]
GMA : maximum non-coplanar energy release rate [16]
RNC : empirical criterion of Richard [17] with $\alpha_1 = 1$ $(K_{Ic} = K_{IIc})$
PIA : principal of independent action [6]

The fracture criteria which were employed to calculate σ_{eq} have been studied in recent work, i.e. [7] and [18]. As has been illustrated in [18], the differences in the multiaxial strength predictions for some particular loading conditions are very subtle compared to experimental errors. This stresses the importance of a high experimental accuracy.

For natural defects, the constant (Y/Z) mainly depends on the shape and the location of the defect. For volume defects, $Y=\sqrt{\pi}$ and for surface defects, $Y=1.12\sqrt{\pi}$. The shape dependency can be taken into consideration for criteria II, III and IV. In the present work, "penny-shaped" cracks (PSC, $Z=\pi/2$) and "through-the-thickness cracks" (TTC, $Z=1$) are discerned.

If the weakest-link principle holds true, the following expression is valid for each test series k on a particular material:

$$\bar{S}_k = S_u F_k \qquad (4)$$

in which $\bar{S}_k$ represents the mean nominal fracture stress of test series k and S_u is the unit strength of unit volume V_u or unit surface A_u. The value of the geometric parameter F_k can be calculated for each test series according to the weakest-link principle for volume or surface defects:

$$F_k = [\frac{V_u}{V_k \Sigma(V)_k}]^{1/m} \quad \text{or} \quad F_k = [\frac{A_u}{A_k \Sigma(A)_k}]^{1/m} \qquad (5)$$

where V_u is the unit volume, V_k the specimen volume of test k and $\Sigma(V)_k$ the stress volume integral. The symbols A_u, A_k and $\Sigma(A)_k$ are defined analogously for surface defects. The value of m is the mean value resulting from all tests on a material. The stress volume integral and the stress surface integral are given by [19]:

$$\Sigma(V)_k = \frac{1}{V} \int_V \frac{1}{4\pi} \int_{B_u} (\frac{\sigma_{eq}}{S_k})^m dB_u \, dV \qquad (6a)$$

and

$$\Sigma(A)_k = \frac{1}{A} \int_A \frac{1}{2\pi} \int_{C_u} (\frac{\sigma_{eq}}{S_k})^m dC_u \, dA \qquad (6b)$$

respectively where B_u is the unit sphere and C_u is the unit circle (both with radius 1) and S_k is the nominal fracture stress.

TEST RESULTS

The results will be discussed for each material separately. The number of specimens of a test series, N, the Weibull modulus and the mean nominal fracture stresses $\bar{S}_k$ are listed in Tables 3-7.

In order to trace the strength-determining defect population of a

material, fractograpy was applied where possible. It should be
mentioned this technique was in most cases carried out only on the
remnants of the uniaxial bend tests. Due to their limited thickness,
the biaxial specimens generally did not reveal a clear fracture plane
morphology.

Alumina I-- Fractography on this material yielded poor results, due
to the large grain size of the material and internal reflections with
optical microscopy. Scanning electron microscopy (SEM) on some of the
specimens, however, showed that the origin of failure was probably
located at the surface. The microstructure is characterized by a wide
grain size distribution with an intercept length varying from 2 - 80
μm. The mean linear intercept length is 7.9 μm and the porosity is 5 %.
    The ROR test produced a somewhat lower value for m. This may be due
to the fact that the roughness of these disks is about 1.0 μm, where it
is 0.3 μm for the other specimens.

TABLE 3--Results of the strength tests on alumina I and alumina II.

		alumina I			alumina II	
TEST	N	$\bar{S}_k$(MPa)	m	N	$\bar{S}_k$(MPa)	m
3P20	40	288	28.5	20	368	8.5
3P40	40	280	27.0			
4P20/40	40	264	22.0	20	291	6.7
BOR	40	288	20.8			
ROR	36	231	12.9	20	269	9.0

Alumina II--The Weibull moduli do not differ significantly for the
tests series on alumina II, thus it is concluded that a single defect
population was responsible for the fracture of the specimens. No
fractographical results were obtained for the reasons as mentioned
before for alumina I. The mean linear intercept length is 3.1 μm, but
large grains up to 40 μm are frequently present. The porosity is 3 %.

Alumina III--The strength tests on alumina III show that the Weibull
modulus is fairly equal for all tests. Fractography indicated that the
specimens failed from defects at the surface, but the defects
themselves could not be discerned due to the large grain size.
The microstructure of the material revealed a quite large porosity
(10 %) in comparison with the other aluminas. The mean linear
intercept length is 10.6 μm.

HIPSIC--Fractography on the remnants of both the uniaxial and
biaxial tests revealed that in both cases sub-surface pores were
strength determining. The microstructure is quantified by a mean linear
intercept length of 2.7 μm and a porosity less than one percent.

TABLE 4--Results of the strength tests on alumina III and HIPSIC.

TEST	alumina III			HIPSIC		
	N	$\overline{S}_k$(MPa)	m	N	$\overline{S}_k$(MPa)	m
3P20	20	285	13.3	20	557	3.6
4P20/40	20	235	10.1	20	484	3.5
BOR	17	360	14.0	20	685	7.8
ROR				20	411	5.6

Ferrite I--The microstructure of the material is rather
inhomogeneous. The porosity of the material is 6 %, where it is less
within the zones where large grains are concentrated. Within the
homogeneous zones the mean linear intercept length is 5.9 μm.
Fractography showed that sub-surface pores were strength determining.

TABLE 5--Results of the strength tests on ferrite I and
ferrite II.

TEST	ferrite I			ferrite II		
	N	$\overline{S}_k$(MPa)	m	N	$\overline{S}_k$(MPa)	m
3P20				20	103	8.6
4P40	18	162	13.7	20	83	6.9
BOR	18	241	14.7	20	124	9.7
ROR	16	164	12.6	20	78	9.6

Ferrite II--The defects responsible for fracture of ferrite II
consisted of large pores located at the surfaces of the specimens. In
low magnification micrographs it was visible that the large pores were
located throughout the material in circular zones, approximately 0.5 mm
in diameter, which possibly represent the shape of original powder
granules. The overall porosity of the material is 25 % and the mean
linear intercept length is 2.9 μm.

All four strength test resulted in similar values for m. The main
difference with ferrite I is that for all tests $\overline{S}_k$ is much lower, which
is presumably due to the large porosity.

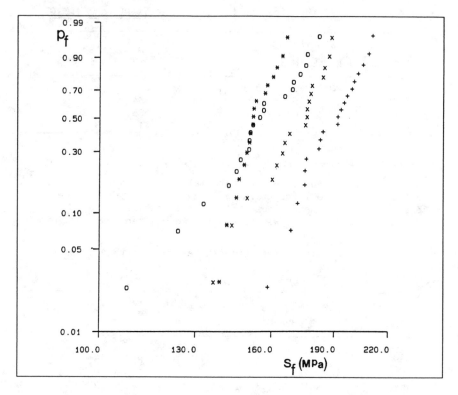

FIG. 2--Weibull diagram of the glass-ceramic. * = ROR test,
O = 4PB test, × = 3PB test at span length 20 mm, + = BOR test.

Barium-titanate--The results of fractography indicate a partial
similarity with the HIPSIC. The strength determining defects were
mainly sub-surface pores. The Weibull moduli are fairly equal, with the
exception of the BOR test which cannot yet be explained. The material
is very fine-grained (mean linear intercept length estimated at .2 μm)
and has a porosity less than one percent.

TABLE 6--Results of the strength tests on barium-titanate and the glass
ceramic.

	Barium-titanate			Glass-ceramic		
TEST	N	$\bar{S}_k$(MPa)	m	N	$\bar{S}_k$(MPa)	m
3P20	19	273	13.9	18	170	13.6
4P20/40	19	259	19.8	20	154	9.5
BOR	20	329	28.8	20	188	13.5
ROR	20	261	18.4	18	153	19.7

Glass-ceramic--The strength-determining defect population of the glass ceramic consisted mainly of large pores, which are almost perfectly round. In some cases, the strength-determining defect was a combination of a pore and machining damage. The porosity of the material is less than one percent. The Weibull diagram of the material is presented in figure 2.

Refractory I--With respect to the results of refractory I, it is immediately noticed that of each bend test $\overline{S}_k$ is rather low in comparison with the other materials. This is not very surprising since the material contains a large porosity. Due the large grain size of both refractories and their variety in composition, fractography was difficult for both materials. Although the origin of fracture was not directly clear, it was believed to be a pore or a crack.

Table 7--Results of the strength tests on refractory I and II.

| | Refractory I | | | Refractory II | | |
TEST	N	$\overline{S}_k$(MPa)	m	N	$\overline{S}_k$(MPa)	m
3P60	30	45	12.4	30	9.7	9.8
4P30/60	30	43	8.9	30	9.1	7.2
BOR50	20	50	8.6	30	19.9	8.1
ROR30/50	20	33	7.1	30	11.3	12.9

Refactory II--All specimens of refractory II failed from the material's intrinsic pores or cracks. The Weibull modulus is fairly equal for all tests, although the ROR test resulted in a somewhat higher value. It should be mentioned that during the mechanical tests, the sound of cracks could be heard before fracture of the specimen. It possibly indicated that microcracks grew during the test, thus violating the weakest-link principle.

ANALYSIS AND DISCUSSION

The calculations of the geometric parameter F_k for each combination of material and test were performed with the aid of the finite element post processor FAILUR [20] both for volume and surface defects. The finite element calculations were done with a mesh such that 1) the outer fiber stresses for each test agreed with the analytical formulae, and 2) the calculations were independent of the number of elements in the finite element mesh.

If for a particular material one of the criteria predicts the strength of test k ideally, equation (5) should hold. Figure 3 represents the measured values of $\overline{S}_k$ of the glass-ceramic as a function of F_k calculated with the criterion III for TTC for volume defects.

Although the points lie on a straight line, this line does not pass the origin which is required according to the standard weakest-link theory.

Similar results were obtained if the values of F_k were calculated according to different criteria. Since this phenomenon was noted for a number of the tested materials, a different fit procedure was applied, for which equation (5) was modified to:

$$\overline{S}_k = S_u F_k + S_r \qquad (7)$$

This equation allows to incorporate possible deviations from the standard theory by means of the parameter S_r. In the present work, it will be referred to as a "size-independent strength". The fit procedure was carried out for all material-fracture criterion combinations, both for volume and surface defects. Thus, each combination yielded estimates for both S_u and S_r, $\hat{S}_u$ and $\hat{S}_r$. Hence, the predicted value of $\overline{S}_k$, $\overline{S}_k^*$ can be calculated using:

$$\overline{S}_k^* = \hat{S}_u F_k + \hat{S}_r \qquad (8)$$

Subsequently, the error in the predicted value of test k from the measured value, ε_k, was defined as:

$$\varepsilon_k = [(\overline{S}_k^* - \overline{S}_k)/\overline{S}_k] \times 100 \% \qquad (9)$$

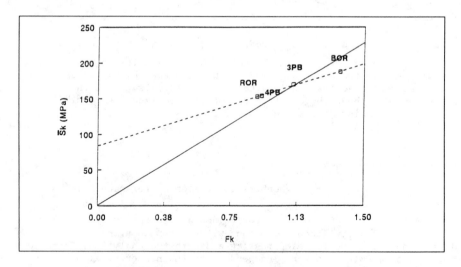

FIG. 3--Mean nominal fracture stresses $\overline{S}_k$ of the glass ceramic as a function of F_k calculated according to the GMA criterion for "through-the-thickness" (TTC) cracks. The dotted line results from fitting the data according to equation (8).

In order to obtain quantitative information about the applicability of a fracture criterion for a range of M tests series on a material, the mean error was defined as:

$$\varepsilon = \sum_{k=1}^{M} |\varepsilon_k| \ / \ M \qquad (10)$$

The calculations of F_k were carried out using the mean value of m for all tests on a material. The results of the fit procedure are listed in Table 8 together with results of the microstructural analysis. For a particular material, the results are given according to the criterion which yielded the smallest value of ε. It should be noted that the differences between the models was in some cases small. The results of the fit procedure for refractory II for all criteria resulted in errors ε larger than 30 %. The material is not taken into account further. These large errors are not very surprising if it is considered that crack growth occurred prior to fracture, which is violating the weakest-link principle.

Table 8--Best strength predictions resulting from all material-
fracture criterion combinations.

Material	defect	$\overline{m}$	criterion	$\varepsilon[\%]$	$\hat{S}_u$ (MPa)	$\hat{S}_r/\hat{S}_u$	P[%]
Alumina I	S	22.0	NSA	1.5	229	0.22	5
Alumina II	V	8.1	COP-TTC	0.2	295	0.08	3
Alumina III	V	12.5	PIA	1.3	344	-0.16	10
HIPSIC	V	5.1	NSA	3.1	257	0.99	<1
Ferrite I	V	13.7	PIA	0.5	177	0.09	6
Ferrite II	V	8.7	NSA	0.8	62	0.37	25
Glass-ceramic	V	14.1	GMA-TTC	0.5	77	1.09	<1
Bariumtitanate	V	17.4	PIA	0.3	93	1.91	<1
Refractory I	V	9.3	PIA	2.9	62	0.08	15

S: prediction based on surface defects
V: prediction based on volume defects
$\overline{m}$: mean value of Weibull modulus resulting from all tests
*: estimated values

It is concluded that with the introduction of the parameter S_r nearly all materials are predicted within two percent of the measured values, which is about equal to the experimental error. With the exception of alumina I, the best results were obtained for volume defects, which is in agreement with fractography, but the differences between the various criteria are in general too small to be significant [21]. For alumina I it was assumed that fracture nucleated mainly from surface damage. For the various materials, different fracture criteria had to be applied and the deviation from "standard" weakest-link is sometimes considerable. The relationship of the fracture criteria with the microstructural parameters such as porosity and G_{mli} is not directly clear.

An important observation is that the value of S_r is highest for the HIPSIC, the barium-titanate and the glass-ceramic, which all have a porosity less than one percent. A reliability analysis on the results revealed that the values of S_r for the HIPSIC, ferrite II and the glass-ceramic differ significantly from zero within 95 % confidence limits and for the barium-titanate within 90 % confidence limits [21]. If it is considered that pores are the main defects from which these materials fail, the application of weakest-link theory may be questionable. An important assumption in the theory is that the number of defects per unit volume is a constant and sufficiently high. Possibly these assumptions are not realistic for these materials. The effect may also be responsible for the higher values of m for the BOR test on the HIPSIC and barium-titanate because the effective volume of the BOR test is small.

CONCLUSIONS

Ten ceramics were tested, both uniaxially and biaxially. The main purpose was the prediction of the test results using a mixed-mode fracture criterion and relating these predictions to the microstructures of the various materials.

The interpretation of the test results showed that for four materials the standard weakest-link theory could not be applied. With the introduction of an additional parameter, the "size-independent strength" S_r, most materials were predicted within two percent according to one of the mixed-mode failure criteria. No single failure criterion predicted the strengths of all materials properly. The differences between the predictions were in some cases relatively small, which stresses the conclusion of earlier work [18] that high experimental accuracy is necessary.

The best-fitting types of defects (i.e. volume, surface) were in agreement with the results from fractography. It should be mentioned that this technique could be applied only for the dark-coloured and fine-grained materials. It is more difficult and elaborate for the lighter materials, especially if the grain size is large with respect to the specimen size.

A relation between the best-fitting criterion and microstructure could not be established. It was attempted to relate the "size-independent strength" to the microstructural parameters. With respect to the latter, it was noted that the most dense materials revealed the largest deviation from weakest-link theory. A possible explanation is provided by the fact that for these materials the number of defects per unit volume is to small, thus violating a basic assumption of the weakest-link theory. Whether this conclusion is true should be verified experimentally by mechanical tests in which the effective volume of the specimen is larger, i.e. a tensile test.

ACKNOWLEDGEMENTS

This work has partly been supported by the Commision for the Innovative Research Program Technical Ceramics (IOP-TK) of the Ministry of Economic Affairs in the Netherlands (IOP-TK research grant 88.B040). Most of the uniaxial bend tests and the K_{Ic} measurements have been carried out the Energy Research Foundation of the Netherlands (ECN).

REFERENCES

[1]   Weibull, W., "Statistical Theory of Strength of Materials",
        Swed. Inst. Eng. Res. Proc., 151, 1-45, 1939.

[2]   Batdorf, S.B. and Crose, J.G., "A Statistical Theory for the
        Fracture of Brittle Structures Subjected to Nonuniform
        Polyaxial Stresses", J. Appl. Mech., 41, 459-461, 1974.

[3]   Batdorf, S.B. and Heinisch, H.L., "Weakest-link Theory
        Reformulated for Arbitrary Fracture Criterion", J. Am. Ceram.
        Soc., 61, 355-358, 1978.

[4]   Evans, A.G., "A General Approach for the Statistical Analysis
        of Multiaxial Fracture", J. Am. Ceram. Soc., 61, 302-308,
        1978.

[5]   Lamon, J. and Evans, A.G., "Statistical Analysis of Bending
        Strengths for Brittle Solids: a Multiaxial Fracture Problem",
        J. Am. Ceram. Soc., 66, 177-182, 1983.

[6]   Stanley, P., Fessler, H. and Sevill, A.D., "An Engineers
        Approach to the Prediction of Failure Probability Prediction of
        Brittle Components", Proc. Brit. Ceram. Soc., 22,453-487, 1973.

[7]   Thiemeier, T., Brückner-Foit, A. and Kölker, H.,"Influence of
        the Fracture Criterion on the Failure Prediction of Ceramics
        Loaded in Biaxial Flexure", J. Am. Ceram. Soc., 74,48-52, 1991.

[8]   de Smet, B.J., Bach, P.W., Scholten, H.F., Dortmans, L.J.M.G.,
        and de With, G., "Weakest-Link Failure Prediction for Ceramics
        III: Uniaxial and Biaxial Bending Tests on Alumina",
        J. Eur. Ceram. Soc., 10, 101-107, 1992.

[9]   Dortmans, L.J.M.G., Morrell, R., and de With, G., "Round Robin
        on Grain Size Measurement for Advanced Technical Ceramics",
        VAMAS report 12, issued through the National Physical
        Laboratory, Teddington, U.K, (to be issued in 1993).

[10]   ENV 623-4. European Standard for Methods of Test for Advanced
        Monolithic Technical Ceramics. General and Textural Properties.
        Part 3: Determination of Grain Size.

[11]   de Smet, B.J. and Bach, P.W., "Fracture Toughness Testing of
        Ceramics", ECN Report ECN-1--91-070, issued through the Energy
        Research Foundation of the Netherlands, Petten, the
        Netherlands, 1991.

[12]   ENV 843-2. European Standard for Methods of Test of Advanced
        Monolithic Technical Ceramics. Mechanical Properties at Room

Temperature. Part 2: Determination of Elastic Moduli. To be
issued in 1993.

[13]   Scholten, H.F., Dortmans, L.J.M.G., de With, G., de Smet, B.J.
and Bach, P.W., "Weakest-Link Failure Prediction for Ceramics
II: Design and Analysis of Uniaxial and Biaxial Bend Tests",
J. Eur. Ceram. Soc., 10, 33-40, 1992.

[14]   Bergman, B., "How to Estimate Weibull Parameters", Brit. Ceram.
Proceedings, Engineering with Ceramics, 2, 175-185, 1987.

[15]   Paris, P.C., and Sih, G.C., "Stress Analysis of Cracks",
Fracture Toughness Testing and Its Applications, ASTM-STP 381.
American Society for Testing and Materials, Philadelphia PA,
30-83, 1965.

[16]   Hellen, T.K., and Blackburn, W.S., "The Calculation of Stress
Intensity Factors for Combined Tensile and Shear Loading",
Int. J. Fract., 10, 305-321, 1974.

[17]   Richard, H.A., "Prediction of Fracture of Cracks Subjected to
Combined Tensile and Shear Loads", VDI Research Report 631/85,
Düsseldorf, Germany, 1985.

[18]   Dortmans, L.J.M.G. and de With, G., "Weakest-Link Failure
Prediction for Ceramics IV: Application of Mixed-Mode Fracture
Criteria for Multiaxial Loading", J. Eur. Ceram. Soc., 10,
109-114, 1992.

[19]   Dortmans, L.J.M.G., Thiemeier, Th., Brückner-Foit, A., and
Smart, J., "WELFEP: a Round Robin for Weakest-Link Finite
Element Postprocessors", J. Eur. Ceram. Soc., 11, 17-22, 1993.

[20]   Dortmans, L.J.M.G. and de With, G., "Weakest-Link Failure
Predictions for Ceramics using Finite Element Post-Processing",
J. Eur. Cer. Soc., 6, 369-374, 1990.

[21]   Scholten, H.F,. "Evaluation of Multiaxial Fracture Models for
Technical Ceramics:, PhD thesis, Eindhoven University of
Technology, Eindhoven, Netherlands, 1993.

Tze-jer Chuang[1] and Stephen F. Duffy[2]

A METHODOLOGY TO PREDICT CREEP LIFE FOR ADVANCED CERAMICS USING CONTINUUM DAMAGE MECHANICS

REFERENCE: Chuang, T.-J. and Duffy, S. F., "A Methodology to Predict Creep Life for Advanced Ceramics Using Continuum Damage Mechanics," Life Prediction Methodologies and Data for Ceramic Materials, ASTM STP 1201, C. R. Brinkman and S. F. Duffy, Eds., American Society for Testing and Materials, Philadelphia, 1994.

ABSTRACT: A methodology is proposed to estimate creep rupture life for advanced ceramics such as continuous fiber reinforced ceramic matrix composites (CFCMC). Based on the premise that the damage pattern takes the form of a heterogeneous distribution of grain boundary cavities in the majority of creep life, a damage parameter is incorporated in various creep strain rate equations. The resulting constitutive equations for creep strain and accumulated damage are cast in terms of stress, and other affinities. It is pointed out that these affinities can be derived from a scalar creep potential in nonequilibrium thermodynamics. The evolutionary laws are formulated based on many micro-mechanical models. The time-dependent reliability or hazard rate for a SiC is then established by damage mechanics with Weibull analysis. A unit cell model is presented for predicting life of a uni-directional CFCMC subjected to a constant far-field stress. A system of coupled first order ordinary differential equations is derived from which the evolution of creep damage can be solved giving the rupture life. It is shown that the stress dependence on the lifetime is very sensitive to the type of damage mechanisms active at the microstructural level.

KEYWORDS: cavity growth, constitutive equation, continuum damage mechanics, creep damage, creep rupture, damage evolution, life prediction

---

In recent years, advanced ceramics such as continuous fiber reinforced ceramic matrix composites have attracted considerable attention due to certain advantages relative to conventional materials (e.g., super-alloys) in structural applications at elevated temperatures. Those advantages include increased strength, enhanced toughness, high creep and corrosion resistance in demanding service environments. The load history of a typical structural member or component, regardless of its application, normally includes a monotonically increasing load regime followed by a long period of sustained load, plus sporadic (perturbated) unsteady or cyclic thermal-mechanical load excursions. Therefore, if advanced ceramics are to be successfully and confidently used in high temperature, load-bearing

---

[1]Physicist, Ceramics Division, National Institute of Standards and Technology, Gaithersburg, MD 20899.

[2]Assistant Professor, Department of Civil Engineering, Cleveland State University, Cleveland, OH 44115.

applications they must survive the initial increasing load regime, **as well as** maintain an acceptable level of reliability during the steady-state portion of the design life. Unfortunately, the lack of a design methodology for assessing service life dominated by creep rupture is one of the major hurdles preventing widespread application of these emerging materials. There are several factors responsible for the present deficiency: (1) since fabrication of advanced ceramics is still in the developmental stage, material parameters are either unknown or constantly being improved; (2) collection of long-term test data under sustained loading conditions is time consuming, and the data collected may be obsolete and irrelevant since the material evolves into a new system due to fabrication improvements; and last but not least, (3) a lack of theories which allow extrapolation of short-term laboratory data to long-term service conditions. The last item is particularly critical for long-term applications since theoretical models are required (in addition to the experimental data necessary for verification purposes) for any material system used in the fabrication of structural components.

This paper focuses on the last issue and presents several methods from which a design engineer can obtain an estimate for the creep rupture lifetime. Since the material under consideration is aimed at applications where loads are of long duration, the design engineer is interested in the material's creep response at low or intermediate levels of sustained loading. It is now well recognized that at the lower end of the applied stress spectrum creep damage usually appears in the form of cavities during the majority of service life. The cavities are distributed in a heterogenous fashion along grain-boundary facets that have orientations coincident with the directions normal to the maximum principal tensile stress. Macroscopic crack growth (i.e., subcritical crack growth) occurs only towards the end of the creep life as a result of microcrack linkage that emerges from the coalescence of cavities growing at the grain boundaries. Under this scenario macroscopic crack growth will be insignificant during a major portion of the life of a structural component. In addition, it seems suitable to adopt continuum damage mechanics concepts such that a damage parameter is incorporated in the constitutive laws in order that the effects of microdefects are represented at the macroscopic level. It should be recognized, however, that if damage is dominated by the single crack growth mode, this approach is not applicable.

The present paper is organized in the following manner. In the section that follows the issue of energy variations during the time-dependent deformation process is discussed within the framework of non-equilibrium thermodynamics. The existence of a scalar-valued creep potential is demonstrated, from which both the creep strain and damage evolutionary laws can be derived. The authors do not present actual formulations for various creep potentials; that effort is reserved for a later date. It is simply pointed out that the constitutive equations that appear here can be obtained from an appropriate formulation of the creep potential. The results are applicable for a general three-dimensional stress states. Next, constitutive laws are presented for a variety of creep damage cases observed in the literature. For the sake of brevity, discussion is limited to only uni-axial formulations. In the section that follows, one of the constitutive equations that incorporate the principles of continuum damage mechanics is used to analyze time-dependent reliability of a monolithic ceramic matrix (i.e. silicon carbide). Here it is shown how probability of failure and hazard rate depends on the applied stress in a component undergoing uniaxial creep. The results will serve as a foundation in extending reliability analysis to multiaxial stress states using finite element methods in conjunction with a unit cell approach. However, in the final section prior to the summary, the constitutive creep laws (including damage) are applied to a class of advanced ceramics, namely continuous

fiber reinforced ceramic matrix composite subject to uniaxial tensile
creep in the fiber direction. By modeling the fiber and matrix as
separate phases connected in parallel, a system of (coupled) ordinary
differential equations of first order for stress, strain and damage is
obtained. The analysis assumes that the constituents are subject to
uniaxial states of stress (i.e, a one dimensional analysis). However,
using this technique it is possible to acquire solutions for stress,
strain and damage states as functions of time. The approach can be used
as a screening technique where a creep rupture lifetime can be estimated
from the time at which damage in the matrix or fiber reaches a critical
value.

THERMODYNAMIC FORMALISM

        Here the existence of a creep (flow) potential Ψ is demonstrated.
Evolutionary laws for both creep strain and damage can be derived from
this scalar valued potential. The approach utilizes an internal state
variable theory originally proposed by Rice [1], and is based on
irreversible thermodynamics. First, consider a continuum element for an
arbitrary material system undergoing creep deformation. The element has
a volume V which represents the initial (undeformed) material state.
Next, subject the volume element to a far-field (Kirchoff) stress tensor
Σ (which results in a total strain tensor E) while maintaining the
volume element at an isothermal temperature T. The first law of
thermodynamics dictates that

$$\dot{U} = V(\Sigma : \dot{E}) + Q \tag{1}$$

where Q is the heat increase per unit time, and U is the total internal
energy. Note that a dot designates a time derivative. The Cauchy
stress tensor $\tilde{T}$ (denoted later as σ_{ij} in terms of index notation) is
related to the Kirchoff stress tensor through the expression

$$\Sigma = [\det(F)] F^{-1} \tilde{T} (F^{-1})^T \tag{2}$$

Here F is the deformation gradient tensor, which is related to the
total strain tensor E by the expression

$$E = \left(\frac{1}{2}\right)(F^T : F - I) \tag{3}$$

Note that I is the second order unity tensor.

        If ζ represents total entropy production rate, then

$$T\dot{\eta} = Q + T\zeta \tag{4}$$

where η is the total entropy of the system. Let creep damage due to
cavitation be one pattern of microstructural rearrangement that affects
system entropy. Furthermore, let ω represent the internal state
variable associated with creep damage. The introduction of an internal
state variable allows the representation in some statistical sense of
the distribution and density of defects locally. The irreversible
process is then considered as a sequence of events where the system
passes from one constrained equilibrium thermodynamic state to another.

This follows the concepts proposed by de Groot and Mazur [2], and represents a reasonable assumption during creep, since the process is slow enough such that gradients representing the thermal and mechanical affinities are mild. The variation of Eq 1 between two neighboring constrained states takes the form

$$\delta U = V\boldsymbol{\Sigma}:\delta\boldsymbol{E} - \boldsymbol{f}\delta\boldsymbol{\omega} + T\delta\eta$$

(5)

which accounts for energy dissipation by creep damage. Writing the total free energy as

$$\phi = \phi(\boldsymbol{E}, T, \boldsymbol{\omega}) \equiv U - T\eta$$

(6)

and the Legendre transformed complimentary energy as

$$\lambda = \lambda(\boldsymbol{\Sigma}, T, \boldsymbol{\omega}) = \boldsymbol{E}:\boldsymbol{\Sigma} - \phi$$

(7)

then by Eq 5 the variation in λ becomes

$$\delta\lambda = \boldsymbol{E}:\delta\boldsymbol{\Sigma} + \left(\frac{1}{V}\right)\boldsymbol{f}\,\delta\boldsymbol{\omega} + \eta\,\delta T$$

(8)

Hence, the total strain is given by $\boldsymbol{E} = \partial\lambda/\partial\boldsymbol{\Sigma}$, and the Kirchoff stress is $\boldsymbol{\Sigma} = \partial\phi/\partial\boldsymbol{E}$. The thermodynamic force (i.e., the affinity) associated with the damage parameter is $\boldsymbol{f} = \partial\lambda/\partial\boldsymbol{\omega} = -\partial\phi/\partial\boldsymbol{\omega}$. Using Eq 3, the Maxwell relation yields

$$\frac{\partial\boldsymbol{E}(\boldsymbol{\Sigma}, T, \boldsymbol{\omega})}{\partial\boldsymbol{\omega}} = \left(\frac{1}{V}\right)\frac{\partial\boldsymbol{f}(\boldsymbol{\Sigma}, T, \boldsymbol{\omega})}{\partial\boldsymbol{\Sigma}}$$

(9)

and the second law of thermodynamics takes the form $\zeta = \boldsymbol{f}\cdot\boldsymbol{\omega}/(TV) \geq 0$. This inequality dictates that dissipative energy consumption must be non-negative. If $\delta\boldsymbol{E}$ denotes the variation in total strain between two neighboring states, then the inelastic part of the total strain is defined as that portion which results from the changes due to damage accumulation at constant stress and temperature, namely

$$(\delta\boldsymbol{E})^P = \frac{\partial\boldsymbol{E}(\boldsymbol{\Sigma}, T, \boldsymbol{\omega})}{\partial\boldsymbol{\omega}}\,\delta\boldsymbol{\omega} = \left(\frac{1}{V}\right)\frac{\partial\boldsymbol{f}(\boldsymbol{\Sigma}, T, \boldsymbol{\omega})}{\partial\boldsymbol{\omega}}\,\delta\boldsymbol{\omega}$$

(10)

Define a creep (or flow) potential $\boldsymbol{\Psi} \equiv (1/V)\int\boldsymbol{\omega}\cdot d\boldsymbol{f}$ such that

$$\dot{\boldsymbol{E}}^P = \frac{\partial\boldsymbol{\Psi}(\boldsymbol{\Sigma}, T, \boldsymbol{\omega})}{\partial\boldsymbol{\Sigma}}$$

(11)

This equation implies that if the creep potential $\boldsymbol{\Psi}$ is known, then the inelastic creep rate can be obtained simply by taking the derivative with respect to stress. In addition, if the potential surface is smooth in stress space, then the inelastic strain rate satisfies the normality condition, i.e., the vector $\dot{\boldsymbol{\varepsilon}}^P$ is the gradient to level surfaces of $\boldsymbol{\Psi}$.

The above derivation is based on a macroscopic (global) description of the materials behavior. On a local scale, creep may result from a collection of separate mechanisms that can be associated with individual internal state variables. Examples include power-law creep within the grain, and cavitation-induced creep at grain-boundaries. Let ω_i be denoted as the local damage in the ith region with volume V_i. Designating $\boldsymbol{\Psi}_i = \boldsymbol{\Psi}_i(\boldsymbol{\sigma}_i, T, \boldsymbol{\omega}_i)$ as the local creep potential associated with V_i, Cocks and Leckie [3] have shown that $\dot{\boldsymbol{\varepsilon}}_i^P = \partial \boldsymbol{\Psi}_i / \partial \boldsymbol{\sigma}$ and

$$\dot{\boldsymbol{\varepsilon}}^P = \left(\frac{1}{V_{total}} \right) \sum V_i \dot{\boldsymbol{\varepsilon}}_i^P \qquad (12)$$

Here $\boldsymbol{\sigma}$ is a local, or microscopic stress tensor. Thus the flow potential can be formulated separately for each individual micro-mechanism (see the subsequent section). An appropriate homogenization process from continuum mechanics permits the summation of individual contribution to arrive at the final result.

CONSTITUTIVE EQUATIONS

In general, explicit expressions for $\boldsymbol{\Psi}$ (from which a creep strain rate can be derived) are dependent on appropriate expressions for the damage parameter $\boldsymbol{\omega}$, which in turn, are functions of the active micro-mechanisms. However, there is no unique way of defining damage. In the case of creep deformation, damage could be defined as a crack density which is equal to the number of cracks per unit volume times the effective cracked volume (e.g., a^3, see Budiansky and O'Connell [4], Rodin and Parks [5], Duva and Huchinson [6], Hasselman, et al. [7]). Alternatively, damage could be defined as the cavity volume fraction along the grain-boundary of a volume element (Cocks and Leckie [3]) or the ratio of cavitated area over total grain-boundary area (Cocks and Ashby [8]). Different authors have pointed out that for a general anisotropic material subject to multi-axial stress states, damage must have a tensorial character. Damage can be quantified either by a fourth or eighth order tensor (see Ju [9], or Chow and Wang [10]). However, for macroscopically isotropic and homogeneous solids subject to uniaxial tensile creep, it is sufficient to characterize damage in terms of fraction of total cavitated grain-boundary areas (Cocks and Ashby [8], Cocks and Leckie [3]). Utilizing this method as a measure of damage, it is possible to develop several constitutive laws for uniaxial creep while simultaneously incorporating continuum damage principles. These constitutive laws can be derived from micro-mechanistic models that have found application to ceramic materials. These micro-mechanistic models include power-law creep due to grain-boundary sliding, lenticular cavity growth in rigid grains, and crack-like cavity growth in rigid as well as elastic grains. Other possible kinetic laws are outlined by Cocks and Ashby [8], where kinetic laws that address dislocational creep, trans-granular hole growth, and grain-boundary reaction with precipitates are presented. These mechanisms are more relevant to creep in metal alloys and will not be discussed here.

Phenomenological Power-law Creep

In the absence of damage, ceramic composites with poly-crystalline fibers and matrix can deform by power-law creep. This assumes that the material is fully dense and subject to very low stress levels. Here the creep potential takes the form:

$$\Psi = \frac{\sigma_0 \dot{\epsilon}_0}{n+1} \left(\frac{\sigma_e}{\sigma_0} \right)^{n+1}$$

(13)

where the reference stress σ_0, the reference strain rate $\dot{\epsilon}_0$, and the stress exponent n are materials constants. Note that Eq 13 is the only instance where a specific form for Ψ is provided. Here σ_e is an equivalent stress defined by $\sigma_e = (3/2) (S_{ij} S_{ij})^{1/2}$ where S_{ij} is the deviatoric stress tensor. Alternatively the equivalent stress can be expressed in terms of the three Cauchy principal stresses $(\sigma_1, \sigma_2, \sigma_3)$ as $\sigma_e = \{(1/2) [(\sigma_1 - \sigma_2)^2 + (\sigma_1 - \sigma_3)^2 + (\sigma_2 - \sigma_3)^2]\}^{1/2}$. Notice that for simple uniaxial tension $\sigma_1 = \sigma$ and $\sigma_2 = \sigma_3 = 0$, thus $\sigma_e = \sigma$. Taking the derivative of Eq 13 with respect to stress yields the empirical power-law relation

$$\dot{\epsilon} = \dot{\epsilon}_0 \left(\frac{\sigma_e}{\sigma_0} \right)^n$$

(14)

For this particular classical formulation, power-law creep due to diffusion (n=1) occurs in the matrix by lattice diffusion (Nabarro [11], Herring [12]), diffusion along grain boundaries (Coble [13]), or through solution-reprecipitation (Chen [14]). Alternatively, diffusion that is accommodated by grain boundary sliding (n=2) results in grain rotation (Ashby and Verrall [15]) or the emergence of new grains (Gifkins [16]). However, literature reviews that focus on the subject of power-law creep in structural ceramics indicate that n ranges from 1 to 8, where in some instances n is reported in excess of 10. It is important to note that most values are consistently higher than 1 or 2, i.e., higher n values than predicted by conventional theories (Cannon and Longdon [17]). This discrepancy points to the need for taking creep damage into account in the formulation of constitutive creep laws. In the following sections, three prominent creep damage mechanisms observed in advanced ceramics are discussed, and constitutive equations for creep and damage rates are derived. These rates are expressed in terms of stress, a damage parameter, temperature and other relevant material properties.

Growth of Lenticular Cavities in Rigid Grains

Consider a lenticular cavity with radius a located at a grain boundary with an area A. The area fraction $(\pi a^2 / A)$ is subject to an applied far field uniaxial stress σ_∞ in a direction normal to A. For growth of a lenticular cavity in a rigid grain, the following conditions must be met: (1) surface diffusivity dominates (i.e., grain-boundary diffusivity is minimal) such that $\Delta = (D_s \delta_s / D_b \delta_b) \gg 1.0$; (2) the applied stress is low in comparison to the capillary (or sintering) stress; and (3) the focus is on the early stage of creep (Chuang, et al. [18]). Furthermore, in order to have the matrix behave rigidly, cavities must be closely spaced, i.e., cavity spacing must be small compared to cavity size a so that material diffused from the cavity surfaces can be distributed **uniformly** along the grain boundary. Fig. 1 is a transmission electron microscope (TEM) photograph taken from a post crept silicon nitride specimen subjected to a tensile stress of 125 MPa at 1350°C for 125 hours. As indicated in the photograph, lenticular cavities were formed at a grain-boundary facet.

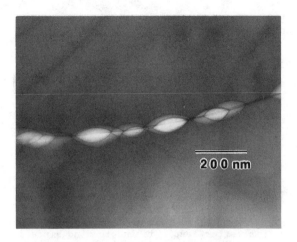

Fig. 1   TEM photograph showing lenticular creep cavities at a grain boundary (courtesy of B.J. Hockey)

Growth rates for lenticular cavities have been solved for by several authors (see Hull and Rimmer [19]; Raj and Ashby [20]; or Speight and Harris [21]). By approximating the cavity shape as spherical and neglecting surface tension effect, the evolution of ω (defined as a^2/b^2 where $2b$ is the cavity spacing) is given by

$$\dot{\omega} = \frac{\Gamma \dot{\epsilon}_0}{\sqrt{\omega}\, ln(1/\omega)} \left(\frac{\sigma_\infty}{\sigma_0} \right) \tag{15}$$

where $\Gamma = 2D_b \delta_b \Omega \sigma_0 / kTb^3 \epsilon_0$ is a material constant depending on atomic volume Ω, geometry, temperature T, and grain-boundary diffusivity, i.e., $D_b \delta_b$. Here damage is represented as a scalar quantity, and damage evolution follows a linear dependence on stress. However, the damage rate depends on the current damage state in a nonlinear fashion. A closed form solution for creep rupture life is obtained by integrating Eq 15, provided that the initial damage state is known and the stress is held constant throughout the load history (Cocks and Ashby [8]).

The creep strain rate induced by lenticular cavity growth assumes that the grain is non-deforming. In essence the strain rate is the result of a mechanism that uniformly distributes material at the grain-boundary ligament in a jacking fashion. The resulting strain rate is given by the expression

$$\dot{\epsilon} = \frac{2\Gamma \dot{\epsilon}_0}{ln(1/\omega)} \left(\frac{b}{d} \right) \left(\frac{\sigma_\infty}{\sigma_0} \right) \tag{16}$$

As was the case for the damage rate, the inelastic strain rate is linearly dependent on stress, and increases in a nonlinear fashion with increasing ω. It should be noted that Eq 16 only represents that portion of the strain rate contributed by cavity growth. The total strain rate is the summation of contributions from all active deformation mechanisms (including elastic deformations).

Growth of Crack-like Cavities in Rigid Grains

When surface diffusion is slow in comparison to grain-boundary diffusion, or when the applied stress is high in comparison to the sintering stress, or during later stages of creep life, cavities grow into crack-like shapes (Chuang, et al. [18]). The shape of the growing crack at steady state conditions is controlled by surface diffusion. The solution obtained by Chuang and Rice [22] indicates a constant crack thickness H is developed, i.e.,

$$H = 2\sqrt{2}\sqrt{1 - \frac{\gamma_b}{2\gamma_s}} \left(\frac{D_s \delta_s \gamma_s \Omega}{\dot{a} k T} \right)^{1/3} \tag{17}$$

Note that a thinning crack will develop with high crack velocity or applied stress. This latter result is interesting since it contradicts the behavior of an elastic crack opening displacement where a larger opening occurs with higher stresses.

The growth of a crack-like cavity in a rigid grain has been considered by Chuang, et al. [18]. The damage rate in the lower stress regime can be cast in the following form:

$$\dot{\omega} = \frac{\prod_1 \sqrt{\omega} \dot{\epsilon}_0}{(1-\omega)^3} \left(\frac{\sigma_\infty}{\sigma_0} \right)^3 \tag{18}$$

where $\prod_1 = D_s \delta_s \Omega \sigma_0^3 / (\sqrt{2} k T b \dot{\epsilon}_0 \gamma_s^2)$. The creep strain rate due to jacking (i.e., the removal of material from the crack surfaces via surface diffusion and the resulting uniform deposition along the grain boundary via grain-boundary diffusion) takes the following form (see Chuang, et al. [18], Cocks and Ashby [8], or Cocks and Leckie [3]):

$$\dot{\epsilon} = \frac{\prod_2 \sqrt{\omega} \dot{\epsilon}_0}{(1-\omega)^3} \left(\frac{\sigma_\infty}{\sigma_0} \right)^2 \tag{19}$$

where $\prod_2 = 4 \prod_1 \gamma_s / d\sigma_0$.

Growth of Crack-like Cavities in Elastic Grains

In the early stages of creep when the damage is minimal, crack-like cavities growing along grain boundaries will not interact since neighboring cracks are remotely located due to large cavity spacings. Here the grains must deform elastically to accommodate material deposited primarily in the near-tip zone. A diffusive crack growth model has been presented by Chuang [23] to address this situation. In this model plane-strain conditions are assumed so that measures are based on a unit thickness. This is the only self-consistent geometry possible, i.e., the only geometry where the crack front will remain straight during growth. Once growth commences for penny shaped cracks, the axi-symmetric properties of the crack are destroyed due to crack interactions. A typical crack-like cavity is shown in Fig. 2 where a TEM photograph depicts a portion of an alumina bend bar near the tensile edge. The alumina bar was subjected to creep deformation under four-

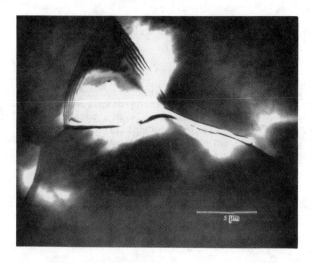

Fig. 2   Diffusional crack growth in alumina at 1720°C, 20 MPa. (After Chuang and Tighe [24])

point bend load conditions at 1720°C. The crack length is approximately 8 micrometers, and the thickness is approximately 125 nanometers. The direction of the applied stress is vertical in this microphotograph.

Solving the integro-differential equation for the unknown stress distribution at the grain boundary, the crack growth rate $\dot{a}$ can be expressed in the following form (Chuang [23]):

$$\dot{a} = V_{min}\left[0.59\frac{K}{K_G}+\sqrt{0.35(\frac{K}{K_G})^2-1}\right]^{12} \tag{20}$$

where $K \equiv \sigma_\infty\sqrt{b}[\tan(\omega/2)]^{1/2}$ and K_G and V_{min} are material constants defined as

$$K_G = \sqrt{\frac{E(2\gamma_s-\gamma_b)}{(1-\nu^2)}} \tag{21}$$

and

$$V_{min} = 8.13\frac{D_s^4\Omega^{7/3}}{kT\gamma_s^2}\left[\frac{E}{(1-\nu^2)D_b\delta_b}\right]^3 \tag{22}$$

The damage parameter ω is defined as the area fraction of cavitated area over total grain-boundary facet ($\omega=a/b$) where b is half the center-to-center crack spacing. Thus $\dot{\omega}=\dot{a}/b$, and the damage rate is

$$\dot{\omega}(\omega,\sigma_\infty) = \left(\frac{V_{min}}{b}\right)\left[0.59\frac{K}{K_G} + \sqrt{0.35\left(\frac{K}{K_G}\right)^2 - 1}\right]^{12} \qquad (23)$$

After the material is deposited non-uniformly in the vicinity of the grain boundary near the crack-tip, the displacement opening at the grain boundary widens, and the inelastic strain rate is approximated by $\dot{\varepsilon} = \langle\delta\rangle/d$ where $\langle\delta\rangle$ is the average opening displacement. Thus $\dot{\delta} = \dot{a}(\delta/x) = (\dot{a}H/L)(-\partial\delta/\partial\hat{x})$, where L is a length parameter defined as $L = (ED_b\delta_b\Omega/\dot{a}kT)^{1/2}$. Examination of the solution of $\langle\delta\rangle$ given by Chuang [23] indicates that $(-\partial\delta/\partial\hat{x}) = 0.5$. Accordingly, the strain rate takes the form $\dot{\varepsilon} = \dot{a}H/(2dL) = \Lambda\dot{a}^{7/6}/d$ or

$$\dot{\varepsilon}(\omega,\sigma_\infty) = \left(\frac{\Lambda}{d}\right)\left(\frac{V_{min}}{b}\right)^{7/6}\left[0.59\frac{K}{K_G} + \sqrt{0.35\left(\frac{K}{K_G}\right)^2 - 1}\right]^{14} \qquad (24)$$

where

$$\Lambda = \left(\frac{\gamma_s^2\Delta^2kT}{E^3D_b\delta_b\Omega}\right)^{1/6} \qquad (25)$$

In this section constitutive creep laws have been derived that are based on a number of well-accepted cavity growth models. In addition the laws also incorporate a damage parameter. In the next section attention is focused on incorporating the constitutive laws for cracklike cavities in elastic grains within the framework of a reliability theory.

CONTINUUM DAMAGE MECHANICS AND RELIABILITY

Advanced ceramic material systems have several mechanical characteristics which must be considered in the design of structural components. Focussing on laminated CFCMC the most deleterious of these characteristics are low strain tolerance and a large variation in ply failure strength specifically in the direction transverse to the fiber. Analyses of components fabricated from these types of advanced ceramics require a departure from the well-entrenched deterministic design philosophies currently utilized for metal and polymer based composites (i.e., the factor-of-safety approach). Although a diminished size effect in the fiber direction has been reported in the literature (see the discussion by Duffy et al. [25]), the bulk strength of a unidirectional-reinforced ply will decrease transverse to the fiber direction as the size of the component increases. For this reason the use of a weakest-link reliability theory is advocated in the design of components fabricated from advanced ceramics.

Reliability is defined as the probability that a component performs its required function adequately for a specified period of time under predetermined (design) conditions. Methods of analysis exist that capture the variability in strength of ceramics as it relates to fast fracture (see Gyekenyesi [26]). However the calculation of an expected lifetime of a ceramic component has been limited to a single statistical analysis based on subcritical crack growth (see Wiederhorn and Fuller [27] for a detailed development). The subcritical crack growth approach establishes relationships among reliability, stress, and time to failure based on principles of fracture mechanics. The analysis combines the

Griffith [28] equation and an empirical crack velocity equation with the underlying assumption that steady growth of a pre-existing flaw is the dominant failure mode. However creep rupture, which is highlighted here, typically entails the nucleation, growth, and coalescence of voids dispersed along grain boundaries. A method to determine an allowable stress for a given component lifetime and reliability is presented for component service life dominated by creep rupture. This is accomplished by combining Weibull analysis with the principles of continuum damage mechanics (which was originally developed by Kachanov [29] to account for tertiary creep and creep rupture of ductile metal alloys).

This effort does not represent the first application of continuum damage mechanics to brittle materials. The observed differences of creep behavior in tension and compression have been addressed through the use of damage mechanics by Krajcinovic [30] for concrete, and Rosenfield et al. [31] for ceramics. In addition, Krajcinovic and Silva [32] explored several fundamental aspects of combining damage mechanics with statistical strength theories for perfectly brittle materials. What is novel here is that the incorporation of damage mechanics within the framework of a weakest link theory allows the computation of reliability for intermediate times less than a component's given life. This effort parallels the work of Cassenti [33], however here creep rupture is highlighted and evolutionary laws for damage reflect microstructural aspects of the material. In Cassenti's work a macroscopic J_2 evolutionary law is utilized without justification.

Ideally, any theory that predicts the behavior of a material should incorporate parameters that are relevant to its microstructure (grain size, void spacing, etc.). In a section that follows a damage model outlined earlier for growth of crack-like cavities in elastic grains is utilized. This model incorporates several microstructural parameters that are known to affect creep deformation and creep rupture in ceramic materials. Here the damage model is applied only to monolithic materials. This would include the constituents of a ceramic matrix composite if the constituents are considered separately. The authors will focus future research efforts on extending the reliability model presented here to continuous fiber reinforced ceramic matrix composites (i.e., the combined or homogenized effects of damage in each constituent). This future effort will require a determination of volume averaged effects of microstructural phenomena reflecting the growth of microdefects in the matrix, along the interface between the matrix and the fiber, and in the fiber.

Theoretical Development

As noted earlier, it is assumed that the evolution of the microdefects represents an irreversible thermodynamic process. On the continuum level this requires the introduction of an internal state variable that serves as a measure of accumulated damage. This analysis is suitable for materials where damage evolution takes place in a distributed manner. Consider a uniaxial test specimen and let A_0 represent the cross-sectional area in an undamaged (or reference) state. Denote A as the current cross-sectional area associated with a damaged state where material defects exist in the cross section (i.e., $A < A_0$). The macroscopic damage associated with this specimen is represented by the scalar

$$\omega = (A_o - A)/A_o \qquad (26)$$

or alternatively by $\psi=1-\omega$, which is referred to as "continuity". The variable ψ represents the fraction of cross-sectional area not occupied by voids. A material is undamaged if $\omega=0$ or $\psi=1$.

For time dependent analysis the rate of change of continuity $d\psi/dt$ (or the damage rate $d\omega/dt$) must be specified. This rate is functionally dependent on stress and the current state of continuity, that is

$$\frac{d\psi}{dt} = f(\sigma,\psi) \tag{27}$$

Note that the damage rate must be monotonically decreasing (i.e., $d\psi/dt<0$) given an applied load. At this point an effective stress can be introduced using the concepts put forth by Lemaitre and Chaboche [34]. A damaged material element subject to a specified uniaxial stress σ_o will exhibit the same strain response (elastic or inelastic) as an undamaged material element with an applied effective stress $\tilde{\sigma}$, i.e.,

$$\tilde{\sigma} = \left(\frac{A_o}{A}\right)\sigma_o = \frac{\sigma_o}{\psi} \tag{28}$$

As Arnold and Kruch [35] point out, this approach has been supported by the results of creep rupture experiments, and damage induced by fatigue loading. A specific kinetic equation is adopted for ψ in the next section that is based on the evolution of defects at the microstructural level.

Now consider that the material element is a monolithic ceramic (or one of the constituents in a ceramic composite, *if* the boundary conditions are identical) with its inherent large scatter in strength. The variation in strength can be suitably characterized by the weakest link theory by using Weibull's [36] statistical distribution function. This is often referred to as Weibull analysis. With Weibull analysis the reliability of a uniaxial specimen is

$$R = \exp\left(-\int_V \left(\frac{\sigma}{\beta}\right)^\alpha dV\right) \tag{29}$$

This assumes that the stress state is homogeneous and that the two-parameter Weibull distribution sufficiently characterizes the material in the failure probability range of interest. Taking σ equal to the net stress defined above and (for simplicity) assuming a unit volume yield the following expression for reliability

$$R = \exp\left[-\left(\frac{\sigma}{\psi\beta}\right)^\alpha\right] \tag{30}$$

Examples of reliability curves with their dependence upon time and a specific form of the damage evolutionary law are presented in the next section.

Now consider the hazard rate function. By definition the hazard rate (or mortality rate) is the instantaneous probability of failure of a component in the time interval $(t, t+\Delta t)$, given that the component has survived to time t. In more general terms, this function yields the failure rate normalized to the number of components left in the

surviving population. This function can be expressed in terms of R, or
the probability of failure P_f, as

$$h(t) = -\frac{dR}{dt}\left(\frac{1}{R}\right) = \frac{dP_f}{dt}\left(\frac{1}{1-P_f}\right) \tag{31}$$

With Eq 30 used to define R, the hazard rate becomes

$$h(t) = -\left(\frac{\sigma}{\psi\beta}\right)^{\alpha}\left(\frac{\alpha}{\psi}\right)\left(\frac{d\psi}{dt}\right) \tag{32}$$

The hazard function can be utilized from a modeling standpoint in one of
two ways. First it can be used graphically as a goodness-of-fit test.
If any of the underlying assumptions or distributions used to construct
Eq 30 (or the explicit form of Eq 27) are invalid, one would obtain a
poor correlation between model prediction of the hazard rate and
experimental data. On the other hand, experimental data can be used to
construct the functional form of the hazard rate and R can be determined
from Eq 31. In a sense this represents an oversimplified curve fitting
exercise. Since it was assumed that the creep rupture failure mechanism
can be modeled by continuum damage mechanics, this effort has followed
the first approach. In this spirit the hazard rate function would be
used to assess the accuracy of the model in comparison to experiment.

The hazard rate is interpreted as follows:

(1)   A decreasing hazard rate indicates component failure
       has been caused by mechanical damage in the pre-
       service stage or defective processing.

(2)   A constant hazard rate indicates failure is caused by
       random factors.

(3)   An increasing hazard rate denotes wear-out of the
       component.

Here it is noted that negative values of α are physically absurd.
Hence Eq 32 yields an increasing hazard rate (recall $d\psi/dt < 0$). This is
compatible with the underlying assumption that creep rupture is the
operative failure mechanism if it is recognized that creep rupture is
strictly a wear-out mechanism. Examples of hazard rate curves based on
a particular form of the evolutionary law for ω are presented in the
section that follows.

Fundamental Implications of the Model

Unfortunately, at the present time there is a lack of experimental
data needed to properly estimate model parameters. Thus an assessment
of the model in comparison to experimental data is reserved for a later
date, and for the examples that follow, model parameters are arbitrarily
chosen for the purpose of illustration. However, the material constants
used here are representative of commercial grade silicon carbide. In
addition, a specific form of the damage rate equation must be adopted.
As noted earlier, the damage evolutionary law for crack-like cavities in
elastic grains (i.e., Eq 23) is adopted here. However, upon studying
the form of Eq 23 it is easily discerned that if the initial value of ω
is zero, the initial rate of change in damage is zero. In other words,
the damage process does not evolve in the absence of damage. Thus an

initial value of $\omega_0 = 0.15$ was adopted for the examples discussed below. Furthermore, specific values for the material constants appearing in Eq 23 are as follows: $V_{min} = 1\times10^{-11}$ m/sec; $b = 1\times10^{-4}$ m; and $K_G = 1.0$ $MPa\sqrt{m}$. In addition, the initial (fast fracture) Weibull parameters are specified as $\alpha = 10$, and $\beta = 700$. For dimensionless R, the Weibull parameter β has units of $stress\cdot(volume)^{1/\alpha}$ and the Weibull modulus α is unitless.

Fig. 3 depicts several time dependent reliability plots as a function of normalized time. Here normalized time is defined as the ratio of real time (t) divided by the time to failure (t_f). In this figure four curves representing applied stresses of σ=400, 425, 450, and 475 MPa are shown. Note that the intercepts along the vertical axis would lie on a fast fracture curve (i.e., the time independent or inert strength curve). In addition as the applied stress is increased, the relative position of the curve shifts downward. For this particular set of material parameters the reliability curves are relatively stable beyond 0.8 normalized time. However, beyond this point reliability drops off rapidly, indicating that the material is becoming unstable.

Fig. 4 represents a family of hazard rate curves. Once again the four curves in this figure represent applied stresses of σ=400, 425, 450, and 475 MPa are shown. The hazard rate curves serve as a more useful tool for the design engineer. Note that there tends to be a well-defined knee in the curves where values of the hazard rate increases rapidly with small increments in normalized time (t/t_f). The design engineer would use these curves to remove a component from service at a normalized time (t/t_f) less than the value at the breakpoint of a given curve.

APPLICATION TO CFCMC

Now consider a CFCMC with a volume fraction of fibers V_f, that is subjected to a constant remote tensile stress Σ_a such that the material creeps. A cross-section of the material under consideration is depicted in Fig. 5(a) where the direction of the applied load is parallel to the

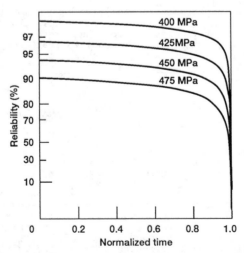

Fig. 3 A family of time dependent reliability curves demonstrating the effect of increasing applied stress.

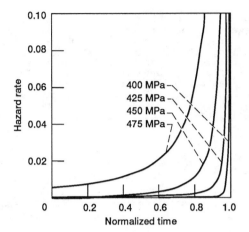

Fig. 4   Family of hazard rate curves corresponding to the reliability curves in Fig. 3.

fiber orientation. Fig. 5(b) is a schematic of the parallel connection model representing the behavior of a unit cell, which is indicated by dashed lines in Fig. 5(a). The model consists of two elements, one representing matrix material, and the other representing the fiber element. Both are connected in parallel and are subjected to far-field (remote) constant stress that gives rise to a creep strain rate. A similar exercise with metal matrix composites (MMC) has been considered by Goto and McLean [37]. A basic assumption of the model is that the fiber is perfectly bonded to the matrix phase so that no displacement discontinuities occur across the fiber/matrix interface. An analysis by Hill, et al., [38] indicated that this condition can be fulfilled by a finite element model representing a similar unit cell approach (material was a SiC/RSBN composite).

As a result of compatibility requirements, the strain rate within fiber and matrix must be equal to the remote creep strain rate. However, the stress states within the two elements will change continuously with time, although the macroscopic (or homogenized) stress remains constant. In continuum mechanics terms, this modeling approach is analogous to the idea of "constraint" cavitation proposed by Dyson [39] and Rice [40]. Denoting the time-dependent stress in the fiber and matrix as σ_f and σ_m respectively, equilibrium conditions require

$$\Sigma_a = \sigma_f V_f + (1 - V_f)\sigma_m \tag{33}$$

Since Σ_a is constant, we have $\dot{\sigma}_f = -(1 - V_f)(\dot{\sigma}_m/V_f)$ or

$$\sigma_f(t) = \epsilon_0\left[E_f + \frac{E_m(1 - V_f)}{V_f}\right] - \left[\frac{1 - V_f}{V_f}\sigma_m(t)\right] \tag{34}$$

where $\epsilon_0 = E(0) = \Sigma_a/[E_m(1 - V_f) + E_f V_f]$ is the initial elastic strain in either matrix or fiber, and the initial stresses in the fiber and matrix are $\sigma_f = E_f \epsilon_0$ and $\sigma_m = E_m \epsilon_0$, respectively. The total strain includes an elastic portion as well as inelastic strains induced by deformation mechanisms discussed previously. In the case of ceramic fibers, results from creep tests on free fiber bundle indicate that fibers usually suffer power-law creep due to grain-boundary sliding in addition to crack damage (see DiCarlo [41], and Tressler [42]). The reader is cautioned that power-law creep in an unconstrained fiber may differ from a fiber that is constrained by the matrix within a CFCMC. Fibers within a CFCMC are usually constrained such that sliding may be difficult to activate. Accordingly, the rate of fiber strain at any time can be expressed as

$$\dot{\epsilon}_f = \frac{\dot{\sigma}_f}{E_f} + A_f \sigma_f^n + F_f(\sigma_f, \omega_f) \qquad (35)$$

where F_f is the strain rate obtained from either Eq 16, Eq 19 or Eq 24, depending on the material and the active damage mechanism. Similarly, the total strain rate in the matrix is

$$\dot{\epsilon}_m = \frac{\dot{\sigma}_m}{E_m} + A_m \sigma_m^N + F_m(\sigma_m, \omega_m) \qquad (36)$$

where, again, F_m is the strain obtained from either Eq 16, Eq 19 or Eq 24. Now with $E = \epsilon_f = \epsilon_m$, from compatibility requirements the following expression

$$\dot{\sigma}_m = \left[\frac{E_m E_f V_f}{E_m + V_f(E_f - E_m)} \right] F(\sigma_m) \qquad (37)$$

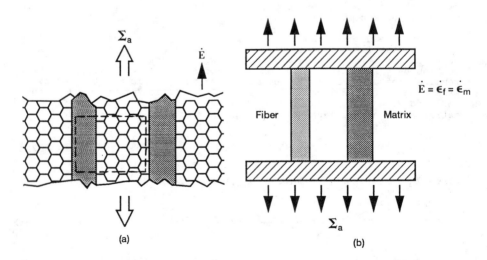

Fig. 5 Modeling of creep in CFCMC: (a) schematic of the material; (b) model of unit cell

is obtained by equating Eq 35 and Eq 36 and solving for $\dot{\sigma}_m$, subject to the initial condition $\sigma_m(0) = E_m \varepsilon_0$. Hence, a numerical algorithm can be developed to solve for the matrix and fiber stresses, the total strain and damage within both the fiber and matrix at every time step. The algorithm is as follows:

(1)   at t=0 $\sigma_m(0)$, $\sigma_f(0)$, $\omega_m(0)$ and $\omega_f(0)$ are known;

(2)   at t= $t_0 + \Delta t$, compute $\sigma_m(t)$ from Eq 37;

(3)   knowing $\sigma_m(t)$, $\sigma_f(t)$ is calculated from Eq 34 and E(t) is calculated from Eq 36;

(4)   similarly, ω can be calculated from Eq 15, Eq 18 or Eq 23;

(5)   steps (2) and (3) are repeated to obtain the stresses, strain and damage values until both ω_f and ω_m reach a critical value;

(6)   the time at which ω_f or ω_m reach a critical value becomes the creep life (t_t) of the material and computations cease.

Fig. 6 depicts a generic plot of the evolution of stress within the fiber and matrix, the local strain, and damage in a typical unit cell. Here it is assumed that the fibers are stiffer ($E_f > E_m$) and creep at a slower rate than the ceramic matrix. Thus the fibers stresses are higher than the ceramic matrix, and damage accumulates more in the fiber resulting in earlier fracture than the matrix. In this way, the effect on the rupture life of material properties (e.g., volume fraction of fibers), and other parameters such as stress and temperature can be studied.

It should be reiterated that this unit cell model assumes a perfect bonding between the two constituents so that the whole body is simply a building block of these unit cells. In the case where local slips between fiber and matrix do occur as indicated by the general phenomena of fiber pull-out in many conventional composites, this model is inapplicable and a finite element approach using some type of interface or interphase elements may have to be implemented. Unlike the unit cell where creep strain is independent of location, it becomes a field parameter and will vary throughout the body.

CONCLUDING REMARKS

A methodology to estimate creep service life based on continuum mechanics concepts has been presented. A numerical algorithm is presented to solve the initial value problem in which stress, strain and damage states can be computed in a time-step fashion, if the constitutive creep laws which dictate the materials behavior are given. In this regard, it is shown by irreversible thermodynamics that a creep flow potential exists from which the evolution of a damage parameter, and inelastic strain rate can be derived. A number of creep damage mechanisms that are most likely active in advanced ceramics are examined, and their corresponding constitutive laws, which incorporate damage, are established. This methodology has potential application to CFCMC (a promising aerospace material) where estimates of creep life can be made when stress, temperature, volume fractions of the constituents, and material properties are known.

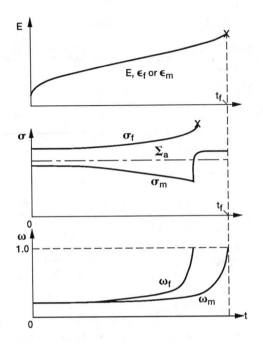

Fig. 6 Schematic plots of stresses, strain and damage developed in a
typical CFCMC as a function of time.

In addition, a time-dependent reliability model was developed by
integrating the principles of continuum damage mechanics within the
framework of Weibull analysis. It was assumed that the failure
processes of subcritical crack growth and creep rupture are separable in
that failure due to the former is a result of preexisting flaws, whereas
creep rupture is characterized by the nucleation, growth, and
coalescence of a new population of natural flaws. The nucleation of new
flaws is a grain boundary phenomenon and an attempt was made to
reconcile the continuum damage formulation to existing material science
models that predict void growth along the grain boundary of
polycrystalline ceramics. However the main objective of this work was
to provide the design engineer with a reliability theory that
incorporates the expected lifetime of a ceramic component undergoing
damage in the creep rupture regime. Several features of the model were
presented in a qualitative fashion, including predictions of both
reliability and hazard rate. The predictive capability of this approach
depends on how well the macroscopic scalar state variable captures the
growth of these grain boundary microdefects. The influence of
microdefects can be measured and used to quantify damage. Density
change, acoustic attenuation, and change in the reliability of the
material by using the concept of effective stress are methods that can
be used to quantify damage. The use of nondestructive evaluation
techniques is currently being explored to accomplish this task.

ACKNOWLEDGEMENTS

This work was partially supported by the HITEMP Program at NASA Lewis Research Center (Program monitor is Dr. John P. Gyekenyesi) under Interagency Agreement No. C-82000-R with NIST. The authors are grateful to Drs. James A. DiCarlo and Stanley R. Levine for suggesting the model problem illustrated in the section of **APPLICATION TO CFCMC**.

REFERENCES

[1]   Rice, J.R., "Inelastic Constitutive Relations for Solids: An Internal Variable Theory and its Application to Metal Plasticity," Journal of the Mechanics and Physics of Solids, Vol. 19, 1971, pp 433-455.

[2]   de Groot, S.R. and Mazur, P., Non-equilibrium Thermodynamics, 2nd Edition, North-Holland Publishing Company, Amsterdam, Netherlands, 1963.

[3]   Cocks, A.C.F. and Leckie, F.A., "Creep Constitutive Equations for Damaged Materials," Advances in Applied Mechanics, Eds. J.W. Hutchinson and T.Y. Wu, Academic Press, Vol. 25, 1987, pp 239-294.

[4]   Budiansky, B. and O'Connell, R.J., "Elastic Moduli of a Cracked Sold," International Journal of Solids and Structures, Vol. 12, January 1976, pp 81-97.

[5]   Rodin, G.J. and Parks, D.M., "Constitutive Models of a Power-law Matrix Containing Aligned Penny-Shaped Cracks," Mechanics of Materials, Vol. 5, No. 3, September, 1986, pp 221-228.

[6]   Duva, J.M. and Hutchinson, J.W., "Constitutive Potentials for Dilutely Voided Nonlinear Materials," Mechanics of Materials, Vol. 3, 1984, pp 41-54.

[7]   Hasselman, D.P.H., Donaldson, K.Y. and Venkateswaran, A., "Observations on the Role of Cracks in the Non-linear Deformation and Fracture Behavior of Polycrystalline Ceramics," in Fracture Mechanics of Ceramics, Vol. 10, Eds. R.C. Bradt, D.P.H. Hasselman, D. Munz, M. Sakai and V. Ya. Shevchenko, Plenum Press, New York, 1992, pp 493-507.

[8]   Cocks, A.C.F. and Ashby, M.F., "On Creep Fracture by Void Growth," Progress in Materials Science, Vol. 27, 1982, pp 189-244.

[9]   Ju, J.W., "A Micromechanical Damage Model for Uniaxially Reinforced Composites Weakened by Interfacial Arc Microcracks," Journal of Applied Mechanics, Vol. 58, December 1991, pp 923-930.

[10]   Chow, C.L. and Wang, J., "An Anisotropic Theory of Continuum Damage Mechanics for Ductile Fracture," Engineering Fracture Mechanics, Vol. 27, May 1987, pp 547-558.

[11]   Nabarro, F.R.N., "Steady-State Diffusional Creep," Philosophical Magazine, Vol. 16, 1967, pp 231-237.

[12]   Herring, C., "Diffusional Viscosity of a Polycrystalline Solid," Journal of Applied Physics, Vol. 21, 1950, pp. 437-445.

[13]   Coble, R.L.,"A Model for Boundary Diffusion Controlled Creep in Polycrystalline Materials," Journal of Applied Physics, Vol. 34, June 1963, pp 1679-1682.

[14]   Chen, C.F., "Creep Behavior of Sialon and Siliconized Silicon Carbide Ceramics," Ph.D. Dissertation, University of Michigan, Ann Arbor, MI, 1987.

[15]   Ashby, M.F. and Verrall, R.A., "Diffusion Accommodated Flow and Superplasticity," Acta Metallurgica, Vol. 21, January 1973, pp 149-163.

[16]   Gifkins, R.C., "Grain Rearrangements during Superplastic Deformation," Metallurgical Transactions, Vol. 13, November 1978, pp 1926-1936.

[17]   Cannon, R.W. and Langdon, T.G., "Review : Creep of Ceramics, Part I: Mechanical Characteristics," Journal of Materials Science, Vol. 18, January 1983,pp 1-50.

[18]   Chuang,T.-J., Kagawa,K.I., Rice J.R., and Sills, L.B., "Overview No.2: Non-equilibrium Models for Diffusive Cavitation of Grain Interfaces," Acta Metallurgica, Vol. 27, February 1979, pp 265-284.

[19]   Hull, D. and Rimmer,D.E., "The Growth of Grain-Boundary Voids under Stress," Philosophical Magazine, Vol.4 [42] 1959, pp 673-687.

[20]   Raj, R. and Ashby, M.F., "Intergranular Fracture at Elevated Temperature," Acta Metallurgica, Vol. 23, June 1975, pp 653-666.

[21]   Speight, M.V. and Harris,J.E., "Kinetics of Stress-Induced Growth of Grain-Boundary Voids," Metal Science Journal, Vol. 1 [1] 1967, pp 83-85.

[22]   Chuang T.-J. and Rice, J.R.,"The Shape of intergranular creep Cracks Growing by Surface Diffusion," Acta Metallurgica, Vol. 21, No. 12, December 1973, pp 1625-1628.

[23]   Chuang, T.-J., "A Diffusive Crack Growth Model for Creep Fracture," Journal of the American Ceramic Society, Vol. 65, No. 2, February 1982, pp 93-103.

[24]   Chuang T.-J. and Tighe, N.J., "Diffusional Crack Growth in Alumina," in Proceedings of the 3rd International Conference on Fundamentals of Fracture, pp 129-132, Ed. P. Neumann, Max-Plank-Institut für Eisenforschung GmbH, Düsseldorf, Germany, June, 1989.

[25]   Duffy, S.F., Palko, J.L., and Gyekenyesi, J.P., "Structural Reliability Analysis of Laminated CMC Components," Journal of Engineering for Gas Turbines and Power, in press.

[26]   Gyekenyesi, J.P., "SCARE - A Post-processor Program to MSC/NASTRAN for Reliability Analysis of Structural Ceramic Components, Journal of Engineering for Gas Turbines and Power, Vol.108, 1986, pp 540-546.

[27]   Wiederhorn S.M., and Fuller, E.R., "Structural Reliability of Ceramic Materials," Material Science and Engineering, Vol. 71, 1985, pp 169-186.

[28]   Griffith, A.A., "The Phenomena of Rupture and Flow in Solids," Philosophy, Transactions of the Royal Society of London, Series A 221, 1921, pp 163-198.

[29]   Kachanov, L.M., "Time of the Rupture Process Under Creep
       Conditions," Izvestia Akademi Nauk SSSR, Otd Tekh. Nauk 8, 1958,
       pp 26-31 (in Russian).

[30]   Krajcinovic, D. "Distributed Damage Theory of Beams in Pure
       Bending," Journal of Applied Mechanics, Vol. 46, 1979, pp 592-596.

[31]   Rosenfield, A.R., Shetty, D.K, and Duckworth, W.H., "Estimating
       Damage Laws from Bend-test Data," Journal of Materials Science,
       Vol. 20, 1985, pp 935-940.

[32]   Krajcinovic, D. and Silva, M.A.G., "Statistical Aspects of the
       Continuous Damage Theory," International Journal of Solids and
       Structures, Vol. 18, 1982, pp 551-562.

[33]   Cassenti, B.N., "Time-Dependent Probabilistic Failure of Coated
       Components," AIAA Journal, Vol. 29, No. 1, January 1991,
       pp 127-134.

[34]   Lemaitre, J. and Chaboche, J.L., Mechanics of Solid Materials,
       Cambridge University Press, 1990.

[35]   Arnold, S.M. and Kruch,S., "Differential Continuum Damage
       Mechanics Models for Creep and Fatigue of Unidirectional Metal
       Matrix Composites," NASA Technical Memorandum 105213,
       November 1991.

[36]   Weibull, W., "A Statistical Theory for the Strength of Materials,"
       Ingneoirs Vetenskaps Akademien Handlinger, No. 151, 1939.

[37]   Goto, S. and McLean, M., "Role of Interfaces in Creep of Fibre-
       reinforced Metal-Matrix Composites - I. Continuous Fibres," Acta
       Metallurgica et Materialia, Vol. 39, No. 2, February, 1991,
       pp 153-164.

[38]   Hill, B.B., Hahn, H.T., Bakis, C.E. and   Duffy, S.F.,
       "Micromechanics-Based Constitutive Relations for Creep of SiC/RSBN
       Composites," in HITEMP Review 1991, NASA CP-10082, 1991, pp. 58-1
       through 58-13.

[39]   Dyson, B. F., "Constrained Creep Cavitation," Metal Science,
       Vol. 10, 1976, pp 349-353.

[40]   Rice, J.R., "Constraints on the Diffusive Cavitation of Isolated
       Grain Boundary Facets in Creeping Polycrystals," Acta
       Metallurgica, Vol. 29, 1981, pp 675-681.

[41]   DiCarlo, J. A. "Creep of Chemically Vapor Deposited SiC Fibers,"
       Journal of Materials Science, Vol. 21, January 1986, pp 217-224.

[42]   Tressler, R.E. and Pysher, D.J., "Mechanical Behavior of High
       Strength Ceramic Fibers at High Temperatures," in Proceedings of
       7th CIMTEC World Ceramic Congress, Ed. P. Vincenzini, Elsevier
       Science Publishers, Amsterdam, Netherlands, 1991.

Luen-Yuan Chao[1] and Dinesh K. Shetty[2]

TIME-DEPENDENT STRENGTH DEGRADATION AND RELIABILITY OF AN ALUMINA CERAMIC SUBJECTED TO BIAXIAL FLEXURE

REFERENCE: Chao, L.-Y. and Shetty, D. K., "Time-Dependent Strength Degradation and Reliability of an Alumina Ceramic Subjected to Biaxial Flexure," Life Prediction Methodologies and Data for Ceramic Materials, ASTM STP 1201, C. R. Brinkman and S. F. Duffy, Eds., American Society for Testing and Materials, Philadelphia, 1994.

ABSTRACT: Fracture stresses of a sintered alumina ceramic were assessed in qualified uniaxial (three- and four-point) and biaxial (uniform pressure-on-disk) flexure tests under inert conditions (dry N_2, 100 MPa/s stressing rate) and in deionized water at a low stressing rate (1 MPa/s). The size and stress-state effects on the inert fracture stresses of the alumina ceramic could be explained by a reliability analysis based on randomly oriented surface flaws and a mixed-mode fracture criterion. The decreased fracture stresses measured in both uniaxial and biaxial flexure tests in water were consistent with subcritical crack growth behavior inferred from dynamic fatigue tests in water. The reliability analysis of the size- and stress-state effects on and time-dependent degradation of fracture stresses included consideration of the statistical uncertainties (90 % confidence bands) of the estimated Weibull (Weibull modulus, m, and characteristic strength, σ_θ) and slow-crack-growth (stress-intensity exponent, N, and critical crack growth rate, V_C) parameters. Results suggested that subcritical crack growth and strength degradation were more severe in biaxial flexure than in uniaxial flexure, which implies an interaction between stress state and subcritical crack growth.

KEYWORDS: reliability analysis, subcritical crack growth, dynamic fatigue, mixed-mode fracture criterion, flexure tests.

[1]Research Associate, Department of Materials Science and Engineering, University of Utah, Salt Lake City, UT 84112.

[2]Professor, Department of Materials Science and Engineering, University of Utah, Salt Lake City, UT 84112.

Fracture stresses of structural ceramics assessed in laboratory tests typically exhibit large scatter reflecting the variability of the severity of the strength-controlling flaws. Fracture strengths are, therefore, often presented in terms of the parameters of a distribution function, most commonly, the Weibull distribution and its parameters, Weibull modulus, m, and characteristic strength, σ_θ [1]. The characteristic strength is affected by the following test variables: (a) surface area or volume of the ceramic specimen under stress, (b) stress gradients in the specimen, (c) stress state in the specimen, and (d) environment and/or temperature of the test specimen. The effects of environment and/or temperature on characteristic strength are manifested as a decrease in the strength with decreasing stressing rate and this is referred to as dynamic fatigue. The Weibull modulus, on the other hand, is independent of the first three variables and only weakly dependent on the stressing rate provided the same flaw population controls the fracture strength under all test conditions.

The dependence of the characteristic strength on specimen size, stress gradient and stress state is treated by extreme-value statistical fracture theories, such as the classical theory of Weibull [2] or the more recent fracture mechanics based theories of Batdorf and Crose [3] and Evans [4]. The two latter theories provide a general theoretical framework for treating multiaxial stress fracture in which specific crack-size distribution, crack orientation and fracture criterion can be employed to predict cumulative probability of fracture of a ceramic under inert conditions. The equivalence of the two latter theories has been demonstrated by Chao and Shetty[5] and a computer code for design of ceramics has been developed by Gyekenyesi [6] using this theoretical approach. Fracture stress measurements on a variety of ceramics in both uniaxial and biaxial tests are generally in agreement with the theoretical predictions [7-9].

The dynamic fatigue of structural ceramics at low temperatures is usually caused by environment-assisted subcritical growth of cracks [10]. At elevated temperatures, the subcritical crack growth may be affected by creep deformation or damage [11]. Evans [10] has developed a general methodology to describe stressing rate dependence of fracture stress based on an empirical relation between subcritical crack growth rate and stress intensity of the crack. This same general methodology has been extended to predict time to failure under sustained stress, i.e. static fatigue, using subcritical crack growth parameters inferred from dynamic fatigue tests [12,13].

It is clear from the above literature review that specimen size and stress-state effects and stressing rate effects on fracture stresses of structural ceramics have been studied quite extensively but independently. There have not been many attempts to combine the two studies in the same ceramic to see if there is significant interaction between stress-state effect and time-dependent strength degradation. Accordingly, the objective of the present study was to see if strength degradation due to slow crack growth in biaxial flexure can be predicted from inert fracture stresses and dynamic fatigue assessed in simple uniaxial tests. A commercial-grade alumina ceramic§, known to exhibit slow crack growth in water, and qualified uniaxial and biaxial flexure tests were employed for this purpose in the present study.

---

§ Grade AD-94, Coors Ceramics, Norman, OK.

MATERIAL AND TEST METHODS

Alumina Ceramic

The alumina ceramic used in this study was liquid-phase sintered and contained nominally 94% of aluminum oxide (Al_2O_3). Two batches of the ceramic were purchased from the same commercial source. The first batch was in the forms of plates (127 x 127 x 5 mm) and rods (50.8 mm in diameter, 76.2 mm long). The plates and rods were made from the same powder lot with identical isostatic pressing and sintering conditions. The cylindrical surfaces of the rods were ground by the manufacturer to ensure the roundness to a maximum deviation of ± 0.0025 mm. The fracture toughness of the ceramic, K_{Ic} = 4.13 ± 0.07 MPa$\sqrt{m}$, was measured using short-bar specimens. The Young's modulus (E) and Poisson's ratio (ν) were measured in uniaxial compression to be 297.2 GPa and 0.23, respectively. Microstructure of this ceramic has been reported in reference 9. The average grain size was about 6 μm.

The second batch of the ceramic was purchased subsequently to examine dynamic fatigue in biaxial stress state. It was in the form of rods with dimensions identical to those of the first batch.

Uniaxial Flexure Tests and Specimens

Beam specimens were cut from plates and initially ground on all faces with a 180 grit diamond wheel. The tension faces were finished by a specific grinding procedure consisting of a sequence of grinding steps designed to minimize strength anisotropy. The final grinding direction was either parallel or transverse to the uniaxial stressing direction. The final dimensions of the specimens were 3 x 4 x 45 mm. The grinding procedure was as follows:

1. The tension surface was first ground with the 180 grit diamond wheel to remove a total thickness of 0.23 mm using 0.0127 mm depth of cut per pass.
2. The specimen was rotated 120° with respect to the initial grinding direction and the surface was ground with a 320 grit diamond wheel to remove a total thickness of 0.1 mm using 0.005 mm depth of cut per pass.
3. The specimen was rotated 120° again and the surface was ground with a 600 grit diamond wheel to the final thickness using 0.005 mm depth of cut per pass.

The edges of the tension faces were rounded and polished with a final finish with 1 μm diamond to minimize edge failures.

Beam specimens were tested in four-point and three-point flexure. The loading fixtures conformed to the design recommended in the military standard [14,15]. The support span was 40 mm in both tests and the loading span was 20 mm in four-point flexure. Tests were conducted on a universal testing machine[¶] . The loading fixtures were calibrated by measuring strains on the specimens. For the four-point flexure test, two strain gauges were mounted at 5 mm on either side of the center on the tension face. Strain values measured from the two locations were within 2%. For the three-point flexure test, one strain gauge was mounted at

---

[¶] Model 1125, Instron Corp., Canton, MA.

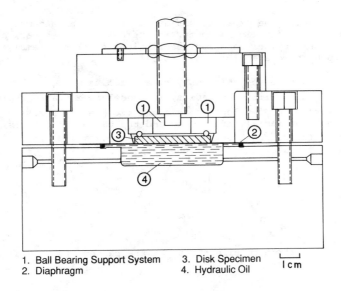

1. Ball Bearing Support System   3. Disk Specimen
2. Diaphragm   4. Hydraulic Oil |——| 1 cm

FIG. 1--Cross Section of the Hydraulic Test Cell Used in Biaxial Flexure
Tests.

the center of the tension face. The thickness normalized strain (εt^2)
was calibrated as a function of load for both four-point and three-point
flexure tests. These calibrations were used to determine the fracture
strain from the fracture load of each specimen and the fracture strains
were converted to fracture stresses using measured value of the Young's
modulus. The fracture stresses estimated from the measured strains were
typically 2.4% and 0.3% higher than the values given by the standard
bending formulae for four-point and three-point flexure, respectively.

Biaxial Flexure Test and Specimen

    Disk specimens, approximately 4.0 mm in thickness, were sliced
from the rods using diamond-impregnated blades. The as-cut specimens
were first ground on both surfaces using the 180 grit diamond wheel to
obtain flat and parallel faces, and the tension faces were then finished
by the grinding procedure described for the beam specimens. The final
thickness of the disk specimens was about 3.175 mm.

    A hydraulic test system featuring uniform pressure loading of
disks was employed in this study to measure fracture stresses in biaxial
flexure. Shetty et al.[16] have used a loading system similar to this
one. Figure 1 shows the cross section of the biaxial test cell. The disk
specimen was supported along its periphery on a ball-bearing support and
transversely loaded by uniform pressure. The specimen support unit
consisted of 40 freely rotating ball bearings (3.175 mm in diameter)
spaced uniformly along a circle, 49.53 mm in diameter. The ball bearings
were used to minimize friction at the contact points. The opposite face
of the disk was transversely loaded by uniform hydraulic pressure. A
brass foil diaphragm (0.025 mm thick) was used to separate the specimen

and hydraulic oil. The pressure was generated from an assembly
consisting of a hydraulic handpump, ram, pressure transducer, and the
universal test machine. Detailed descriptions of this assembly and
operating procedure are given in ref.[9].

A disk specimen of thickness, t, and radius, r_2, freely supported
along a concentric circle of radius, r_1, and loaded transversely by
uniform pressure, p, develops an axisymmetric biaxial stress
distribution. The radial stress, σ_r, and the tangential stress, σ_t, are
functions of radial position, r, only, and their variations, as given by
plate theory [17,18], are as follows :

$$\sigma_r = \sigma_b [1 - \alpha (\frac{r}{r_1})^2]$$ (1)

$$\sigma_t = \sigma_b [1 - \beta (\frac{r}{r_1})^2]$$ (2)

where σ_b, the maximum tensile stress at the center of the disk, is given
as:

$$\sigma_b = \frac{3pr_1^2}{8t^2} \left[2(1-v) + (1+3v)(\frac{r_2}{r_1})^2 - 4(1+v)(\frac{r_2}{r_1})^2 \ln(\frac{r_2}{r_1}) \right] + \frac{(3+v)p}{4(1-v)}$$ (3)

α and β are parameters that determine the stress gradients and are given
as:

$$\alpha = \frac{3p(3+v)r_1^2}{8t^2\sigma_b}$$ (4)

$$\beta = \frac{3p(1+3v)r_1^2}{8t^2\sigma_b}$$ (5)

The biaxial loading system was calibrated by measuring strains on
disk specimens under load. A strain gauge was mounted at the center of
the tension surface of a disk specimen and the thickness normalized
strain (εt^2) was measured as a function of the applied pressure (p).
Calibration was established with three specimens varying slightly in
thickness. The measured strains (εt^2) showed good reproducibility for
the three specimens. The agreement between the measured values and the
theoretical values calculated from Eq. 3 was excellent (measured strains
were typically only 1.5% lower than the predicted values). The strain
calibration was used to calculate the fracture stress from the fracture
pressure of each disk specimen using the Young's modulus assessed in
uniaxial compression.

Test Environments

Two types of environment were employed to assess fracture stress
distributions in inert and slow-crack growth enhanced conditions. For
the inert condition, specimens were tested in dry N_2 and a stressing

rate of 100 MPa/s. For the subcritical-crack-growth enhanced condition, specimens were tested in deionized water and a stressing rate of 1 MPa/s. Both types of tests were conducted at room temperature.

For the dynamic fatigue experiments, specimens were tested in deionized water with various stressing rates. The stressing rates selected were 0.02, 0.1, 10 and 100 MPa/s.

RESULTS

Fracture Stresses in Inert Condition

The probability of fracture for each specimen in a test group was defined by the following rank statistics:

$$F = \frac{(i-0.5)}{N} \qquad (6)$$

where i is the rank of a specimen in increasing order of fracture stress and N is the sample size. The "best fit" strength distribution for single flaw population was determined for each set of strength data using the maximum likelihood method [19] and expressed in terms of the two-parameter Weibull distribution function:

$$F = 1-\exp\left[-\left(\frac{\sigma}{\sigma_\theta}\right)^m\right] \qquad (7)$$

where m and σ_θ are the Weibull modulus and the characteristic strength of the distribution, respectively. Figure 2 shows the linearized Weibull plots of the three sets of fracture stress data obtained in inert condition. Table 1 lists the values of the two parameters determined for the three fracture stress distributions. For the three-point and the four-point flexure tests, the differences between the average fracture stresses of specimens with grinding directions parallel and transverse to the stressing direction were less than 1.5%.

The statistical uncertainty of the experimental data and of the distribution parameters must be quantified in order to accurately assess size and stress-state effects on fracture stresses. The parametric bootstrap technique described by Johnson and Tucker [20] was used to determine the 90% confidence bands and intervals on the measured stress distributions and the estimated Weibull parameters, respectively. In the parametric bootstrap technique, the best fit Weibull function for a given set of fracture stress data (Eq. 7) is first determined using the maximum likelihood method. The best fit Weibull distribution is then used as a base distribution to generate simulated fracture stresses by randomly selecting fracture probabilities and corresponding stress values from the base distribution. Specifically, the 'linear congruential method' [21] was used in this study to generate pseudo-random numbers on a computer. Parameters used in the generator were selected according to the rules developed by Knuth [22]. Integer numbers were randomly generated in the range (0, 121500); each number was normalized by 121500 to define a fracture probability, F, according to Eq. 6. Corresponding to each fracture probability, a simulated fracture stress was calculated from the base stress distribution function. The

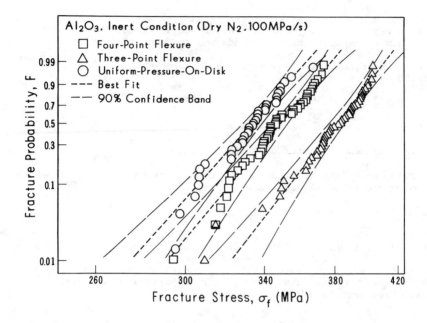

FIG. 2--Linearized Weibull Plots of the Fracture Stresses Measured In
Different Tests in Inert Condition

Table 1--Weibull Parameters Estimated for Strength Distributions
Measured in Different Tests in Inert Condition.

Parameter	Four-Point Flexure		Three-Point Flexure		Biaxial Flexure	
	Best Fit	90% confidence interval	Best Fit	90% confidence interval	Best Fit	90% confidence interval
m	23.77	20.48 ~ 30.11	25.43	21.91 ~ 32.21	22.25	18.51 ~ 28.50
σ_θ (MPa)	353.4	349.4 ~ 356.9	385.9	381.8 ~ 389.5	338.8	333.7 ~ 342.9

simulated fracture stresses in a constant sample size determined a
simulated stress distribution for which the Weibull parameters were
again estimated using the maximum likelihood method (Note: the sample
sizes for the three-point, four-point and biaxial flexure tests were 48,
48, 35 in inert condition and 35, 33, 33 in slow-crack growth enhanced
condition). Totally, 1000 simulated distributions were generated for
each test group using this procedure. The 1000 m and σ_θ values were
ordered according to their magnitudes from the lowest to the highest.

The 90% confidence intervals on the parameters were defined by the 51st and 950th values of the array. In a similar manner, at any given fracture probability, the corresponding 1000 fracture stress values taken from the simulated distributions were ordered to determine the 90% confidence limits on the fracture stress. The confidence band of the distribution was thus generated by linking the 90% limits at all fracture probabilities. Table 1 lists the 90% confidence intervals on the Weibull parameters for the three test data sets. It can be noted that the 90% intervals on the Weibull moduli overlapped in all tests. This was taken to imply that variations of the Weibull moduli estimated from the different tests could be attributed to statistical error due to limited sample size. On the other hand, the 90% confidence intervals on σ_θ did not overlap. Thus, the differences were significant and reflected size and/or stress-stats effects on fracture stresses. The 90% confidence bands on the fracture stress distributions are also plotted in Fig. 2 as the dashed curves. The uncertainty in the fracture stress estimates is larger at low and high fracture probability regimes but smaller in the intermediate fracture probability regime.

Fracture Stresses in Water and 1 MPa/s Stressing Rate

Figure 3 shows the linearized Weibull plots of the fracture stresses measured in the three tests in water. Also shown in the figure are the best Weibull fits and the 90% confidence bands of the distributions. Table 2 lists the values of the best-fit Weibull parameters and their 90% confidence intervals. The 90% confidence intervals on the Weibull moduli again overlapped, thus implying that the differences could be due to sampling error only. However, significant strength degradation and modest increase in Weibull moduli could be noted for all three tests when compared with the results obtained in inert condition (Table 1).

Dynamic Fatigue in Four-Point Flexure

Fifteen specimens were tested at each stressing rate. The maximum likelihood method was used to fit Weibull distribution to each group. The median fracture stresses (i.e. σ_f at F = 0.5) taken from the fitted Weibull distributions are plotted as a function of stressing rate in Figure 4. The 90% confidence intervals on these median fracture stresses were calculated using the 'bootstrap method' described earlier. The 90% confidence intervals are shown in the figure as the error bars. The median fracture stress and its 90% confidence intervals for the stress distribution assessed in water at 1 MPa/s are also plotted in the figure. The inert median fracture stress is plotted in the figure as a shaded band with its width indicating its confidence interval. It can be noted that despite the small sample size the error in these median fracture stresses due to statistical uncertainty in sampling is small. This is in agreement with the shape of the confidence bands of the Weibull distributions illustrated in Figs. 2 and 3.

The dynamic fatigue data were combined with the inert fracture stresses to infer a crack growth rate (V) - stress intensity (K_I) relation describing subcritical crack growth. The following power function is commonly used to describe a single-stage slow crack growth behavior [23]:

$$V = V_c \left(\frac{K_I}{K_{IC}}\right)^N \qquad (8)$$

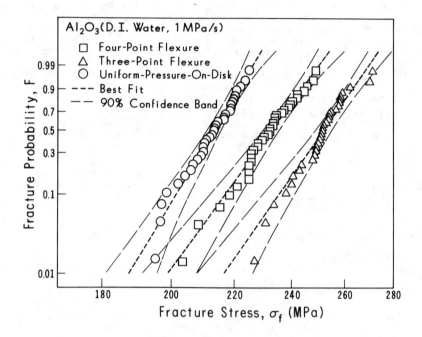

FIG. 3--Linearized Weibull Plots of the Fracture Stresses Measured in
Three Different Flexure Tests in Water

Table 2--Weibull Parameters for Strength Distributions Estimated in
Three Different Flexure Tests in Water and 1 MPa/s Stressing Rate
Condition.

Parameter	Four-Point Flexure		Three-Point Flexure		Biaxial Flexure	
	Best Fit	90% confidence interval	Best Fit	90% confidence interval	Best Fit	90% confidence interval
m	26.70	22.21 ~ 34.20	27.83	23.05 ~ 36.43	32.66	26.95 ~ 42.15
σ_θ (MPa)	236.3	233.3 ~ 238.7	255.6	252.6 ~ 258.0	215.9	213.8 ~ 217.7

where K_{IC} is the fracture toughness and N and V_C are two empirical
parameters described as crack growth exponent and critical crack
velocity, respectively. Equation 8 can be integrated to express the

time-dependent fracture stress (σ_f) as a function of stressing rate ($\dot{\sigma}$),

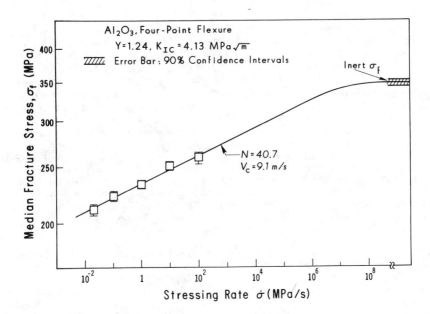

FIG. 4--Stressing Rate Dependence of the Median Fracture Stress Assessed
in Four-Point Flexure in Water and the Best Fit Dynamic Fatigue Relation
Based on Single-Stage V-K$_I$ Relation.

initial crack size (a$_i$) and critical crack size (a$_c$) for a given
fracture probability, F :

$$\sigma_{f(F)} = \dot{\sigma} \left[\frac{2K_{IC}^N}{V_C Y_0^N (N-2)} \left(a_{i(F)}^{1-(N/2)} - a_{c(F)}^{1-(N/2)} \right) \frac{N+1}{\dot{\sigma}^N} \right]^{1/(N+1)} \qquad (9)$$

The initial crack size, a$_i$, was estimated from the inert fracture
stress, σ_I, using the following relation :

$$a_{i(F)} = \frac{K_{IC}^2}{\sigma_{I(F)}^2 Y_0^2} \qquad (10)$$

where Y_0 is a fracture mechanics parameter which incorporates the
influences of crack shape, free surface and stress gradient effects on
the stress intensity factor. At a given stressing rate, Eq. 9 can be
used to calculate the time-dependent fracture stress (σ_f) for an initial
crack size (a$_i$) estimated from the inert fracture stress (σ_I) using the
slow crack growth parameters N and V$_C$. The following procedure was used.
A pair of N and V$_C$ values was initially assumed (the N and V$_C$ values
obtained by fitting the linearized fatigue data with least square linear
regression method provided good starting values). Equation 9 was then

used first to calculate a tentative value for σ_f for an assumed small crack length increment, $a = a_i + \Delta a$. The stress intensity factor at that crack size was calculated using $K_I = \sigma_f Y_0 \sqrt{a}$. The resulting K_I was compared to the critical stress intensity factor, K_{IC}. This procedure was used repetitively with gradual increments in 'a' until the calculated K_I reached K_{IC}. The final crack length was the critical crack length (a_c) and the corresponding stress was the fracture stress (σ_f). Using this calculation scheme, stressing rate dependence of fracture stress could be generated for any given pair of values of N and V_C. Various combinations of N and V_C were used to predict dynamic fatigue relation and compared with the experimental data. The best fit parameters were defined by the criterion that the sum of the variance in stress was a minimum. The solid line in Fig. 4 represents the best fit dynamic fatigue relation determined by this procedure. As can be seen, the best fit dynamic fatigue relation predicts a strength saturation at the inert strength level at a stressing rate of $\sim$ 2×10^7 MPa/s. The best fit values of N and V_C were determined to be 40.7 and 9.1 m/s, respectively.

Several features of the above calculation should be noted here. First, Eq. 9 was derived from Eq. 8 based on the assumption that there was no threshold stress intensity factor (K_{It}) required to initiate crack extension. Second, it was assumed that Y_0 remained constant ($\sim$1.24 [24]) during slow crack growth. In principle, variation of Y_0 can also be included in the calculation using a specific fracture mechanics model. Third, the method used for calculating the best fit relation is different from and superior to the traditional method where least-square linear regression is applied to the linearized data (i.e. the coordinates of the stress and stressing rate on a logarithmic scale).

ANALYSIS OF THE STRENGTH DATA

The objective of the analysis was to see if the fracture stress distributions assessed in three-point and biaxial flexure in slow crack growth conditions could be predicted from the inert fracture stress distribution and the subcritical crack growth behavior assessed in four-point flexure. The analysis was done in two stages. First, the inert fracture stress distribution measured in four-point flexure was used as a basis to predict inert fracture stress distributions in three-point and biaxial flexure using statistical fracture formulations of size and size and stress-state effects, respectively. Analytical formulations of the size and stress-state effects on inert fracture stress distributions have been discussed in detail in Ref. 9. A brief review is provided in the following for the sake of completeness. In the second stage of the analysis, fracture stress distributions in slow crack growth conditions in three-point and biaxial flexure were predicted from the respective inert fracture stress distributions predicted in the first stage using the slow crack growth parameters assessed from Fig. 4.

Predicted fracture probabilities are presented in two forms. In the first form, predictions based on the best fit Weibull parameters (Weibull modulus and characteristic strength) of the base four-point distribution are presented as single Weibull distributions. In the second form, an upper and lower bound predictions were established by taking into account the statistical uncertainties of the measured

Weibull parameters of the base distribution and the slow-crack-growth parameters. These prediction bands were established at the the 90% confidence level. In all cases, a prediction was considered satisfactory if the 90% confidence band of the measured distribution (shown in Figs. 2 and 3) overlapped the 90% confidence prediction band at any stress level.

Size and Stress-State Effects on Inert Fracture Stresses

Fractographic examinations revealed that fracture of all specimens initiated from the surfaces placed in tension [9]. For surface flaw fracture, the theory of Batdorf and Crose [3] gives the following equation for probability of fracture:

$$F = 1 - \exp\left[-\int_A \int_0^{\sigma_h} \frac{\Omega}{2\pi} \frac{dN(\sigma_c)}{d\sigma_c} d\sigma_c \, dA\right] \tag{11}$$

In Equation 11, σ_c is the critical normal or effective stress of a crack, σ_h is an upper limit that σ_c can achieve and is a function of the fracture criterion. σ_h can be greater than the applied principal stress, σ, for some strong shear-sensitive fracture criteria. Ω is a solid angle in the principal stress space that encloses all the normals to crack planes so that an effective stress, σ_c, which is a function of the applied stress, σ, and crack orientation, satisfies the fracture criterion. $N(\sigma_c)$ is a crack-size distribution function characterizing the surface flaws and in this study it was assumed to be the following [3]:

$$N(\sigma_c) = \bar{k} \, \sigma_c^{\ m} \tag{12}$$

where $\bar{k}$ and m are the scale and shape parameters in analogy with the Weibull parameters. Application of Eq. 11 to the four-point, three-point and biaxial flexure specimens was discussed in detail in a previous paper [9]. Three factors that affect the formulation in Eq. 11 are briefly reviewed here. First, based on fractographic examination and fracture patterns of the tested biaxial disk specimens the planes of the fracture initiating cracks were found to be normal to the tension face and the crack-orientation distribution was determined to be random in the two-dimensional plane. Second, the fracture initiating flaws were assumed to be half-penny shaped surface cracks. Third, a noncoplanar strain energy release rate fracture criterion [25,26] was used in this study. For a semi-circular surface crack, this fracture criterion reduces to [9]:

$$\frac{\sigma_n}{\sigma_c} + \left[\frac{2\,\tau}{C\,(2-\nu)\,\sigma_c}\right]^2 = 1 \tag{13}$$

where σ_n and τ are the normal and shear stress components on the crack plane and C is a parameter indicating the shear sensitivity of the criterion.

Figure 5 shows the linearized Weibull plots of the fracture stresses measured in the three-point and the biaxial flexure tests in

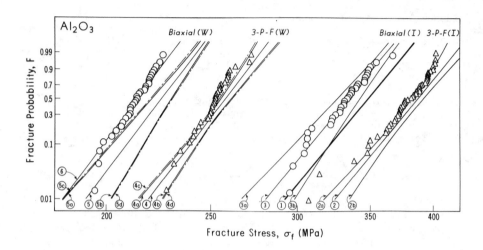

Fig. 5--Weibull Plots of Fracture Stresses and Predictions for Three-
Point and Biaxial Flexure Tests in Inert and Slow-Crack-Growth
Conditions.

inert and slow crack growth conditions. The heavy solid line (line 1)
represents the Weibull function fitted to the inert fracture stresses
assessed in four-point flexure and used as the base distribution. The
fracture probabilities in the inert three-point flexure (denoted as 3-P-
F(I)) were predicted using Eq. 11. The ratio of the fracture stresses in
three-point and four-point flexure at any given fracture probability is
a function of specimen geometries and the Weibull modulus [9]:

$$\frac{\sigma_3 (F)}{\sigma_4 (F)} = f_1 (\text{geometries, } m) \qquad (14)$$

The light solid line 2 represents the prediction of fracture
probabilities in inert three-point flexure based on the best fit m and
σ_θ values of the fracture stress distribution in inert four-point
flexure (i.e. line 1) using Eq. 14. The prediction slightly
underestimated the measured fracture probabilities. The upper and lower
bound predictions were constructed using the following procedure. First,
the 1000 simulated distributions (i.e. 1000 sets of (m, σ_θ)) for the
inert four-point flexure as described earlier were recalled; predictions
for the inert three-point flexure were made based on each of them using
Eq. 14. Next, at a given fracture probability, the corresponding 1000
fracture stress values taken from the predicted distributions were
ordered to determine the 90% confidence limits. The prediction band for
the inert three-point flexure test was thus constructed by linking the
90% confidence limits at all fracture probabilities. This is indicated
by the curved lines 2a and 2b in Fig. 5. The Weibull distribution fitted
to the measured fracture stresses in three-point flexure lay within the
two prediction bounds; thus, it was considered a satisfactory
prediction.

The fracture probabilities in the inert biaxial flexure test (denoted as Biaxial(I) in Fig. 5) were predicted using Eq. 11 by accounting for both size and stress-state effects. The ratio of the fracture stresses in biaxial and four-point flexure is a function of specimen geometries, the Weibull modulus and the fracture criterion (i.e. the value of C, the shear-sensitivity parameter in Eq. 13) [9]:

$$\frac{\sigma_b(F)}{\sigma_4(F)} = f_2(\text{geometries, m, fracture criterion}) \qquad (15)$$

According to Eq. 15, predictions based on the best fit m and σ_θ values of line 1 depend on the C value. Mixed-mode fracture toughness envelopes for polycrystalline alumina ceramics subjected to combined mode I-mode II loading have been experimentally measured by Singh and Shetty[27] using diametral compression and by Suresh et al.[28] using asymmetric four-point bending techniques. Equation 13 with C values of 2.0 and 1.0 adequately described the respective fracture toughness data. For the alumina ceramic used in this study, it has been demonstrated in the previous paper [9] that the C value corresponding to the best fit predictions to the inert biaxial strength data was 1.0. Predictions based on C = 1 value are shown in the figure as the straight line 3. The upper- and lower-bound fracture probability predictions for the same C value were constructed in a manner similar to that described for the three-point flexure with Eq. 15 replacing Eq. 14. These bounds are shown as the curved lines 3a and 3b.

The prediction bands for the inert three-point and the inert biaxial flexure tests have a common characteristic. The stress corresponding to the minimum separation of the confidence bands, i.e. the 'pinch points', are located at the same fracture stress level (~ 365 MPa). This is the same stress level at which the base distribution has its highest confidence. In other words, confidence intervals translate vertically during predictions; the highest confidence of a prediction is maintained at the same stress level where data for predictions were obtained.

Analysis of Time-Dependent Fracture Stresses

Fracture probabilities for the three-point flexure in water (denoted as 3-P-F(W) in Fig. 5) were predicted from the respective inert fracture stress distribution predicted in the previous stage (i.e. line 2). For a given fracture probability, the time-dependent fracture stress was calculated from the corresponding inert fracture stress on line 2 using Eq. 9 with the inferred slow crack growth parameters (N = 40.7 and V_C = 9.1 m/s) using the calculation procedure described in the dynamic fatigue section. Calculations were repeated for a range of fracture probabilities. These predictions are indicated by the solid line identified as line 4 in Fig. 5. The slope of line 4 indicated a Weibull modulus equal to 25.6. This value is consistent with the following expected relation between Weibull moduli of fracture stress distributions in slow-crack growth (m(SCG)) and inert (m(inert)) conditions [23]:

$$m(\text{SCG}) = \left(\frac{N+1}{N-2}\right) m(\text{inert}) \qquad (16)$$

with N = 40.7 and m(inert) = 23.77. The prediction of line 4, however, slightly underestimated the measured fracture probabilities. The upper and lower bounds of the prediction were constructed in two steps. In the first step, only the statistical uncertainties of the Weibull parameters of the base distribution were taken into account. Two prediction bounds for the fracture probabilities were constructed from the 1000 pairs of σ_θ and m values corresponding to the inert condition, curved lines 2a and 2b, using the methodology discussed before (i.e. Eq. 9). These are shown in the figure as curved lines 4a and 4b. In the second step, the statistical uncertainties of the subcritical crack growth parameters were also considered. This involved an error analysis of these parameters as described in the following.

Errors in the estimated values of N and V_C were assessed using a procedure similar to that employed in the error analysis of the Weibull parameters. The procedure consisted of randomly selecting a median fracture stress within the 90% confidence interval at each stressing rate in water (Fig. 4) and fitting Eq. 9 with best fit values of N and V_C. This procedure was repeated 100 times to obtain 100 sets of N and V_C values. Predicted fracture stress for the three-point flexure in water, at a given fracture probability, were then calculated using Eq. 9, where the Weibull (m, σ_θ) and subcritical crack growth (N, V_C) parameters in the calculation were selected by randomly pairing the previous 1000 predicted Weibull distributions (i.e. distributions used to obtain curves 2a and 2b) and the 100 dynamic fatigue relations. 1000 fracture stresses were calculated using this procedure at a given fracture probability, and their 90% confidence interval was determined. The resultant confidence band is shown in Fig. 5 as 4c and 4d. The prediction of the degraded strength in three-point flexure is considered satisfactory since the 90% confidence band of the measured distribution overlapped the final prediction band at all stress levels.

Fracture probabilities in the biaxial flexure tests in water (denoted as Biaxial(W)) were predicted from the predicted inert fracture stress distributions (straight line 3 and curved lines 3a, 3b) using Eq. 9 and the same calculation procedure. The predictions are shown as the solid line 5 obtained by considering the best fit values of m, σ_θ, N and V_C, the curved lines 5a, 5b obtained by considering the uncertainties of m and σ_θ and best fit values of N and V_C and curved lines 5c, 5d obtained by considering the uncertainties of m, σ_θ, N and V_C. The Weibull modulus of line 5 is the same as line 4. It is apparent that even after taking into account the statistical uncertainties of both the Weibull and the subcritical crack growth parameters the measured fracture probabilities are greater than predictions. The degraded biaxial fracture stress distribution and majority of its 90% confidence band fell outside the prediction bounds and, therefore, are not in agreement with the predictions.

DISCUSSION

The discrepancy noted above between the measured and the predicted fracture stresses in biaxial flexure tests in water warranted further study because of its important implications for reliability analysis. The discrepancy implied that strength degradation in biaxial flexure was intrinsically more severe than that observed in uniaxial flexure. This suggested an interaction between stress-state and subcritical crack growth. Specifically, the results suggested that subcritical crack growth and, therefore, strength degradation were more severe in biaxial flexure as compared to uniaxial flexure. To verify this implication, dynamic fatigue tests were conducted in biaxial flexure in water.

The study of dynamic fatigue behavior in the biaxial flexure test was complicated by the lack of availability of the ceramic material of the first batch. Specifically, all of the ceramic purchased in the form of rods in the first batch had been used in the two biaxial test series. For the biaxial tests required for the dynamic fatigue study it was necessary to purchase additional rods and prepare new disk specimens. Unfortunately, the alumina ceramic in the second batch turned out to be not exactly identical to the ceramic in the first batch in terms of its fracture stresses. This is evident from the inert fracture stress and the stressing rate dependence of the median fracture stresses plotted in Figure 6. Approximately, 10 disk specimens each from the new batch were tested in the inert condition and at each of the stressing rates in the water tests. The best fit Weibull distributions were determined for each group. The median fracture stresses taken from the best fit distributions and their 90% confidence intervals are plotted as a function of stressing rate. For comparison, the median fracture stresses and their 90 % confidence intervals of the inert fracture stresses and the fracture stresses in water at 1 MPa/s of the first batch ceramic are also plotted in the figure. Fracture stresses of the disk specimens from the new batch were consistently higher than the fracture stresses of the first batch in both the inert and the water environments. The differences in the fracture stresses of the disk specimens from the two batches could likely come from one of two sources : (1) difference in fracture toughness from batch to batch, and/or (2) different flaw populations or flaw-size distributions between the two batches. In this study, fractures of all specimens were initiated by surface flaws produced in the surface finishing. Since all specimens were machined and surface finished identically, it was reasonable to assume that the higher strength of the second batch was due to a higher fracture toughness, K_{IC}. This conclusion was also supported by the fact that the 90% confidence interval on m for the inert fracture stress distribution of the second batch of disk specimens, 24.01 $\Leftrightarrow$ 36.11, overlapped significantly with the corresponding interval of m for the biaxial flexure test for the first batch (see Table 1). This was taken to imply that the flaw populations were very likely identical in the two batches. On the other hand, the 90% confidence interval on the median fracture stress, σ_M, for the second batch ceramic was determined to be 337.7 $\Leftrightarrow$ 355.1 MPa. This interval barely overlapped the corresponding interval of σ_M for the inert biaxial fracture stress of the first batch (328.2 $\Leftrightarrow$ 337.9 MPa). Therefore, the difference between the σ_M values of the two batches was considered significant and reflected the difference in the fracture toughness of the two batches. The fracture toughness of the second batch ceramic was, therefore, estimated from the ratio of the two inert median fracture stresses and the fracture toughness of the first batch ceramic assuming the same flaw-size distribution:

$$K_{IC\,2} = K_{IC\,1} \left(\frac{\sigma_{M\,2}}{\sigma_{M\,1}} \right) \tag{16}$$

$K_{IC\,2}$ was calculated to be 4.30 MPa$\sqrt{m}$.

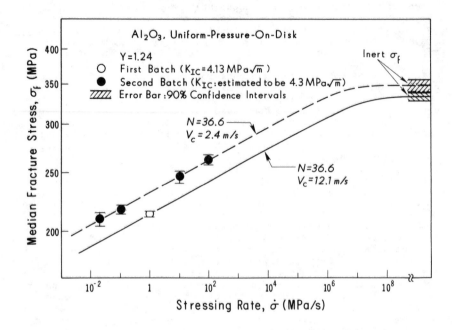

FIG. 6--Stressing Rate Dependence of the Median Fracture Stress Assessed
in Biaxial Flexure in Water and the Best Fit Dynamic Fatigue
Relations Based on Single-Stage V-K_I Relation.

The dashed line in Fig. 6 represents the best fit of the dynamic
fatigue relation, Eq. 9, to the fracture stress data of the second batch
obtained by minimizing the total variance of the measured fracture
stresses. The best fit slow crack growth parameters calculated were N =
36.6 and V_C = 2.4 m/s. A similar dynamic fatigue analysis could not be
performed for the first batch in view of the limited data in water. In
Eq. 8, the slow crack growth exponent, N, is generally considered a
function of the material and test environment. It is not clear at this
point, however, whether there is any dependence of N on fracture
toughness. It was assumed in the following analysis that N value did not
change with fracture toughness but the critical velocity, V_C, was
different for the two batches. Thus, the N value determined from the
dynamic fatigue relation of the second batch ceramic was used for the
first batch. Only the value of V_C was determined to be 12.1 m/s to
achieve agreement with the measured fracture stress in water at 1 MPa/s.
This is shown by the solid line in Fig. 6.

Table 3 summarizes the best fit values of the subcritical crack
growth parameters, N and V_C, assessed from the four-point and inferred
for the biaxial dynamic fatigue test data of the first batch. It can be
noted that the N value was lower and V_C was higher for biaxial flexure
as compared to uniaxial flexure. A lower N value and a higher V_C value
predict greater degradation of fracture stress in dynamic fatigue. Thus,

Table 3--<u>Best Fit Values and 90% Confidence Intervals on N and V_C Determined From Four-Point and Biaxial Dynamic Fatigue Tests.</u>

| Parameter | Four-Point Flexure | | Biaxial Flexure | |
	Best Fit	90% confidence interval	Best Fit	90% confidence interval
N	40.7	36.8 ~ 44.6	36.6	34.1 ~ 40.1
V_C (m/s)	9.1	2.2 ~ 43.0	12.1	-

the above trend in N and V_C noted for the dynamic fatigue in biaxial flexure relative to uniaxial flexure can explain the discrepancy noted before in the prediction of the fracture stresses in biaxial flexure in water. Fracture stress distribution of the biaxial flexure in water was, therefore, recalculated based on line 3 and the subcritical crack growth parameters obtained from biaxial dynamic fatigue. The results are shown in Fig. 5 as solid line 6. The good fit of line 6 was, of course, expected since the slow crack growth parameters used in this prediction were based on the fracture stresses measured in biaxial flexure in water at 1 MPa/s.

An interesting observation to be made here is with respect to the influences of the subcritical crack growth parameters, N and V_C, and their uncertainties on the predictions of strength in water. Although the parameters N and V_C strongly influence strength degradation via Eq. 9, the uncertainties in N and V_C do not lead to widely separated prediction band. This is evident, for example, from a comparison of the prediction bounds 5c and 5d relative to the prediction bounds 5a and 5b. This small enlargement of the prediction band when the uncertainties in N and V_C are taken into account is due to the fact that uncertainties in N and V_C are not independent but coupled. Thus, a high value of N in the 90% confidence interval is coupled with a high value of V_C and vice versa. This coupling tends to negate or minimize the strong influences N and V_C separately have on the degraded strength in water. For this very reason, the N and V_C values obtained from dynamic fatigue in biaxial flexure predict lower strength in water (line 6) as compared to the predictions based on N and V_C from the four-point bend tests (lines 5c and 5d), even though nominally there is no significant difference between the two sets of the slow crack growth parameters.

A single-stage $V-K_I$ relation was inferred in this study because it was adequate to describe the stressing rate dependence of the fracture stress of the alumina. Often, ceramics do exhibit multistage $V-K_I$ relation. In principle, dynamic fatigue tests can still be used to estimate the slow-crack-growth parameters of a multistage $V-K_I$ relation [10]. However, to do this, fracture stresses should be measured over a much wider stressing rate range. Due to the limitations of the universal testing machine and the relatively low strength of the alumina, the highest stressing rate used in this study was 100 MPa/s. Existence of an intermediate stage II in the V- K_I relation can only be ascertained by testing specimens at still higher stressing rates [10]. If dynamic fatigue tests are to be used for precisely estimating the slow-crack growth parameters as part of an overall design methodology, fracture stresses must be assessed over a much wider range of stressing rates.

Finally, in the present study, it was assumed that cracks undergoing subcritical growth were subjected to pure mode I loading (Eq. 10). Under mixed-mode loading conditions, studies have shown that cracks kink during subcritical growth[29]. A precise assessment of subcritical crack growth must take into consideration the changing crack geometry and its effects on crack driving force.

CONCLUSIONS

1. Size and stress-state effects on the inert fracture stresses of the alumina ceramic were in agreement with a reliability analysis based on randomly oriented surface flaws and a mixed-mode fracture criterion.

2. The measured fracture stresses of the alumina in three-point bending in water were within the prediction bounds based on the inert fracture stresses measured in four-point bending and the subcritical-crack-growth parameters assessed from dynamic fatigue tests in four-point bending when the uncertainties in both the Weibull (m and σ_θ) and the subcritical crack growth (N and V_C) parameters were taken into account.

3. The fracture stresses of the alumina in biaxial flexure in water were less than the prediction bounds based on four-point bend inert and dynamic fatigue tests. The increased strength degradation in biaxial flexure was consistent with the subcritical crack growth parameters assessed directly from dynamic fatigue in biaxial flexure.

4. The strength degradation assessed in biaxial flexure was significantly greater than that in four-point bending even though the subcritical crack growth parameters, N and V_C, assessed in the two test series were not significantly different. This arises from a counteracting interaction between the uncertainties of the two subcritical crack growth parameters.

ACKNOWLEDGEMENTS

This paper is based on research supported by NASA Lewis Research Center under grant no. NAG-3-789 at the University of Utah. Discussions with Dr. John Gyekenyesi and Noel Nemeth of NASA Lewis Research Center, Dr. William Tucker of General Electric Company and Dr. John Cuccio of Allied-Signal Aerospace Company are gratefully acknowledged.

REFERENCES

[1] Weibull, W., "A Statistical Distribution of Wide Applicability," Journal of Applied Mechanics, Vol.18, No.3, September 1951, pp 293-297.

[2] Weibull, W., "A Statistical Theory of the Strength of Materials," Ingenioersvetenskapsakademiens Handlingar, No.151, 1939, pp 5-45.

[3] Batdorf, S. B. and Crose, J. G., "A Statistical Theory for the Fracture of Brittle Structures Subjected to Nonuniform Polyaxial Stresses," Journal of Applied Mechanics, Vol.41, No.2, 1974, pp 459-464.

[4] Evans, A. G., "A General Approach for the Statistical Analysis of Multiaxial Fracture," Journal of the American Ceramic Society, Vol.61, No.7-8, July 1978, pp 302-308.

[5] Chao, L. Y. and Shetty, D. K., "Equivalence of Physically Based Statistical Fracture Theories for Reliability Analysis of Structural Ceramics in Multiaxial Loading," Journal of the American Ceramic Society, Vol.73, No.7, July 1990, pp 1917-1921.

[6] Gyekenyesi, J. P., "SCARE: A Postprocessor Program to MSC/NASTRAN for Reliability Analysis of Structural Ceramic Components," Journal of Engineering for Gas Turbines and Power, Vol.108, No., 1986, pp 540-546.

[7] Giovan, M. N. and Sines, G., "Biaxial and Uniaxial Data for Statistical Comparisons of a Ceramic's Strength," Journal of the American Ceramic Society, Vol.62, No.10, October 1979, pp 510-515.

[8] Shetty, D. K., Rosenfield, A. R. and Duckworth, W. H., "Statistical Analysis of Size and Stress-State Effects on the Strength of An Alumina Ceramic," Methods for Assessing the Structural Reliability of Brittle Materials, ASTM STP 844, S. W. Freiman and C. M. Hudson, Eds., American Society for Testing and Materials, Philadelphia, 1984, pp 57-80.

[9] Chao, L. Y. and Shetty, D. K., "Reliability Analysis of Structural Ceramics Subjected to Biaxial Flexure," Journal of the American Ceramic Society, Vol.74, No.2, February 1991, pp 333-344.

[10] Evans, A. G., "Slow Crack Growth in Brittle Materials Under Dynamic Loading Conditions," International Journal of Fracture, Vol.10, No.2, June 1974, pp 251-259.

[11] Lange, F. F., "Interrelations Between Creep and Slow Crack Growth for Tensile Loading Conditions," International Journal of Fracture, Vol.12, No.5, 1976, pp 739-744.

[12] Jakus, L., Coyne, D. C. and Ritter, J. E., Jr., "Analysis of Fatigue Data for Lifetime Predictions for Ceramic Materials," Journal of Materials Science, Vol.13, 1978, pp 2071-2080.

[13] Fett, T. and Munz, D., "Lifetime Prediction for Hot-Pressed Silicon Nitride at High Temperatures," *Methods for Assessing the Structural Reliability of Brittle Materials*, ASTM STP 844, S. W. Freiman and C. M. Hudson, Eds., American Society for Testing and Materials, Philadelphia, 1984, pp 154-176.

[14] "Flexure Strength of High Performance Ceramics at Ambient Temperature," MIL-STD-1942 (MR), U.S. Department of Defense, Nov.,1983.

[15] Baratta, F. I., "Requirements for Flexure Testing of Brittle Materials", *Methods for Assessing the Structural Reliability of Brittle Materials*, ASTM STP 844, S. W. Freiman and C. M. Hudson, Eds., American Society for Testing and Materials, Philadelphia, 1984, pp 194-222.

[16] Shetty, D. K., Rosenfield, A. R., Duckworth, W. H. and Held, P. R., "A Biaxial-Flexure Test for Evaluating Ceramic Strengths," *Journal of the American Ceramic Society*, Vol.66, No.1, January 1983, pp 36-42.

[17] Szilard, R., *Theory and Analysis of Plates, Classical and Numerical Methods*. Prentice-Hall, Englewood Cliffs, N.J. 1974, pp 628.

[18] Field, J. E., Gorham, D. A., Hagan, J. T., Mathewson, M. J., Swain, M. V. and Van Der Zwaag, S., "Liquid Jet Impact and Damage Assessment for Brittle Solids", *Proceedings of the 5th International Conference on Rain Erosion and Allied Phenomena*, Cambridge, England, September 1979.

[19] Pai, S. S., Gyekenyesi, J. P., "Calculation of the Weibull Strength Parameters and the Batdorf Flaw Density Constants for Volume and Surface Flaw Induced Fracture in Ceramics"; in *Proceedings of the 3rd International Symposium on Ceramic Materials and Components for Engines*, Las Vegas, Nevada,, 1988. American Ceramic Society, Westerville, OH, 1989.

[20] Johnson, C. A., Tucker, W. T., "Advanced Statistical Concepts of Fracture in Brittle Materials," *Engineered Materials Handbook, Vol.4, Ceramics and Glasses*, ASM International, The Materials Information Society, 1991, pp 709-715.

[21] Press, W. H., Flannery, B. P., Teukolsky, S. A. and Vetterling, W. T., "Chapter 7 Random Numbers," *Numerical Recipes in C*, Cambridge University Press, New York, 1990.

[22] Knuth, D. E., *Seminumerical Algorithms*, 2nd ed., Vol.2 of the Art of Computer Programming, Addison-Wesley, 1981.

[23] Wiederhorn, S. M., "Subcritical Crack Growth in Ceramics," *Fracture Mechanics of Ceramics*, Vol.2, Edited by Bradt, R. C., Hasselman, D. P. H. and Lang, F. F., Plenum, New York, 1974, pp 613-646.

[24] Scott, P. M. and Thorpe, T. W., "A Critical Review of Crack Tip Stress Intensity Factors for Semi-elliptic Cracks," _Fatigue of Engineering Materials and Structures_, Vol.4, No.4, 1981, pp 291-309.

[25] Palaniswamy, A. and Knauss, W. G., "On the Problem of Crack Extension in Brittle Solids under General Loading," _Mechanics Today_, Vol.4, 1978, pp 87-148.

[26] Shetty, D. K., "Mixed-Mode Fracture Criteria for Reliability Analysis and Design with Structural Ceramics," _Journal of Engineering for Gas Turbines and Power_, Vol.109, 1987, pp 282-289.

[27] Singh, D. and Shetty, D. K., "Fracture Toughness of Polycrystalline Ceramics in Combined Mode I and Mode II Loading," _Journal of the American Ceramic Society_, Vol.72, No.1, January 1989, pp 78-84.

[28] Suresh, S., Shih, C. F., Morrone, A. and O'Dowd, N. P., "Mixed-Mode Fracture Toughness of Ceramic Materials," _Journal of the American Ceramic Society_, Vol.73, No.5, January 1990, pp 1257-67.

[29] Singh, D. and Shetty, D. K., "Subcritical Crack Growth in Soda-Lime Glass in Combined Mode I and Mode II Loading," _Journal of the American Ceramic Society_, Vol.73, No.12, December 1990, pp 3597-3606.

C. A. Johnson,[1] and William T. Tucker[1]

WEIBULL ESTIMATORS FOR POOLED FRACTURE DATA

REFERENCE: Johnson, C. A. and Tucker, W. T., "Weibull Estimators for Pooled Fracture Data," Life Prediction Methodologies and Data for Ceramic Materials, ASTM STP 1201, C. R. Brinkman and S. F. Duffy, Eds., American Society for Testing and Materials, Philadelphia, 1994.

ABSTRACT: An estimator is a method or algorithm to analyze fracture data and estimate useful quantities such as distribution parameters and predicted component strengths. There are advantages in efficiency and model validation that typically result from combining or pooling of fracture data from multiple specimen sizes and geometries. Three types of information are contained in pooled data sets: variability in strength within subgroups, dependence of strength on specimen size, and dependence of strength on loading geometry. Efficient pooled estimators extract information of all three types from the data to yield the best overall estimates of distribution parameters and fracture strengths. Two Weibull estimators for pooled fracture data are derived and discussed. One is based on linear regression and the other on the maximum likelihood technique. The pooled estimators are extensions of conventional linear regression and maximum likelihood estimators. The estimators are derived for the two-parameter, size-scaled, uniaxial Weibull distribution function. It has been shown, however, that these estimators are also valid for multiaxial Weibull models. The pooled estimators are demonstrated using strength data from sintered SiC tested in six different combinations of specimen size and bending configuration.

KEYWORDS: Fracture strength, pooled data, maximum likelihood, linear regression

INTRODUCTION

An estimator is a method or algorithm to analyze fracture data and estimate useful quantities. These quantities may include both point estimates such as distribution parameters and predicted strengths, as well as estimates of intervals such as confidence and tolerance bounds. Conventional Weibull estimators are limited to analysis of strength data taken on a single

---

1. Ceramist and statistician, respectively, General Electric Company, Corporate Research and Development, Schenectady, NY 12301.

specimen size and loading geometry. Methods used for such estimators include linear regression, maximum likelihood, moments methods, etc. There are several potential advantages in measuring and pooling fracture strengths from multiple specimen sizes and loading geometries, including:

- The validity of size-scaling aspects of the Weibull model can be tested.

- Point estimates can often be calculated with smaller confidence intervals.

- Fracture data may already exist from a different specimen size and geometry that could be combined with current data.

- Initial component test data can be combined with laboratory specimen fracture data to determine self-consistency and to update point and interval estimates.

In our studies of combined data,* it has been useful to categorize problems according to the type of loading and the presence or absence of size scaling. The following four "classes" of problems have been considered:

- Class I: Uniform tensile stress and single specimen size

- Class II: Uniform tensile stress and multiple specimen sizes

- Class III: Common load factor (k) and multiple specimen sizes

- Class IV: Differing load factors and multiple specimen sizes

In each, it has been assumed that the Weibull distribution with a size-scaling term (volume, area or edge length) adequately describes the probabilistic nature of strength to failure. Class I problems of strength/probability estimation are the easiest to analyze but the least useful for practical applications. On the other hand, Class IV problems are the most useful of the four but, accordingly, are the most difficult to properly analyze. The progression of increasing complexity from Class I through Class IV problems has helped in the development of flexible and rigorous Class IV estimators.

The two adjustable parameters of the Weibull distribution are derived from fracture data using a Weibull estimator. For the first three classes of problems described above, methods of estimating the parameters from fracture data are readily available in the literature. However, only a few methods have been developed for estimating parameters in Class IV problems [1,2].

This paper describes two Weibull estimators developed for Class IV problems of pooled fracture data. The first is based on linear regression techniques and the second on maximum likelihood. Both require successive approximation methods and are therefore best suited for computer analysis. For various reasons reviewed later in this paper, the maximum likelihood estimator is preferred over the linear regression estimator for most applications.

As background for derivation and demonstration of the pooled estimators, the next two sections review the size-scaled Weibull distribution function and a useful example of Class IV

* Research sponsored by the U.S. Department of Energy, Assistant Secretary for Conservation and Renewable Energy, Office of Transportation Technologies, as part of the Ceramic Technology Project of the Materials Development Program under contract DE-AC05-84OR21400 with Martin Marietta Energy Systems, Inc.

fracture data taken at six different combinations of specimen size and loading geometry. The two sections that follow then derive, demonstrate and discuss the linear regression and maximum likelihood estimators for pooled data. A final section summarizes and compares the two estimators.

THE WEIBULL DISTRIBUTION FUNCTION

Quantitative descriptions of the statistical nature of fracture in brittle materials rely on a distribution function to define the cumulative probability of failure, P, as a function of all dependent variables that influence failure such as stress level, component size and geometry, etc. The majority of distribution functions considered for this purpose are based on the weakest link concept which simply states that the entire body will fail when the stress at any flaw is sufficient for unstable crack propagation of that flaw. Examples of applicable weakest link distribution functions include the Weibull distribution [3,4] and extreme value distributions [5].

The Weibull distribution has become very popular in descriptions of ceramic fracture for a number of reasons: The function is mathematically simple and easy to manipulate; it is a weakest link distribution with the proper "limiting conditions"; it is one of the three extreme value distributions; but most important, the distribution has been successful in describing a great deal of fracture data representing many different materials from a large number of investigations. In many respects, the Weibull distribution is to weakest link problems what the Gaussian distribution is to problems where the central limit theorem applies. For these reasons, this paper will consider only the Weibull distribution.

In his 1939 paper [3], Weibull introduced both a uniaxial model of probabilistic failure and a related multiaxial model that treats some aspects of failure produced by multiaxial stresses. His uniaxial model is given by

$$P = 1 - \exp\left[-\int_V \left(\frac{\sigma}{\sigma_0}\right)^m dV\right] \tag{1}$$

where P is the cumulative probability of failure, σ is the stress of a unit volume dV at the time of failure, m is the Weibull modulus, and σ_0 is a normalizing parameter. (Note that the symbol F is commonly employed for cumulatives in the statistical literature. P, however, is more popular in the materials literature for the cumulative probability of failure.) If failure is caused by flaws that are distributed randomly through the volume of a specimen or component, then the integration is carried out over all elements of volume, dV. If the flaws are restricted to surfaces (such as machining flaws), then the integration must be carried out over all surface elements, dA.

For the general case where stress is a function of position in the body, the integration can be carried out to yield

$$P = 1 - \exp\left[-kV\left(\frac{\sigma_{max}}{\sigma_0}\right)^m\right] \tag{2}$$

where k is a dimensionless "load factor" or "structure factor" and σ_{max} is the maximum stress in the structure at the time of failure. k is then defined as:

$$k = \frac{1}{V}\int_V \left(\frac{\sigma}{\sigma_{max}}\right)^m dV \tag{3}$$

where V is the total volume covered by the integration. For the case of uniform uniaxial tension, k is unity; for all other loading geometries, k is a function of m and, when evaluated, is always less than unity. The product of k times V is often termed the "effective volume" or "stressed volume" and, as the term implies, is the volume of material that is effectively under uniform uniaxial tension.

If the uniaxial Weibull model described above is valid for a given material, then the two adjustable parameters, m and σ_0, are material constants. The values of the Weibull parameters are estimated from experimental fracture strengths using an "estimator" such as linear regression or maximum likelihood [6]. By standard convention, estimates of these parameters from analysis of data are known as $\hat{m}$ and $\hat{\sigma}_0$

In the simplest practical situation, Equation 2 can then be used to estimate the probability of failure of a component of interest where the maximum stress, the component size and the load factor are known for that structure. Conversely, Equation 2 can be solved for σ_{max} to estimate the maximum stress that can be survived in the component at a specified probability of failure (specified failure rate).

EXAMPLE OF CLASS IV STRENGTH DATA

In order to demonstrate the derivation and application of Weibull estimators for Class IV problems, strength tests were performed on specimens of GE boron-doped sintered SiC. Testing was done in six bending configurations. The A, B and C specimen geometries of the MIL-STD-1942MR (virtually identical to ASTM Standard C1161, adopted in September 1990) were tested in both 3-point and 4-point bending using rolling pin fixture designs similar to those discussed in the testing standard. Specimen A is defined to have a cross-section of 1.5 x 2.0 mm and is tested on a 20 mm outer span (10 mm inner span when 4-point testing is done). Specimen B doubles every dimension of specimen A and specimen C doubles those dimensions yet again. The minimum and maximum specimen volumes therefore vary by a factor of 64 and the areas by a factor of 16. The difference between 3-point and 4-point bending contributes another factor of approximately 10 to the range of effective volumes and areas. Therefore, the SiC data span a range of 500-1000 in effective volume and a range of 100-200 in effective area.

Specimen preparation of the sintered SiC included: isopressing of submicron spray-dried beta SiC powder (with 0.5 weight percent boron and 0.3 weight percent carbon used as sintering additives) into billets with green dimensions of approximately 10 x 60 x 200 mm; sintering at approximately 2100 C in helium to a density of approximately 96 percent of theoretical; and slicing and surface grinding of specimens according to MIL-STD-1942MR. Specimens were always removed from the billets such that the center of the specimen thickness corresponded with the center of the billet thickness. Eight billets of SiC were used where all were prepared and sintered side by side.

Six of the eight billets were used for the size-scaling study and two for a specimen-location study. In the size-scaling study, each billet was cut into 6 type A, 6 type B and 6 type C specimens for a total of 108 specimens. They were divided such that 18 specimens were tested in each of the six testing configurations. Each group of 18 contained 3 specimens from each of the billets. Testing was done carefully to minimize testing errors and to maintain information to allow detection of billet-to-billet variations, etc. The remaining two billets were cut into a total of 30 type B specimens for testing in 4-point bending only. Location-within-billet was recorded for each of the 30 specimens to allow a search for any dependence of strength on location.

All 138 bend specimens were loaded to failure at room temperature using loading rates that resulted in times-to-failure from 30 seconds to one minute. One 4-point A specimen was accidently destroyed before testing due to rapid, uncontrolled preloading. Analysis of the resulting strengths using analysis of variance techniques indicated no significant billet-to-billet variation and no significant position-to-position variation. All specimens are therefore considered to have a single distribution of flaws. All further discussions will group the specimens by size and testing geometry with no reference to billet.

The results of analyzing the specimens one group at a time are included in (Table 1). The Weibull parameters listed in (Table 1) are the output of conventional maximum likelihood analysis. Group-to-group variations in the two Weibull parameters are typical of statistical sampling error on data sets of this size. Therefore, the variation in Weibull modulus, for instance, is not believed to indicate that different groups contained different fundamental flaw distributions. As discussed in the next paragraph, failure in these specimens is controlled by a distribution of surface flaws. In order for the exponent of Equation 2 to be dimensionless, the dimensions of $\hat{\sigma}_0$ must account for the surface area term in the exponential. Therefore, the dimensions of $\hat{\sigma}_0$ are a function of the Weibull modulus and are ($MPa \cdot mm^{2/m}$).

TABLE 1 -- Analysis of sintered SiC data treated one group at a time

Geometry	Number	Av Str (Mpa)	Std Dev (MPa)	$\hat{m}$	$\hat{\sigma}_0$ (MPa $\cdot$ mm$^{2/m}$)
3-pt A	18	388.11	33.36	14.57	430.99
4-pt A	17	312.85	34.83	9.43	459.24
3-pt B	18	350.96	31.12	12.20	449.93
4-pt B	48	303.31	24.17	14.28	430.28
3-pt C	18	325.65	21.69	16.39	418.83
4-pt C	18	283.78	22.35	14.48	441.08

Fractography in the form of low power (50X) stereo microscopy was used to identify the location of each fracture initiating flaw and to determine whether it was a surface, sub-surface, or edge-related flaw. Approximately 70 percent of the fracture origins were clearly associated with the as-ground surfaces, 20 percent with chamfered edges, and 10 percent with sub-surface microstructural flaws. Those specimens that failed from sub-surface and edge-related flaws were uniformly spread through the spectrum of strengths within each group and showed no tendency to be stronger or weaker than the average. Due to the dominance of surface-related fracture origins observed in this data set, all Class IV analyses and all references to effective size discussed below assume that strength was controlled by surface flaws.

Data such as the SiC flexure strengths that contain multiple active flaw distributions should be analyzed using an estimator that can simultaneously derive the Weibull parameters of each independent flaw distribution. This can be done for Class I data sets using censored data analysis techniques, but the pooled estimators described in this paper are not derived here for use with censored data. (A derivation of the maximum likelihood estimator for pooled and censored data is included in an associated paper [7].) Nevertheless, it is important to emphasize the critical role of fractography in probabilistic failure analysis. An incorrect assumption concerning the "dimensionality" (volume, area, or edge length) of the strength controlling flaws can result in very large errors in predictions of component strength, especially if the prediction of component strength involves an extrapolation in size from the experimental fracture data.

In addition to the six values of $\hat{m}$ in (Table 1), the strength as a function of specimen size can be used to derive a seventh independent $\hat{m}$. This is illustrated in Figure 1 where the average fracture strength is plotted versus the effective area on a log-log scale. The error bars for each data point indicate the standard deviation of strengths within each group. Strength behavior consistent with the Weibull two-parameter distribution will result in a linear relationship on Figure 1. The slope of the line is the negative inverse of the Weibull modulus; therefore a regression line through the data can be used to generate yet another $\hat{m}$. This seventh estimate adds to the confusion of determining the "best" estimate of the true Weibull modulus. Should the seven estimates be averaged? Should they be weighted according to the number of specimens contributing to each estimate? The Class IV estimators described in the next two sections are more rigorous and statistically efficient approaches to extracting the useful information from data sets such as the SiC data.

LINEAR REGRESSION ESTIMATOR FOR POOLED FRACTURE DATA

The conventional Weibull linear regression estimator for data sets with a single specimen size and geometry involves regression of a transformed probability term ($\ln \ln \{1/(1-P)\}$) on a transformed fracture strength term ($\ln \sigma$). These are the axes used on Weibull probability plots such as Figure 2. When fracture data from a material with an underlying Weibull strength distribution are plotted on such axes, the data tend to fall on a straight line.

The strength of each specimen is a directly measured quantity, but the associated probabilities of failure, P, must be deduced indirectly using ranking statistics. For a group of N strength measurements, this is done by ordering the strengths from smallest to largest and assigning a ranking number, i, to each with $i = 1$ for the weakest and $i = N$ for the strongest. The probabilities of failure can then be assigned by a relationship such as:

$$P_i = (i\text{-}0.5)/N \tag{4}$$

Several other relationships are available to assign probabilities, but the one defined in Equation 4 is preferred because it yields small statistical bias errors in the resulting Weibull parameters from Class I data sets.

When linear regression is used to regress probability on strength, estimates of the two Weibull parameters are found that yield the smallest possible sum-of-squared deviations between the values of probability from ranking and the corresponding values predicted by the distribution function. Included on Figure 2 are the 48 strengths of the 4-point B specimens and the associated estimates of Weibull parameters. The small deviations of the data points from the straight line are typical of data sets of this size and should not be reason to question the validity of the distribution in describing the data.

When pooled fracture data from different specimen sizes and geometries are available, conventional linear regression techniques cannot be used directly. An additional transformation is needed to either transform all data to a common specimen size or to a common probability of failure. These transformations are easily derived from Equation 2 at constant fracture strength. Consider two specimens, A and B, with identical fracture strengths but with different sizes and/or geometries such that they have different effective volumes, kV. The specimens are fabricated from the same material and, therefore, are assumed to have the same Weibull parameters. The probabilities of failure for the two specimens can be described using Equation 2 as follows:

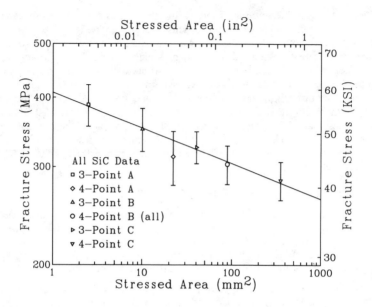

Figure 1 -   Dependence of strength on specimen size for boron-doped, sintered beta SiC
tested in six configurations of specimen size and testing geometry.   Data
points are average strengths; intervals indicate one standard deviation.

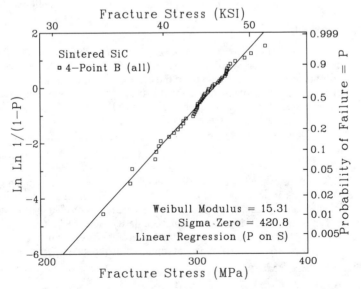

Figure 2 -   Weibull probability plot for 48 bend specimens of 4-point B geometry.
Weibull parameters estimated by regressing probability on fracture stress.

$$P_A = 1 - \exp\left[-k_A \, V_A \left(\frac{\sigma}{\sigma_0}\right)^m\right] \tag{5a}$$

$$P_B = 1 - \exp\left[-k_B \, V_B \left(\frac{\sigma}{\sigma_0}\right)^m\right] \tag{5b}$$

For simplicity, the subscript "max" from Equation 2 has been dropped from the stress in the numerator of the exponential, but the term still represents the maximum stress in the structure at the time of failure. Equation 5 can be rearranged to:

$$\ln(1 - P_A) = -k_A \, V_A \left(\frac{\sigma}{\sigma_0}\right)^m \tag{6a}$$

$$\ln(1 - P_B) = -k_B \, V_B \left(\frac{\sigma}{\sigma_0}\right)^m \tag{6b}$$

Dividing Equation 6a by 6b at constant fracture stress then yields:

$$\frac{\ln(1 - P_A)}{\ln(1 - P_B)} = \frac{k_A \, V_A}{k_B \, V_B} \tag{7}$$

By assigning A as the untransformed (as-tested) state and B as the transformed state, Equation 7 can be used to transform Class IV fracture data to either a common specimen size or to a common probability of failure. To transform all data to a common probability of failure, Equation 7 can be solved for $k_B \, V_B$; any arbitrary probability, P_B, can be chosen; and the effective volume for each observed fracture strength can be transformed to that which would be necessary to result in the chosen failure probability. When failure is controlled by surface flaws, a similar relationship is easily derived in terms of stressed area instead of stressed volume.

The transformation is illustrated graphically on Figure 3 where the 4-point C strength data are included both in the untransformed state (open data points) and in the transformed state (solid data points). In this case, all data were transformed to a common probability of failure of 0.5, the median. As mentioned above, failure in this data set is dominated by surface flaws, therefore, the parallel version of Equation 7 involving stressed area is used and the abscissa of Figure 3 is stressed area. Note that the untransformed data are all located at a single value of stressed area corresponding to the effective area of the 4-point C specimen geometry, whereas the transformed data points are plotted at varying values of stressed area. Linear regression through the untransformed data points would clearly be meaningless, but regression of the transformed data results in useful estimates of the two Weibull parameters. The linear regression line through the solid points of Figure 3 can be used to estimate the Weibull modulus, $\hat{m}$, since the slope of the line is $-1/\hat{m}$. The position of the line can be used to estimate the second parameter of the Weibull distribution, $\hat{\sigma}_0$. There is a whole family of parallel straight lines representing all probabilities of failure (quantiles) from zero to unity.

In classical linear regression, the dependent variable is regressed on the independent variable, where the dependent variable is assigned as the one expected to contain the majority of scatter due to random variations. When regression is performed on axes such as Figure 3, there is no clear resolution of which axis should be defined as independent. It is important to note that, for fracture data from a single specimen size and testing configuration (Class I problem) such as that of Figure 3, regression of the area on the fracture stress is mathematically identical to conventional analysis on Weibull probability plots where probability is regressed on fracture stress. The regression of Figure 3 was done by regressing area on strength (A on S), and the resulting parameters are included on the graph.

The same concept of transformation to a common probability of failure can be used to transform and analyze Class IV sets of pooled fracture data. This was done for the six specimen sizes and geometries of the SiC data and is included graphically as Figure 4. To reduce confusion, only the transformed data are included. The corresponding effective areas of the untransformed data can be seen on Figure 1.

Application of this linear regression estimator to Class IV pooled data becomes an iterative problem, unlike the Class I example of Figure 3. The iterative nature in Class IV problems arises because the effective areas (or volumes or lengths) of the specimens are functions of $\hat{m}$ for all loading configurations except uniform tension. Therefore, the positions (and the relative positions) of both the transformed and untransformed points on Figure 4 will shift horizontally as $\hat{m}$ is varied. The linear regression estimator, therefore, begins by assuming an arbitrary $\hat{m}$. The estimator then transforms the data and performs linear regression to determine an updated $\hat{m}$. $\hat{m}$ is iteratively calculated and updated until the assumed and calculated m's agree. An efficient algorithm for this process has been programmed that requires only four to eight iterations to determine $\hat{m}$ to greater than five significant digits for most data sets that have been analyzed.

The Weibull parameters shown on Figure 4 result when this iterative algorithm is carried out for the set of 137 strengths of SiC and when the stressed area is regressed on fracture strength (A on S). The line included on Figure 4 is the position of the 0.5 quantile behavior as predicted by the estimated parameters. It should be noted that the Weibull parameters that result from this estimator are independent of the choice of failure probability to which all data is transformed.

Most of the 137 data points including five of the six subgroups tend to visually agree with the regression line. This qualitatively confirms that the size-scaling assumption of the Weibull distribution is justified and that the magnitude of the size-scaling effect on strength is self-consistent with the variability in strength within groups. The exception to the visual agreement is the 4-point A data where 12 of the 17 strengths seem to fall at a disturbing distance below the regression line. Reevaluation of specimen preparation techniques, testing procedures, fractography and data analysis have not found a source for the apparent discrepancy. It is possible that a deviation of this magnitude is common or even expected for data sets of this size due to statistical sampling error. Continuing work on goodness-of-fit tests for Class IV data sets is currently addressing this question [8].

MAXIMUM LIKELIHOOD ESTIMATOR FOR POOLED FRACTURE DATA

The maximum likelihood approach for estimating the adjustable parameters of probability distributions is well-developed and well-respected by the statistical community. The underlying logic of the approach, however, is more abstract than that of linear regression. Possibly for that reason, many people in the materials community are uncomfortable with the approach as it is applied to the Weibull distribution. The first portion of this section will describe a simple graphical representation of the maximum likelihood technique that may help in visualizing the logic and approach of the method, as well as help in later discussions and derivations of the maximum likelihood estimator for pooled data sets.

The graphical representation will be discussed using an example of nine test specimens, all of a single specimen size and geometry (Class I problem), where the resulting strengths were found to vary from approximately 450 to 820 MPa. The nine strengths are clearly marked on the stress axes of the two graphs of Figure 5. In each of the two graphs, a probability density,

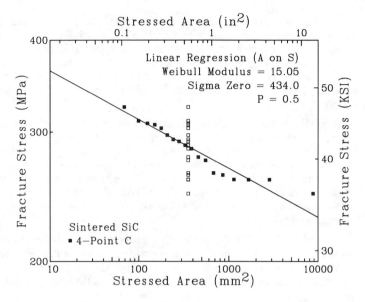

Figure 3 - Dependence of strength on specimen size for 18 4-point C specimens illustrating both untransformed data (open symbols) and data transformed to a common probability of failure of 0.5 (solid symbols).

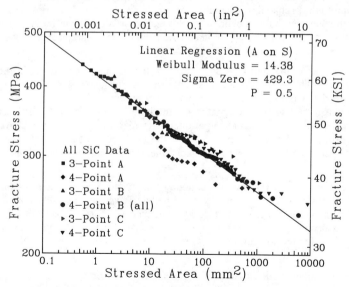

Figure 4 - Strength versus specimen size for pooled linear regression analysis of all 137 strengths of SiC. Data points are shifted in effective area to that which should cause failure at the observed strength and at a 0.5 probability of failure.

f, is plotted versus fracture stress on linear axes. The probability, P, used in most of the previous equations of this paper was the cumulative probability. The probability density is merely the derivative of the cumulative probability with respect to stress, $dP/d\sigma$. The curves shown on the two graphs are examples of Weibull distributions with different values of m and σ_0. In all cases, the area under a probability density curve must be exactly unity. Therefore, the height of the peak of the curve is generally different for different choices of Weibull parameters, as is the case in Figure 5.

Maximum likelihood techniques are iterative in that they repeatedly "guess" the values of the Weibull parameters. The two graphs of Figure 5 represent two such iterative guesses of parameters. First, consider the upper graph of Figure 5. The parameters of this graph are clearly a rather poor guess--for instance all nine observed strengths fall to the right of the peak of the curve. Since the peak of the true probability density curve corresponds to the most likely strength observed in a large number of observations, these parameters seem to be far from correct. Nevertheless, we continue consideration of this choice of parameters in terms of the maximum likelihood technique.

A quantity called the "likelihood" is defined as the product of the probability densities of each of the observed strengths. To help in visualizing the magnitude of the likelihood, an arrow has been drawn at the probability density for each of the nine fracture strengths. The likelihood for this choice of parameters would then be the product of the nine indicated densities in the top graph.

Next consider the bottom graph of Figure 5. The shape and position of the curve "appear" to be a much better fit to the nine fracture strengths. The previous statement is very qualitative, however. The likelihood turns out to be an excellent quantitative measure. Again, the nine arrows point to the probability densities corresponding to the nine fracture strengths, and the product of the densities is the likelihood for this choice of Weibull parameters. A simple inspection is sufficient to conclude that the likelihood from the lower graph is greater than the upper graph.

It has been proven in the statistical literature that the parameters resulting in the maximum value of the likelihood are very good estimates of the true parameters. In fact, in many respects they are the optimum choice of estimated parameters. The mathematics used in derivation of a maximum likelihood estimator and the iterative process used in executing the estimator are merely the means of finding parameters to maximize the likelihood. As might be expected and as seen below, the maximum of the likelihood is found by taking partial derivatives with respect to the unknowns and simultaneously setting those quantities equal to zero.

Derivation of the Class IV maximum likelihood estimator for pooled fracture data parallels the derivation of the Class I estimator as published by Trustrum and Jayatilaka [6], which in turn is based on maximum likelihood derivations such as those of Lawless [9], Mann, Schafer, and Singpurwalla [10] and Nelson [11]. Unlike Class I derivations, the Class IV estimator must account for the loading factor, k, which is generally a function of the Weibull modulus.

Due to simplifications made possible in the derivation, maximum likelihood estimators typically find the maximum of the "log likelihood" instead of the likelihood itself. The log likelihood, l, is defined as the sum of the logs of the probability densities, f, for each observed specimen. It is easy to show that the maximum of the log likelihood corresponds to the maximum of the likelihood. The probability density of Equation 2 is:

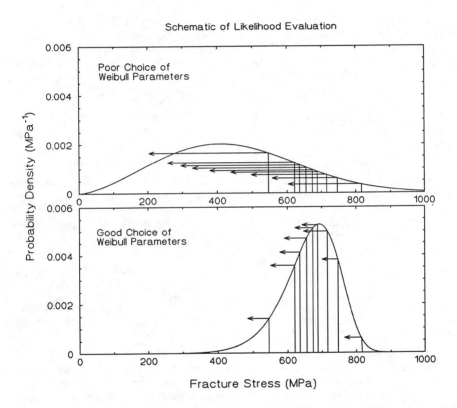

Figure 5 - Schematic of probability density curves for two choices of Weibull parameters to illustrate the maximum likelihood approach.

$$f = \frac{dP}{d\sigma} = \frac{mkV}{\sigma_0^m} \sigma^{m-1} \exp\left[-kV\left(\frac{\sigma}{\sigma_0}\right)^m\right] \tag{8}$$

Again for simplicity, the subscript "max" from Equation 2 has been dropped from the stress in the numerator of the exponential, but the term still represents the maximum stress in the structure at the time of failure. The log likelihood, l, for a group of n strength measurements is then:

$$l \equiv \sum_{i=1}^{n} \ln(f_i) = n \ln(m) - nm \ln(\sigma_0) + \sum_{i=1}^{n} \ln(k_i V_i) + (m-1) \sum_{i=1}^{n} \ln(\sigma_i) - \sum_{i=1}^{n} k_i V_i \left(\frac{\sigma_i}{\sigma_0}\right)^m \tag{9}$$

Partial derivatives of Equation 9 are then taken with respect to the two Weibull parameters, m and σ_0. The derivative with respect to m must account for the functional dependence of k_i on m. Each of these partial derivatives is then set equal to zero to find the combination of the two parameters that leads to the maximum in the log likelihood. The two resulting equations can be combined such that σ_0 drops out leaving the relationship:

$$0 = \frac{n}{m} + \sum_{i=1}^{n} \frac{dk_i/dm}{k_i} + \sum_{i=1}^{n} \ln(\sigma_i) - \frac{n \sum_{i=1}^{n} \left[k_i V_i \sigma_i^m \ln(\sigma_i) + V_i \sigma_i^m \, dk_i/dm \right]}{\sum_{i=1}^{n} V_i k_i \sigma_i^m} \tag{10}$$

The maximum likelihood estimate of the Weibull modulus is then evaluated by iteratively determining the value of $\hat{m}$ that satisfies Equation 10. Under usual conditions, only a single value of $\hat{m}$ satisfies Equation 10 for any given data set. After $\hat{m}$ is known, the second Weibull parameter can be determined without iteration by rearranging the partial of Equation 8 with respect to σ_0 that was earlier equated to zero:

$$0 = n \sigma_0^m - \sum_{i=1}^{n} k_i V_i \sigma_i^m \tag{11}$$

In order to evaluate the maximum likelihood estimates of the Weibull parameters, the dependence of both k and dk/dm must be known as a function of $\hat{m}$ for every specimen geometry tested.

An efficient algorithm for the evaluation of maximum likelihood estimates for Class IV problems has been developed that requires only five to ten iterations to determine $\hat{m}$ and $\hat{\sigma}_0$ to approximately five significant digits for most data sets that have been analyzed. The 137 strengths of SiC described above result in Weibull parameters of $\hat{m} = 14.22$ and $\hat{\sigma}_0 = 433.1$. Again, these estimates assume that failure is controlled by a distribution of surface flaws. Therefore, the dimensions of $\hat{\sigma}_0$ are a function of the Weibull modulus and are (MPa · mm$^{2/m}$).

SUMMARY

The previous two sections described and demonstrated estimators capable of pooling fracture data from multiple specimen sizes and geometries and estimating the adjustable parameters of the two-parameter Weibull distribution. The two estimators are based on linear regression and maximum likelihood respectively. Each estimator is capable of extracting information from the three sources contained in a pooled data set: variability in strength within subgroups, dependence of strength on specimen size, and dependence of strength on loading geometry. The estimators were derived for the uniaxial form of the Weibull distribution with a uniaxial effective volume (kV). The estimators are equally applicable, however, to multiaxial forms of the Weibull distribution by using a parallel concept of a multiaxial effective volume (IV) as defined in an associated paper [12]. The methodology of [12] coupled with techniques that allow pooling of censored data from multiple flaw populations have been integrated and are described in [7].

In the present paper, the uniaxial forms of the two estimators were used to analyze a pooled data set of 137 SiC bend strengths and demonstrated that the resulting parameter estimates were similar but not identical to each other. For linear regression, the parameters were $\hat{m}$ = 14.38 and $\hat{\sigma}_0$ = 429.3 MPa · $mm^{2/m}$; for maximum likelihood they were $\hat{m}$ = 14.22 and $\hat{\sigma}_0$ = 433.1 MPa · $mm^{2/m}$.

While both estimators are able to analyze pooled fracture data, they are not equal in overall capability. The maximum likelihood estimator is preferred for a number of reasons including greater compatibility with censored data analysis [11], smaller bias errors [13], and higher statistical efficiency [14].

Bias in an estimator is a consistent offset error that results in estimates that tend to be either too large or too small. Monte Carlo simulation studies of these two estimators [13] have shown that point estimates from both estimators contain bias error, but the degree of bias error in the maximum likelihood estimates is typically much smaller than that from corresponding linear regression estimates for most problems of pooled fracture data analysis that have been analyzed.

Quantitative measures of the efficiency in extracting information from pooled fracture data requires a method of calculating confidence and tolerance bounds on estimates. Two methods for calculating bounds have been developed for these pooled estimators and reported elsewhere [15,16]. One involves bootstrap simulation and the other the likelihood ratio method. Resulting bounds from the two methods are in good agreement. Studies of the efficiency of the two estimators for pooled data [14] have shown that the maximum likelihood estimator consistently results in smaller confidence bounds and is, therefore, more efficient in using available information. The difference in efficiency is substantial. It was found that the linear regression estimator typically used only 50 to 60 percent of the information that was extracted by the maximum likelihood estimator.

In conclusion, both the linear regression and the maximum likelihood estimators are capable of pooling data from multiple specimen sizes and geometries. The maximum likelihood estimator is strongly preferred, however, due to its strengths in censored data analysis, small bias errors and high statistical efficiency.

REFERENCES

[1]   C.A. Johnson and S. Prochazka, "Investigation of Ceramics for High Temperature Turbine Component," Final Report, Contract N62269-76-0243, June, 1977.

[2]   S.B. Batdorf and G. Sines, "Combining Data for Improved Weibull Parameter Estimation," *J. Am. Ceram. Soc.*, 63, 214-218, (1980).

[3]   W. Weibull, "A Statistical Theory of the Strength of Materials," *Royal Swedish Academy of Eng. Sci. Proc.*, 151, 1-45, 1939.

[4]   W. Weibull, "A Statistical Distribution Function of Wide Applicability," *J. App. Mech.*, 18, 293-297, 1951.

[5]   E.J. Gumbel, Statistics of Extremes, Columbia Univ. Press, 1958.

[6]   K. Trustrum and A. De S. Jayatilaka, "On Estimating the Weibull Modulus for a Brittle Material," *J. Mater. Sci.,* 14, 1080-1084, 1979.

[7]   J.S. Cuccio, A.D. Peralta, J.Z. Song, P.J. Brehm, C.A. Johnson, W.T. Tucker, and H.T. Fang, "Probabilistic Methods for Ceramic Component Design and Implications for Standards," Life Prediction Methodologies and Data for Ceramic Materials, ASTM STP 1201, C.R. Brinkman and S.F. Duffy, Eds., American Society for Testing of Materials, Philadelphia, 1993.

[8]   C.A. Johnson and W.T. Tucker, "Advanced Statistical Concepts of Fracture in Brittle Materials", in *Ceramic Technology for Advanced Heat Engines Project, Semiannual Progress Report for October 1991 Through March 1992,* ORNL/TM-12133, pp 247-268, September 1992.

[9]   J.F. Lawless, Statistical Models and Methods for Lifetime Data, John Wiley and Sons, NY, 1982.

[10] N.R. Mann, R. E. Schafer, and N. D. Singpurwalla, Methods for Statistical Analysis of Reliability and Life Data, John Wiley & Sons, New York, 1974.

[11] W. Nelson, Applied Life Data Analysis, John Wiley & Sons, New York, 1982.

[12] W.T. Tucker and C.A. Johnson, "The Multiaxial Equivalent of Stressed Volume," Life Prediction Methodologies and Data for Ceramic Materials, ASTM STP 1201, C.R. Brinkman and S.F. Duffy, Eds., American Society for Testing of Materials, Philadelphia, 1993.

[13] C.A. Johnson and W.T. Tucker, "Advanced Statistical Concepts of Fracture in Brittle Materials," in *Ceramic Technology for Advanced Heat Engines Project Semiannual Progress Report for October 1989 Through March 1990,* ORNL/TM-11586, pp 298-316, September 1990.

[14] C.A. Johnson and W.T. Tucker, "Advanced Statistical Concepts of Fracture in Brittle Materials," in *Ceramic Technology for Advanced Heat Engines Project, Semiannual Progress Report for October 1990 through March 1991,* ORNL/TM-11859, pp 268-282, July 1991.

[15] C.A. Johnson and W.T. Tucker, " Advanced Statistical Concepts of Fracture in Brittle Materials," Engineered Materials Handbook, Volume 4, Ceramics and Glasses, pp 709-715, ASM International, 1992.

[16] C.A. Johnson and W.T. Tucker, "Statistical Procedures for Estimating Component Strengths and Associated Confidence Bounds," in *Proceeding of the Annual Automotive Technology Development Contractors' Coordination Meeting, 1991,* 385-395, SAE Publication P-256, June 1992.

William T. Tucker,[1] and Curtis A. Johnson [1]

THE MULTIAXIAL EQUIVALENT OF STRESSED VOLUME

REFERENCE: Tucker, W. T. and Johnson, C. A., "The Multiaxial Equivalent of Stressed Volume," Life Prediction Methodologies and Data for Ceramic Materials, ASTM STP 1201, C. R. Brinkman and S. F. Duffy, Eds., American Society for Testing and Materials, Philadelphia, 1994.

ABSTRACT: A comprehensive probabilistic fracture analysis methodology should allow data to be combined or pooled from multiple specimen sizes and geometries such that all available fracture data can be integrated into strength/probability estimates and such that size-scaling aspects of the model can be tested. Moreover, this capability must, among other things, allow for multiaxial stress states, and be be capable of providing a measure of the statistical uncertainty in estimates of strength and/or probability including confidence and tolerance bounds. Previous efforts of the authors have produced estimators for combined data, confidence and tolerance bounds on estimates, bias and bias correction of estimates and measures of statistical efficiency of the estimators. These developments have been based on Weibull's uniaxial model involving k, a dimensionless "load factor" or "structure factor."

The results presented in this paper show that similar techniques are applicable to more comprehensive models of multiaxial failure. Building on previous work by the first author, reported elsewhere, and generalizing results recently reported in the literature, it is shown that the Batdorf-Heinisch's (B–H) flaw density distribution and the Lamon-Evans' (L-E) elemental strength approaches to weakest-link fracture statistics for multiaxial loading give equivalent probability predictions for equivalent failure criteria. A generalization is also given detailing necessary and sufficient conditions for the B–H and L-E approaches to be equivalent. This allows a general size factor to be defined that simultaneously takes into account geometry, loading, and multiaxial stresses. This general size factor replaces k in the estimators for combined data, confidence and tolerance bounds on estimates, bias and bias correction of estimates and measures of statistical efficiency previously developed by the authors. It can be applied on an elemental basis or to a component structure as a whole, and this has consequences in determining probability predictions. Also employing these results, it is indicated how a contradictory conclusion, also reported in the literature, was reached. All in all, there appear to be no obvious roadblocks in incorporating the effects of multiaxial stresses into current analysis

---

1. Statistician, and ceramist, respectively, General Electric Company, Corporate Research and Development, Schenectady, NY 12301.

methods. Moreover, the form of the general size factor indicates that it may be possible to obtain generalizations that cover time and/or temperature effects. Thus, the equivalence of the B-H and L-E formulations has broad reaching implications.

KEYWORDS: load factor, structure factor, multiaxial failure modes, multiaxial stresses

---

Quantitative descriptions of the statistical nature of fracture in brittle materials rely on a distribution function to define the cumulative probability of failure, P, as a function of all dependent variables that influence failure such as the stress level, the component size and geometry, etc. The majority of distribution functions considered for this purpose are based on the weakest link concept which simply states that the entire body will fail when the stress at any defect is sufficient for unstable crack propagation of that defect. The Weibull distribution has become very popular in descriptions of ceramic fracture for a number of reasons: The function is mathematically simple and easy to manipulate; it is a weakest link distribution with the proper "limiting conditions"; it is one of the three extreme value distributions; but most important, the distribution has been successful in describing a great deal of fracture data representing many different materials from a large number of investigations.

In his 1939 paper, Weibull [1] introduced both a uniaxial model of probabilistic failure and a related multiaxial model that treats some aspects of failure produced by multiaxial stresses. His uniaxial model is given by

$$P = 1 - \exp\left[-\int^V \left(\frac{\sigma}{\sigma_0}\right)^m dV\right] \qquad (1)$$

where P is the cumulative probability of failure, σ is the stress (MPa) of a unit volume dV (mm^3) at the time of failure, m is the Weibull modulus, and σ_0 $(MPa\,(mm)^{3/m})$ is a normalizing parameter. (In passing, we note that the symbol F is more commonly employed for cumulatives in the statistical literature.) If failure is caused by defects that are distributed randomly throughout the volume of a specimen or component, then the integration is carried out over all elements of volume, dV, as indicated by V (mm^3). If the defects are restricted to surfaces (such as machining defects), then the integration is carried out over all surface elements, dA (mm^2).

More generally, when stress is a function of position in the body, the integration can be carried out to yield

$$P = 1 - \exp\left[-kV\left(\frac{\sigma_{max}}{\sigma_0}\right)^m\right] \qquad (2)$$

where k is a dimensionless "load factor" or "structure factor" and σ_{max} is the maximum stress in the structure at the time of failure. For the case of uniform uniaxial tension, k is unity and Eq 2 reduces to the form of the two-parameter Weibull distribution commonly encountered in the statistical literature. For all other loading geometries, k is a function of m and, when evaluated, is always less than unity. The product of k times V is often termed the "effective volume" and as the term implies, is the volume of material that is effectively under uniform uniaxial tension.

If the uniaxial Weibull model described above is valid for a given material, then the two adjustable parameters, m and σ_0, are material constants. The value of the Weibull parameters are estimated from experimental fracture strengths. In the simplest practical situation, Eq 2 can then be used to estimate the probability of failure of a component of interest where the maximum stress, the component size and the load factor are known for that structure. Conversely, Eq 2 can be solved for σ_{max} to estimate the maximum stress that can be survived in the component at a specified probability of failure (specified failure rate).

Unfortunately, the procedure outlined above is oversimplified. In most real-life situations one or more complications are present that require a more complex analysis. Among such complications are: multiple active flaw populations, multiaxial stress states, censored and/or proof-tested data, in-service modification of flaws by oxidation, slow crack growth, etc., directional anisotropy of strength, and gradients in properties (from center to near-surface). If not accounted for, the presence of any of these will reduce the accuracy of strength/probability estimates in unpredictable and often nonconservative ways. Many of the recent efforts in fracture statistics have addressed complications such as these. While progress has been made, many difficult problems remain. In particular, progress is needed in approaches that allow more than one complication to be present simultaneously.

Also a comprehensive probabilistic fracture methodology should at least meet the further requirements: The analysis should allow data to be combined or pooled from multiple specimen sizes and geometries such that all available fracture data can be integrated into strength/probability estimates and such that size-scaling aspects of the model can be tested, and be capable of providing a measure of the statistical uncertainty in estimates of strength/probability (confidence and tolerance bounds). Previous efforts of the authors have given results that meet these requirements and yielded estimators for combined data, confidence and tolerance bounds on estimates, bias and bias correction of estimates, and measures of statistical efficiency of the estimators. All of these approaches and developments have been based on the uniaxial model of Eq 2. The remainder of this paper goes on to show the mathematical equivalence of the two most important multiaxial approaches to probabilistic strength analysis and then demonstrates how multiaxial approaches are compatible with the previously developed techniques. [2] Thus, the form of Eq 2 and methods based on it are quite general.

MULTIAXIAL METHODS

Possibly the most straightforward manner to handle multiaxial stress states is to employ the principal of independent action (PIA) assumption in which the three principal stresses are assumed to act as three independent uniaxial stresses. This results in three independent probabilities of survival which multiplied together give the overall probability of survival.

2. Research sponsored by the U.S. Department of Energy, Assistant Secretary for Conservation and Renewable Energy, Office of Transportation Technologies, as part of the Ceramic Technology Project of the Materials Development Program under contract DE-AC05-84OR21400 with Martin Marietta Energy Systems, Inc.

Batdorf [2] has shown that the pure independence assumption can be either conservative or nonconservative when compared to more refined approaches that take the multiaxial stress state into account, but is usually nonconservative in practice. As a result of the nonconservativeness, the PIA method is usually not employed. The more refined approach employed by Batdorf [2] is equivalent to assuming a failure criterion given by an effective stress defined by the loading and geometry of the crack. Failure of a particular crack occurs when σ_e (the effective stress) $\geq \sigma_c$, the critical stress of the crack. The critical stress is defined as that remote stress applied normal to the crack plane that will just result in failure of that particular crack. Most writers since Weibull use these concepts in evaluating the probability of component failure by employing limiting arguments analogous to Weibull's ([1], pages 20-21) in which a "crack density" is defined. However, this approach is questionable from a mathematical viewpoint. Weibull realized this and, indeed, it was not his first approach to the problem. Weibull's original argument, leading to Eq 1 was in terms of the (cumulative) hazard function or what Weibull called the risk of rupture. The risk of rupture approach is based only on the assumptions of a weakest link model and statistical independence of the individual links (the independence assumption can be generalized). Also the hazard function approach can be put into a limiting argument in a straightforward way. The hazard function approach will be followed herein. For future reference, note that the (uniaxial) Weibull model is given by the maximum principal stress criterion, while Weibull's [1] multiaxial model (employed by Batdorf in [2]) is given by the normal stress criterion.

The two most promising approaches to extending Weibull to cover multiaxial conditions are:

- Batdorf's crack density, due to Batdorf and co-workers and culminating in Batdorf and Heinisch [3], and
- Evans' elemental strength, due to Evans and co-workers as delineated in Lamon and Evans [4] and further explained in Lamon [5].

(Herein, the Batdorf-Heinisch approach will be referred to as the B-H approach and the Lamon-Evans as L-E.) Both of these approaches have earlier developments which are superseded by the above references. The main difference between the two approaches is in their orientation. The B-H approach in effect calculates the fraction of cracks for a given infinitesimal critical stress ($\equiv$ strength) range that would lead to failure based on crack orientation and the relationship to σ_e, i.e., the fraction for which the load, σ_e, is greater than the critical stress, σ_c ($\equiv$ strength). The L-E approach in effect calculates the fraction of strengths ($\equiv$ critical stresses) for a given infinitesimal orientation range that would lead to failure as a result of that strength being less than the critical equivalent stress (S_E, to be defined). Viewed in this context, one would expect that under a common set of assumptions both would lead to the same failure probabilities and, indeed, this is the case. In the remainder of this section, a proof of the equivalence of the B-H and L-E approaches is given, followed by results of analytical/numerical studies aimed at resolving a discrepancy between B-H and L-E as reported by Lamon [5] in re-study of fracture data from Shetty et al. [6].

Other investigators have also made progress in this area. Recently, Chao and Shetty [7] have shown by numerical methods the equivalence of the B-H and L-E approaches for two specific failure criteria. Furgiuele and Lamberti [8] show the equivalence mathematically, but their overall results are not as general as those shown herein. Also results of She, Landes, Boulet, and Stoneking [9] can be employed to show the equivalence as well. The mathematical equivalence was first shown by the authors in [10] (and improved in [11]).

The Equivalence of B-H and L-E

In the development, to the extent possible, the notation and setup common to B-H and L-E are followed. Thus a single internal failure mode with size scaling given by the volume of material under consideration is considered. To begin, consider the L-E approach where the kernel of the hazard is given by:

$$\int_0^s g(S)dS = \int_0^s \frac{1}{2\pi} \int_0^{\pi/2} \int_0^{2\pi} g(S_E)dS_E \cos\phi \, d\phi d\psi \tag{3}$$

with

$$g(S)dS = \frac{1}{2\pi} \int_0^{\pi/2} \int_0^{2\pi} g(S_E)dS_E \cos\phi \, d\phi d\psi \tag{4}$$

from Equations 1, 7, and 10 of [4]. In the L-E approach Eq 3 is integrated over the volume in

$$P_f = P_f(s) = 1 - \exp\left[-\int_V dv \int_0^s g(S)dS\right] \tag{5}$$

(cf. Equation 1 of [4]) to obtain the probability of failure at strength s. This results since the integration in Eq 3 is over those strengths that would produce failure. Lamon and Evans then make assumptions that give

$$g(S_E)dS_E = mS_1^{m-1}S_0^{-m}dS_1 F^m \tag{6}$$

(cf. Equations 9-12 of [4]) so as to evaluate Eq 4 and, ultimately, Eq 5. In deriving Eq 6, Lamon and Evans take

$$g(S_E) = mS_E^{m-1}S_0^{-m} \tag{7}$$

with

$$S_E = S_1 F\left[\frac{S_2}{S_1}, \frac{S_3}{S_1}, \phi, \psi\right] = S_1 F[.]. \tag{8}$$

(Note that the exponent, m, in Eq 6 is given incorrectly in [4], Equation 12 as $m+1$; this situation is corrected by Lamon in [12].) Since the upper limit of integration on the left hand side in Eq 3 is the strength at which failure would just occur for the particular loading geometry, failure criterion, and orientation, $sF[.]$ is the value of the failure criterion, S_E, which would just produce failure. Also note in Eq 8, $F[.]$ is not a function of S_1. This allows Eq 6 to be derived from Eqs 7 and 8, since $dS_E = dS_1 F[.]$. Finally, assume that all integrand functions are such that the order of integration can be interchanged (at will). This is certainly true for the Weibull crack density given by Eq 6 and the failure criterion employed by Lamon and Evans, and since a generalization of Eq 6 will be indicated in what follows, the assumption is made.

Lamon and Evans employ both Cartesian and spherical local coordinate systems in which σ_1 is associated with x, σ_2 is associated with y, and σ_3 is associated with z in order to produce Eq 4.

The angle ψ is measured from the x-axis in the (x,y)-plane and the angle ϕ is measured from the (x,y)-plane toward the z-axis, and are used to also define the local direction cosines. (See Figure 3 in [4].) The strength S_i is associated with the principal stress σ_i, $i = 1,2,3$. This setup will be important in considering the B-H approach, as well. Indeed both S_E and σ_e are functions of the σ_i.

In demonstrating the equivalence, it must be shown that the hazard functions resulting from each of the two approaches are equivalent. (Actually the two hazards only need be equal almost everywhere, a.e., in a measure theoretical sense. Generally, the a.e. designation will be dropped in what follows, since practically it will be of little consequence.) This follows since there is a one-to-one relationship between a probability distribution and its hazard function. Now the L-E hazard is

$$\int_V dv \int_0^s g(S)dS = \int_V dv \int_0^s \frac{1}{2\pi} \int_0^{\pi/2} \int_0^{2\pi} g(S_E)dS_E \cos\phi \, d\phi d\psi$$

$$= \frac{1}{2\pi} \int_0^{\pi/2} \int_0^{2\pi} \left[\int_V \left[\int_0^s g(S_E)dS_E \right] dv \right] \cos\phi \, d\phi d\psi$$

$$= \frac{1}{2\pi} \int_0^{\pi/2} \int_0^{2\pi} \left[\int_V \left[\int_0^s mS_1^{m-1} S_0^{-m} dS_1 F^m \right] dv \right] \cos\phi \, d\phi d\psi$$

$$= \frac{1}{2\pi} \int_0^{\pi/2} \int_0^{2\pi} \left[\int_V \left[(s/S_0)^m F^m \right] dv \right] \cos\phi \, d\phi d\psi. \tag{9}$$

Eq 9 follows by interchanging the order of integration as indicated and the results given in Eqs 3-8. However, as indicated, the product of sF in the inner integrand of Eq 9 is the failure criterion, S_E, evaluated at the stress ($\equiv$ strength) that would just produce failure. Thus another way to express Eq 9 is

$$\int_V dv \int_0^s g(S)dS = \frac{1}{2\pi} \int_0^{\pi/2} \int_0^{2\pi} \left[\int_V \left[(S_E/S_0)^m \right] dv \right] \cos\phi \, d\phi d\psi. \tag{10}$$

The reason for expressing the L-E hazard in both of the forms of Eqs 9 and 10 will become clear shortly.

The hazard kernel in the B-H approach is given from Equation 3 in [3], which is

$$P_f = 1 - \exp\left[-\int_V dv \int_0^\infty d\sigma_c \left[\frac{dN(\sigma_c)}{d\sigma_c} \right] \frac{\Omega(\Sigma, \sigma_c)}{4\pi} \right], \tag{11}$$

where $N(\sigma_e)$ denotes the critical crack density, Σ is the applied stress state throughout an elemental volume, and Ω is the solid angle within which a crack will fail (cf. [3]). Thus in order to prove the conjecture of equivalence of B-H and L-E it must be shown that

$$\int_V dv \int_0^\infty d\sigma_c \left[\frac{dN(\sigma_c)}{d\sigma_c} \right] \frac{\Omega(\Sigma, \sigma_c)}{4\pi} \equiv \frac{1}{2\pi} \int_0^{\pi/2} \int_0^{2\pi} \left[\int_V \left[(S_E/S_0)^m \right] dv \right] \cos\phi \, d\phi d\psi. \tag{12}$$

To do this, expand Ω, by noting that failure occurs if and only if $\sigma_e \geq \sigma_c$, to yield

$$\Omega(\Sigma,\sigma_c) = 2\int_0^{\pi/2}\int_0^{2\pi} I(\Sigma,\sigma_c)\cos\phi\, d\phi\, d\psi \tag{13}$$

where

$$I(\Sigma,\sigma_c) = \begin{cases} 1, & \sigma_e \geq \sigma_c \text{ at the given } (\phi,\psi) \text{ and particular location} \\ 0, & \text{otherwise.} \end{cases} \tag{14}$$

Eqs 13 and 14 follow from the definitions given by Batdorf and Heinisch [3] associated with their Equations 1-4 (see their Eqs 6 and 7 as well). Also note that Eqs 13 and 14 employ the spherical coordinate system of Lamon and Evans [4], discussed earlier. As is made clear in Batdorf and Heinisch [3], Ω is a function of both the fracture criterion selected and the applied stress state. Thus I must be a function of both as well. This is reflected in the definition given by Eq 14. Since the coordinate system denotes the orientation of the plane of a potential failure producing crack, I must be a function of (ϕ,ψ) as shown in Eq 14. Now the B-H hazard is

$$\int_V dv \int_0^\infty d\sigma_c \left[\frac{dN(\sigma_c)}{d\sigma_c}\right]\frac{\Omega(\Sigma,\sigma_c)}{4\pi} = \int_V dv \int_0^\infty f(\sigma_c)d\sigma_c \frac{1}{2\pi}\int_0^{\pi/2}\int_0^{2\pi} I(\Sigma,\sigma_c)\cos\phi\, d\phi\, d\psi$$

$$= \frac{1}{2\pi}\int_0^{\pi/2}\int_0^{2\pi}\left[\int_V\left(\int_0^\infty f(\sigma_c)d\sigma_c I(\Sigma,\sigma_c)\right)dv\right]\cos\phi\, d\phi\, d\psi \tag{15}$$

where $f(\sigma_c)$ denotes $dN(\sigma_c)/d\sigma_c$, a substitution for $\Omega(\Sigma,\sigma_c)$ via Eq 13 has been made, and the order of integration has been interchanged. Moreover, $I(\Sigma,\sigma_c) = 0$, always, when $\sigma_c > \sigma_e$, and is one otherwise, so the upper limit of integration of the innermost integral of Eq 15 is σ_e and the integrand in this range is $f(\sigma_c)$. Thus

$$\int_V dv \int_0^\infty d\sigma_c \left[\frac{dN(\sigma_c)}{d\sigma_c}\right]\frac{\Omega(\Sigma,\sigma_c)}{4\pi} = \frac{1}{2\pi}\int_0^{\pi/2}\int_0^{2\pi}\left[\int_V\left(\int_0^{\sigma_e} f(\sigma_c)d\sigma_c\right)dv\right]\cos\phi\, d\phi\, d\psi. \tag{16}$$

Now set

$$f(\sigma_c) = \left[\frac{dN(\sigma_c)}{d\sigma_c}\right] = m\sigma_c^{m-1}K \tag{17}$$

which follows from Equation 9 in [3]. Then the B-H hazard is given by

$$\int_V dv \int_0^\infty d\sigma_c \left[\frac{dN(\sigma_c)}{d\sigma_c}\right]\frac{\Omega(\Sigma,\sigma_c)}{4\pi} = \frac{1}{2\pi}\int_0^{\pi/2}\int_0^{2\pi}\left[\int_V\left(\int_0^{\sigma_e} m\sigma_c^{m-1}Kd\sigma_c\right)dv\right]\cos\phi\, d\phi\, d\psi$$

$$= \frac{1}{2\pi}\int_0^{\pi/2}\int_0^{2\pi}\left[\int_V \sigma_e^m K dv\right]\cos\phi\, d\phi\, d\psi, \tag{18}$$

upon carrying out the innermost integration.

Since $\cos\phi$ is nonnegative in the domain of integration of Eqs 10 and 18, the innermost integrands of each must be identically equal (a.e.), if the L-E and B-H hazards are to be equivalent. This implies that

$$S_0^{-m} = K, \tag{19}$$

and

$$S_E \equiv \sigma_e. \tag{20}$$

Requiring that Eq 19 hold in order that the L-E and B-H approaches be equivalent is very reasonable; the relationship given in Eq 19 is just notational. Also, requiring that $S_E \equiv \sigma_e$ in Eq 20 just requires equality of the failure criterion in both approaches so as to yield overall equality. This is obvious and rational. The only question may come with regard to the assumption allowing for the interchange of the order of integration. As to this, the Weibull crack density and the usual failure criteria result in this interchange property and since any $g(S)$ and/or $f(\sigma_c)$ must be nonnegative, one is only requiring that some (minor) technical mathematical conditions hold for the integrands and failure criteria.

With these assumptions, a more general result can be shown. Let **G** denote the innermost integrand of Eq 10 and **F** denote the innermost integrand of Eq 18. Then for general **G** and **F**, the two approaches are equivalent if and only if **G**≡**F** (a.e.). Thus, in a very general sense the B-H and L-E approaches are equivalent.

There are further considerations associated with the equivalence of the L-E and B-H methods. Note that σ_e can be re-defined as

$$\sigma_e = \sigma_M F[\Sigma,\phi,\psi] = \sigma_M F[.] \tag{21}$$

where σ_M denotes the maximum value of σ_e occurring in any elemental volume. In this form σ_e is analogous to S_E defined by Eq 8. Indeed, under tensile loading Eq 21 reduces to Eq 8. If it assumed that any failure criterion employed is bounded, then it is always possible to define σ_e via Eq 21. This assumption is implicit in the L-E development and for comparison purposes is made now. Eq 21 generalizes the L-E definition of the failure criterion. Careful consideration indicates that the definition given by Eq 21 is more appropriate than that given by Lamon and Evans [4]. For example, Chao and Shetty [7] indicate that σ_M can be greater than the applied maximum principal stress for some fracture criteria with a strong shear sensitivity; also see Nemeth, Manderscheid, and Gyekenyesi [13]. By employing the definition of Eq 21 for σ_e, it is straightforward to show that Eq 18 becomes

$$\int_V dv \int_0^\infty d\sigma_c \left[\frac{dN(\sigma_c)}{d\sigma_c} \right] \frac{\Omega(\Sigma,\sigma_c)}{4\pi} = \frac{1}{2\pi} \int_0^{\pi/2} \int_0^{2\pi} \left[\int_V \left(\sigma_M^m K F^m \right) dv \right] \cos\phi \, d\phi d\psi. \tag{22}$$

Taking the B-H formulation as in Eq 22 and that of the L-E as in Eq 9, it is clear that the two approaches are equivalent, if and only if, Eq 19 holds,

$$s = \sigma_M, \tag{23}$$

and F is equivalent in both approaches. Having Eq 23 hold and the two F's be equivalent replaces Eq 20 and indicates more clearly the nature of the equivalency of the two approaches. That is, failure occurs on a boundary where the critical equivalent stress, S_E, equals the maximum value of the effective stress, σ_e. While any other value of effective stress that is in a one-to-one and onto relationship to σ_M could be employed in Eq 21, the maximum value is the obvious choice for mathematical considerations and is consistent with the physical mechanism of failure. In any event, the key is to employ a consistent usage of a failure stress, e.g., one that is measured at the same place and conditions for the test specimen. In many cases it will be simplest to employ the observed failure stress as the factor, σ_M.

A Comparison of Results from B-H and L-E

The equivalence of the B-H and L-E approaches has been challenged by Lamon [5], who further claims that the L-E approach is superior to the B-H approach. Lamon [5] compared results of Shetty et al. [6] employing the B-H approach to results of analyzing the same data via the L-E approach. In both [5] and [6] the analysis of 4-point bending data was employed to predict the results of 3-point bending and pressurized disk tests. The L-E approach successfully extrapolated to both the 3-point bending and pressurized disk results, whereas the B-H approach only extrapolated to the 3-point bending results; thus Lamon's conclusion. Given the equivalence of the two methods, they should have given comparable results--not exact since the L-E approach as applied employed a different failure criterion than the shear sensitive criterion employed by Shetty et al. (Shetty et al. also studied the normal stress criterion.) But the shear sensitive fracture criteria employed in each study are comparable in their shear sensitivity, so that the end results should have been more comparable. The more detailed study reported herein is in response to a request by a referee for clarification on this issue during the review process. While the study of the difference between the L-E and B-H approaches could seem to be minor-- they are equivalent--[5] is the only citation known to the authors in which a difference between the approaches is reported. Thus the apparent difference needs to be resolved. The further new results given herein aim to do this.

Perusal of Lamon [5] indicates that he employed "initial" data of Shetty et al. [14] for the shear sensitive failure mode studied. He then compared to results of Shetty et al. [14] for the Weibull approach and to Shetty et al. [6] for the B-H approach. (Apparently Lamon did not carry out any analyses in the B-H manner but just employed published results of Shetty et al. Moreover, the pressurized disk results of [14] were only employed for comparison to the prediction from the 4-point bending results.) The 3- and 4-point data appear to be the same in both the Shetty et al. papers. However, the shear sensitive data are distinctly different and this difference favors Lamon's conclusion. Unfortunately (and apparently resolving the discrepancy) it was reported in an earlier version of this paper and elsewhere that by employing the stress distribution equations for the pressure disk loading from Shetty et al. [14], the coplanar strain energy release rate failure criterion from Shetty et al. [6], and numerically integrating, in essentially the L-E manner via Eqs 9 or 10, a Weibull failure probability plot nearly equal to that obtained by Lamon [5] for the L-E approach (and the L-E failure criterion) was obtained. However, this is not the case. As an aside, all of the combinations of failure criteria and loading geometries given in [6] have been analytically integrated in the L-E manner under the Shetty et al. assumptions and duplicated exactly the Shetty et al. results (Equations 11-20 obtained in the B-H manner by Shetty et al. [6]). Thus, there is a discrepancy between the two analyses of the pressure disk fracture data. Our further studies indicate how this could have come about.

The further analyses begin by developing bounds for the subject failure criteria. The development of the bounds is based on the use of Eq 22 to express P_f as

$$P_f = 1 - \exp\left[-IV\sigma_M^m/\sigma_0^m\right], \tag{24}$$

where

$$I = I(m) = \frac{1}{2\pi V}\int_0^{\pi/2}\int_0^{2\pi}\left[\int_V(\sigma_e/\sigma_M)^m dv\right]\cos\phi\, d\phi\, d\psi, \tag{25}$$

and K has been written as $1/\sigma_0$. All other quantities in Eqs 24 and 25 are as previously defined. As pointed out earlier, use of σ_M is natural from a mathematical viewpoint, but that sometimes other values can be used as well. In the developments that follow (cf. (33)) the observed failure stresses, or a multiple that is not a function of any of the variables of integration, will be employed. With this background, it is clear that bounds only need be developed for the σ_e employed by Lamon and Shetty et al.

The normal stress criterion plays a central role in the development of bounds. While the normal stress criterion is not a function of the shear stress, τ, in the plane of the possible failure producing crack, it is sensitive to the shear stresses within the structure. For a given location and orientation, since σ_1, σ_2, and σ_3 are invariant, it follows that the value of σ (the normal stress to the plane) varies as τ varies through the relationship

$$\sigma^2 + \tau^2 = (\sigma_1 l_1)^2 + (\sigma_2 l_2)^2 + (\sigma_3 l_3)^2, \tag{26}$$

where the l_i denote the direction cosines of the given orientation as measured from the principal axes of the σ_i, $i = 1,2,3$. Thus, if Eq 26 is held constant, i.e., only values of the σ_i that produce the same value for the sum, $\sigma^2 + \tau^2$, are considered, and the failure criterion is only a function of σ, then the event of failure occurs as τ varies. There can be combinations of σ_1, σ_2, and σ_3 for which Eq 26 is constant and failure occurs and combinations for which it does not. Clearly, the normal stress failure criterion could be expressed as a function of τ and the left hand side of Eq 26 as well. Put another way, the normal stress criterion is just a function of l_i and σ_i, with $i = 1,2,3$. However, of those criteria that are direct functions of σ and τ, and are based on energy or ultimate strength considerations, the normal stress criterion is the least shear sensitive. In this sense it serves as an upper bound on the probability of failure for these types of shear sensitive failure criteria. For example, the coplanar strain energy release rate criterion is greater than or equal to the normal stress criterion or $\sqrt{\sigma^2 + \tau^2} \geq \sigma$. Moreover, the maximum strain energy release rate criterion employed by Lamon is given by

$$\sigma_e = (\sigma^4 + 6\sigma^2\tau^2 + \tau^4)^{1/4}$$
$$= \sqrt{\sigma^2 + \tau^2}\left[1 + \frac{4\sigma^2\tau^2}{(\sigma^2 + \tau^2)^2}\right]^{1/4}. \tag{27}$$

Now it is straightforward to show that $(\sigma^2 + \tau^2)^2 \geq 4\sigma^2\tau^2$, so that

$$(\sigma_e)_{\text{Max. Strain}} \leq [2]^{1/4}(\sigma_e)_{\text{Cop. Strain}} \cdot \qquad (28)$$

Finally, we have

$$[2]^{1/4}(\sigma_e)_{\text{Cop. Strain}} \geq (\sigma_e)_{\text{Max. Strain}} \geq (\sigma_e)_{\text{Cop. Strain}} \geq (\sigma_e)_{\text{Nor. Stress}} \qquad (29)$$

since $[1 + 4\sigma^2\tau^2/(\sigma^2 + \tau^2)^2]^{1/4} \geq 1$. In hindsight Eqs 27, 28, and 29 are obvious; earlier they were not. (Along these lines see Figure 2 of [3].)

Eq. 29 implies that if Lamon [5] had employed the coplanar strain energy release rate criterion in the L-E method, then the probability plot for this failure criterion would have plotted to the right of that for the maximum strain energy release rate criterion in his Figure 7 and would have, possibly, plotted on or near the pressure disk data from Shetty et al. [6]. Moreover, the normal stress criterion would have plotted even further to the right. Also if Shetty et al. [6] had employed the maximum strain energy release rate criterion, it would have produced a plot to the left of the lower bound given by Eq. 20 (cf. Eq 34 ff.) in their Figure 4, which is just to the left of the curve for the 4-point bending data. This shows clearly that the treatments of Lamon and Shetty et al. differ in some way.

In order to more completely address the difference between the Lamon and Shetty et al. methods, bounds for the coplanar strain energy release rate criterion were then developed. The coplanar strain energy release rate criterion applied to the pressure disk loading is given by

$$(\sigma_e)_{\text{Cop. Strain}} = \sqrt{(\sigma_t l_1)^2 + (\sigma_r l_2)^2} \,, \qquad (30)$$

where $l_1 = \cos\phi\cos\psi$ and $l_2 = \cos\phi\sin\psi$ are the direction cosines from the local principal axes. Also

$$\sigma_r = \sigma_u[1 - \alpha(r/r_1)^2] \qquad (31)$$

and

$$\sigma_t = \sigma_u[1 - \beta(r/r_1)^2] \,, \qquad (32)$$

where σ_u is the observed failure stress produced at the center of the disk, and α, β, r (*mm*), and r_1 (*mm*) are defined in [6] and [14]. The definitions, setup, and numerical values given in [14] are employed herein. (There are some typographical errors in [6] coupled with a rounding of certain physical dimensions that exacerbate the difference between the Lamon and Shetty et al. results.) Also σ_r will be taken to be zero when Eq 31 would give a negative value for certain $r \leq r_1$. With this setup

$$(\sigma_e)_{\text{Cop. Strain}} = \sigma_u\cos\phi\sqrt{\tilde{\sigma}_t^2\cos^2\psi + \tilde{\sigma}_r^2\sin^2\psi} \,, \qquad (33)$$

where $\tilde{\sigma}_r = [1 - \alpha(r/r_1)^2]$ and $\tilde{\sigma}_t = [1 - \beta(r/r_1)^2]$, follows from Eqs 30-32. From the fact that $\tilde{\sigma}_t \geq \tilde{\sigma}_r \geq 0$, bounds on the portion of Eq 33 involving the square root are given by

$$\tilde{\sigma}_t \geq \sqrt{\tilde{\sigma}_t^2 \cos^2\psi + \tilde{\sigma}_r^2 \sin^2\psi} \geq \tilde{\sigma}_r. \tag{34}$$

In view of Eq 33, and since $\tilde{\sigma}_t$ and $\tilde{\sigma}_r$ are only functions of r and this corresponds to a size integration, it follows that the bounds given by Eq 34 can be evaluated in closed form by use of Eq 25. Also note in view of Eq 33 that σ_u factors out of the integrations. It replaces σ_M in Eq 24: The term $\cos\phi\sqrt{\tilde{\sigma}_t^2 \cos^2\psi + \tilde{\sigma}_r^2 \sin^2\psi}$ is analogous to the term F in the L-E formulation. The result of carrying out the closed form integrations associated with Eqs 33 and 34 yields a lower bound for the product IV equal to the lower bound given by Eq. 20 of [6] and is given by

$$\frac{\sqrt{\pi}\,\Gamma(m/2+1)}{\Gamma(m/2+1+1/2)}\pi r_1^2\left[\frac{1}{\alpha 2(m+1)}\right]. \tag{35}$$

The upper bound obtained by the same method is

$$\frac{\sqrt{\pi}\,\Gamma(m/2+1)}{\Gamma(m/2+1+1/2)}\pi r_1^2\left[\frac{1}{\beta 2(m+1)}\right][1-(1-\beta)^{m+1}], \tag{36}$$

and differs from that given by Eq. 20 of [6] by the term $[1-(1-\beta)^{m+1}]$. Therefore the upper bound given by this method is less than or equal to the upper bound given by Eq. 20 of [6] and is thus a tighter bound than that given by Eq. 20. However, in many cases the two upper bounds are numerically close. All in all, the bounds given by Eq. 20 of [6] are quite good. Similarly bounds for the normal stress criterion applied to the pressure disk loading can be obtained. The relationship to Eq. 15 of [6] is analogous to the relationship of the bounds developed herein to Eq. 20 just discussed.

As a final step in studying the discrepancy between the Lamon and Shetty et al. evaluations, two dimensional integrations of the normal stress and coplanar strain energy release rate failure criteria and numerical evaluation of the various bounds were carried out. A two dimensional integration will suffice in view of Eq 33 and its counterpart for the normal stress failure criterion. As an aside, it will require a triple integration to evaluate the failure criterion employed by Lamon. Programming to carry out the required triple integration is under development. The double integration is based on a double application of a Gaussian type numerical integration rule (see Chapter 25 of [15]) with 28 points employed for each individual integration. Also all computations were done in double precision on a VAX 6000. The earlier integrations that were in error were based on use of a trapezoidal rule and Simpson's rule. (It is not clear how the erroneous results were obtained, but it was probably not due to the integration rules employed.)

To begin the numerical evaluations and to compare with the results of Shetty et al., the physical constants of [6] were employed to evaluate the various bounds and actual solutions for the product IV (mm^3) for the normal stress and coplanar strain energy release rate criteria. These results were then used to make probability plots as shown in Figure 4 of [6]. In doing this the intercept of the line for 4-point bending data was estimated from Figure 4. This intercept changes for the various plots and is given by $(m\ln\sigma_0 - \ln IV)$--the $\ln IV$ term is what actually changes for the various cases. The line that was estimated in this way is given by

$$y = 23.8x - 140.3816, \text{ for 4-point bending}, \tag{37}$$

where x denotes the abscissa and y denotes the ordinate of Figure 4 of [6]. With the exception of the line denoting Eq. 15 in Figure 4 all of the Shetty et al. results were duplicated. Also in view of Eq 29 the location of the line for Eq. 15, [6], is questionable. The upper bound from Eq. 15, [6], computed by the method just described falls about half way between the upper and lower bounds from Eq. 20, [6]. Similarly, the lower bound from Eq. 15, [6], falls about half way between the lower bound from Eq. 20, [6], and the line from the 4-point bending data, [6]. The actual solution for the coplanar strain energy release rate failure criterion produces a line essentially equal to that from the upper bound of Eq. 15, [6]. The actual solution for the normal stress criterion produces a line near to that from the lower bound of Eq. 20, ([6]). (See Table 1 to follow.) These results are self consistent. Thus the plotting of Eq. 15 in Figure 4 of [6] appears to be incorrect. However, the overall implications of Figure 4 therein are correct.

The values of IV that produced these results (from the constants of [6]) are shown in Table 1,

TABLE 1

IV Values (mm^3) for the Two Failure Criteria with Various Loads and Bounds

Weibull Modulus	Normal Stress Criterion						Coplanar Strain Criterion				
	3-Pt. Bend.	4-Pt. Bend.	Pressurized Disk				Pressurized Disk				
			Lower Bound	Actual	Upper Bound		Lower Bound	Actual	Upper Bound		
					(11)	Eq. 15			(11)	Eq. 20	
Constants: Ref. 4											
1.00	50.0000	79.6875	231.728	289.725	347.731	449.496	272.998	349.006	409.661	529.551	
23.8	.265391	4.01569	5.01367	6.98280	9.72532	9.72532	6.98402	9.75159	13.5473	13.5473	
33.3	.137980	2.86611	3.07812	4.28706	5.97082	5.97082	4.30580	6.00757	8.35221	8.35221	
Constants: Ref. 6											
1.00	49.7865	79.3750	134.226	181.256	228.236	260.367	158.132	222.276	268.885	306.737	
23.8	.263732	4.01403	2.90413	4.04473	5.63330	5.63330	4.04543	5.64852	7.84716	7.84716	
33.3	.137118	2.86525	1.78298	2.48324	3.45854	3.45854	2.49409	3.47983	4.83794	4.83794	
Constants: Ref. 6--$r_1 = 23.0/2$											
1.00	49.7865	79.3750	79.7950	116.481	153.068	154.783	94.0062	145.178	180.329	182.349	
23.8	.263732	4.01403	1.72645	2.40451	3.34889	3.34889	2.40493	3.35794	4.66498	4.66498	
33.3	.137118	2.86525	1.05994	1.47624	2.05604	2.05604	1.48269	2.06869	2.87606	2.87606	

along with solutions for the constants given in [14], and solutions for the constants of [14], but with the value of r_1 taken as 23.0/2 mm. The value of 23.0 mm comes about by subtracting the diameter (2.4 mm) of the ball bearings used in the support ring of the pressure apparatus from the diameter of the circle (25.4 mm) on which the ball bearings rested in the supporting ring (cf. [14] for details). This is a crude attempt to correct for any possible residual stresses that may have been produced by the loading of the pressure disk by the ring of ball bearings and, thus, are not accounted for by Eqs 31 and 32. Clearly, this latter solution set will have a stress pattern that is generally below those shown for each of σ_t and σ_r in Figure 2 of [14] and is employed only for illustrative purposes to indicate the sensitiveness of the solutions to the stress state. The maximum value of 33.3 for the Weibull modulus was the largest that the subroutines which were employed to evaluate the gamma function could accomodate. Once the value of the Weibull modulus takes on values of 20 or so or larger the sensitiveness to changes in its value is greatly lessened. Also there are somewhat significant changes between the constants of [6] and those of [14]. However, with the constants of [14] (but letting r_1 vary) and the Weibull modulus in an acceptable range, IV seems to be most sensitive to changes in the stresses, σ_t and σ_r. The normal stress solution with the constants of [14] with $r_1 = 23.0/2$ mm gives a value of 2.40451 mm^3 for IV for the pressurized disk loading. This solution would plot close to the solution that Lamon obtained in [5] but would still be to the left of Lamon's curve. Thus it may well be that the manner in which the principal stress field was determined by Lamon has produced the difference between his results and those of Shetty et al.; [5] does not indicate how the principal stress field is computed. Also the numerical integration must be carried out carefully. In this vein evaluation of the failure criterion appears to be crucial (and this may be the source of our earlier erroneous results). Moreover, the solution that Lamon should have obtained even with the constants of [14] and $r_1 = 23.0/2$ mm would be greater than the coplanar strain solution of 3.35794 mm^3 for the pressurized disk loading. In this situation in view of Eq 29 the maximum strain energy release rate failure criterion should not produce a line about half way between those for the 4-point and 3-point bending case.

In summary Lamon's solution appears questionable unless the stress solutions given by Eqs 31 and 32 are in large error. However, this could be the case due to the effect of residual stresses that these equations do not take into account. Further study of this issue seems warranted.

CONCLUSIONS

The equivalence of the L-E and B-H approaches has a number of consequences and implications. Possibly the most important is the fact that current analysis technology is applicable to the situation of multiaxial stress as shown by Eqs 24 and 25. One could employ CARES [13], for example, to numerically determine $I(m)$ which would replace $k(m)$ in all analyses techniques. All in all, there appear to be no obvious roadblocks in incorporating the effects of multiaxial stresses into analysis methods for combining data, determining confidence limits, etc. Moreover, since all failure criteria considered to date are continuous, the equivalence covers all current models. Among other things, this implies that the statement sometimes made, that the B-H approach is phenomenological whereas the L-E is not, is just not true. The approaches are equivalent and, as such, share the same properties. Also, Eq 25 indicates that it may be possible to obtain generalizations that cover time, temperature effects. The key is to generalize the approach leading to Eq 25 by adding another dimension, say time, and then integrate over this dimension in a setup analogous to Eq 16. Clearly, this implies that a suitable failure criterion, σ_e, can be formulated that covers the time effect; study of this is left for future work.

REFERENCES

[1]   W. Weibull, "A Statistical Theory of the Strength of Materials," Royal Swedish Academy of Eng. Sci. Proc., 151, 1-45, 1939.

[2]   S.B. Batdorf, "Some Approximate Treatments of Fracture Statistics for Polyaxial Tension," Inter. Jour. of Fracture, 13, 1, 5-11, 1977.

[3]   S.B. Batdorf and H.L. Heinisch, "Weakest Link Theory Reformulated for Arbitrary Fracture Criterion," J. Am. Ceram. Soc., 61, 7-8, 355-358, 1978.

[4]   J. Lamon and A.G. Evans, "Statistical Analysis of Bending Strengths for Brittle Solids: A Multiaxial Fracture Problem," J. Am. Ceram. Soc., 66, 3, 177-182, 1983.

[5]   J. Lamon, "Statistical Approaches to Failure for Ceramic Reliability Assessment," J. Am. Ceram. Soc., 71, 2, 106-12, 1988.

[6]   D.K. Shetty, A.R. Rosenfield, and W.H. Duckworth, "Statistical Analysis of Size and Stress State Effects on the Strength of an Alumina Ceramic," Methods for Assessing the Structural Reliability of Brittle Materials, ASTM STP 844, S.W. Freiman and C.M. Hudson, Eds., American Society for Testing Materials, Philadelphia, 57-80, 1984.

[7]   L. Chao and D.K. Shetty, "Equivalence of Physically Based Statistical Fracture Theories for Reliability Analysis of Ceramics in Multiaxial Loading," J. Am. Ceram. Soc., 73, 7, 1917-21, 1990.

[8]   F.M. Furgiuele and A. Lamberti, On the Equivalence of Two Weakest-Link Fracture Statistics Formulations," Inter. J. of Fracture, 51, R15-R20, 1991.

[9]   S. She, J.D. Landes, J.A.M. Boulet, and J.E. Stoneking, "Statistical Theory for Predicting the Failure of Brittle Materials," J. of Appl. Mech., 58, 43-49, 1991.

[10] C.A. Johnson and W.T. Tucker, "Advanced Statistical Concepts of Fracture in Brittle Materials," Ceramic Technology for Advanced Heat Engines Project Semiannual Progress Report for October 1989 Through March 1990, ORNL/TM-11586, Pages 298-316, September 1990.

[11] C.A. Johnson and W.T. Tucker, "Advanced Statistical Concepts of Fracture in Brittle Materials," Ceramic Technology for Advanced Heat Engines Project Semiannual Progress Report for April 1990 Through September 1990, ORNL/TM-11719, Pages 273-285, December 1990.

[12] J. Lamon, "Statistical Analysis of fracture of Silicon Nitride (RBSN) Using the Short Span Bending Technique," Gas Turbine Conference and Exhibit, Houston, TX, March 18-21, ASME, 85-GT-151, 8 pp, 1985.

[13] Nemeth, N.N., J.M. Manderscheid, and J.P. Gyekenyesi, "Ceramics Analysis and Reliability Evaluation of Structures (CARES)," NASA Technical Paper 2916, August 1990.

[14] D.K. Shetty, A.R. Rosenfield, W.H. Duckworth, and P.R. Held, "A Biaxial-Flexure Test for Evaluating Ceramic Strengths," J. Am. Ceram. Soc., 66, 1, 36-42, 1983.

[15] M. Abramowitz and I.A. Stegun, Eds., Handbook of Mathematical Functions with Formulas, Graphs, and Mathematical Tables, National Bureau of Standards, Applied Mathematics Series, 55, Washington, D.C., 1964.

James Margetson[1]

DETERMINATION OF DEFECT DISTRIBUTIONS FOR USE IN FAILURE THEORIES OF LOAD BEARING CERAMICS

REFERENCE: Margetson, J., "Determination of Defect Distributions for Use in Failure Theories of Load Bearing Ceramics," *Life Prediction Methodologies and Data for Ceramic Materials*, ASTM STP 1201, C. R. Brinkman and S. F. Duffy, Eds., American Society for Testing and Materials, Philadelphia, 1994.

ABSTRACT: Probabilistic failure theories have been developed for predicting the structural reliability of complex components subjected to conditions of multiaxial stress. Embodied in these theories are material failure characteristics which describe the statistical variation of defects within the volume and over the surface of the material. In this paper a method is presented for analysing the strength data. It is shown how the respective volume and surface defect distributions can be analysed to yield the strength parameters required by the various probabilistic multiaxial failure stress theories.

KEYWORDS: probability, strength, ceramic, characterisation, Weibull, failure, fracture, probabilistic design, defects

---

For structures made from load bearing ceramics radically new design concepts are required. The conventional method of designing with certainty, assuming a factor of safety, no longer applies. In these materials flaws are randomly distributed both in severity and orientation so failure does not always take place at the point of maximum stress. The distribution of surface and volume defects will also have different failure characteristics. The combination of flaw severity and stress orientation determines the point of failure. The engineering design of components manufactured from these materials requires the use of probabilistic design techniques. These procedures have to be used at the material characterisation, theoretical failure prediction and experimental reliability evaluation stages of the component design. Failure will depend on the material microstructure and hence there will be different failure mechanisms associated with different materials. The various theories used to predict the component failure will need to reflect this requirement.

---

[1]Director, Defence Research Consultancy Limited, c/o Royal Ordnance plc, Westcott, Aylesbury, Bucks HP18 ONZ, U.K.

The strength analysis of brittle components is usually based on a uni-modal failure model where it is assumed that within a material there are surface and volume flaws which are randomly distributed and that failure will originate from one of these defects. In this paper a method is presented for analysing strength data. It is shown how the respective volume and surface defect distributions can be analysed to yield the strength parameters required by the various probabilistic multiaxial failure stress theories.

FAILURE PROBABILITY RELATIONSHIPS

For a ceramic component subject to conditions of multiaxial stress the failure probability relationships for failures originating from volume and surface defects are respectively given by [1, 2, 3]

$$P_f^{(V)} = 1 - \exp\left\{-\int_V \left(\frac{\sigma_{ev}}{\sigma_{ov}}\right)^m \frac{dV}{V}\right\} \tag{1}$$

$$P_f^{(A)} = 1 - \exp\left\{-\int_A \left(\frac{\sigma_{ea}}{\sigma_{oa}}\right)^m \frac{dA}{a}\right\} \tag{2}$$

In the above relationships σ_{ev} and σ_{ea} are effective stress functions which depend on the stress state, defect geometry and failure mechanism. The quantities σ_{ov} and σ_{oa} are characteristic strengths which are respectively associated with the unit volume 'v' and unit area 'a'. The quantity m is the Weibull modulus and is an inverse measure of the variability of the strength of the material.

Expressions for the effective stress functions have been derived for various ceramic failure theories [1, 2, 3]. For the "principle of independent action" (PIA) tensile failure stress model [3]

$$\sigma_{ev} = \sigma_{ea} = \left\{\sum_{k=1}^3 \sigma_k^m\right\}^{\frac{1}{m}} \tag{3}$$

where σ_k, k = 1,2,3 are the principal stress components. For the Batdorf-Heinisch probabilistic fracture mechanics models [2]

$$\sigma_{ev} = \left\{\frac{1}{4\pi} \int_o^{2\pi} \int_o^\pi \sigma_c^m \sin\alpha \, d\alpha \, d\beta\right\}^{\frac{1}{m}} \tag{4}$$

$$\sigma_{ea} = \left\{\frac{1}{2\pi} \int_o^{2\pi} \sigma_c^m \, d\alpha\right\}^{\frac{1}{m}} \tag{5}$$

A common form of the critical stress component σ_c which has been derived for many of the Batdorf-Heinisch failure models is [2]

$$\sigma_c = A\,\sigma_n + B\,\sqrt{\sigma_n^2 + C\,\tau_n^2} \qquad (6)$$

Values of the constants A, B and C are given in Table 1 for the maximum stress and maximum energy failure criteria. In Table 1 GC and PSC respectively denote Griffiths Crack and Penny Shaped Crack defects. The quantities σ_n and τ_n are the normal and shear stress components

$$\sigma_n = \sigma_1 l^2 + \sigma_2 m^2 + \sigma_3 n^2 \qquad (7)$$

$$\tau_n^2 = \sigma_1^2 l^2 + \sigma_2^2 m^2 + \sigma_3^2 n^2 - \sigma_n^2 \qquad (8)$$

acting on a crack with direction cosines relative to the principal stress axes given by

$$\left.\begin{array}{l} l = \cos\alpha \\ m = \sin\alpha\,\cos\beta \\ n = \sin\alpha\,\sin\beta \end{array}\right\} \qquad (9)$$

TABLE 1--Constants for various Batdorf-Heinisch failure criteria.

Criterion	$\sigma_c = A\sigma_n + B\sqrt{\sigma_n^2 + C\tau_n^2}$		
	A	B	C
Max. Stress (GC)	0.5	0.5	1.0
Max. Stress (PSC)	0.5	0.5	$1/(1-\upsilon/2)^2$
Max. Energy (GC)	0.0	1.0	1.0
Max. Energy (PSC)	0.0	1.0	$1/(1-\upsilon/2)^2$

STRENGTH CHARACTERISATION ANALYSIS

It is apparent from the previous section that before the failure probability of a component can be predicted it is first necessary to determine the material failure characteristics σ_{ov}, σ_{oa} and m.

The material failure characteristics of a ceramic material are frequently determined from an analysis of the fracture data derived from a beam subjected to flexure. When considering the fracture of a long slender beam the stress state is uniaxial and non-uniform. The stress state in a beam can be expressed by the equations

$$\sigma_1 = \frac{M(x)\,y}{I} \qquad (10)$$

$$\sigma_2 = 0 \qquad (11)$$

$$\sigma_3 = 0 \qquad (12)$$

where $M(x)$ is the bending moment at a section x, y is the distance from the neutral axis and I is the second moment of area.

For this simplified stress state the failure probability expressions given by equations (1) and (2) can be evaluated and expressed in the compact form

$$P_f^V\,(\sigma_b) = 1 - \exp\left\{-\left(\frac{\sigma_b}{\sigma_{ov}}\right)^m S_V(m)\right\} \qquad (13)$$

$$P_f^A\,(\sigma_b) = 1 - \exp\left\{-\left(\frac{\sigma_b}{\sigma_{oa}}\right)^m S_A(m)\right\} \qquad (14)$$

Here σ_b is the maximum bending stress

$$\sigma_b = \frac{6M}{bd^2} \qquad (15)$$

where b and d respectively denote the breadth and depth of the beam. Expressions for $S_V(m)$ and $S_A(m)$ for various bend test configurations are given by

1. Pure bending

$$S_V(m) = \frac{V_b}{V}\left(\frac{1}{2\,(m+1)}\right) I(m) \qquad (16)$$

$$S_A(m) = \frac{A_b}{a}\left(1 + \frac{\lambda m}{1 + \lambda}\right)\left(\frac{1}{2\,(m+1)}\right) J(m) \qquad (17)$$

2. Four point loading

$$S_V(m) = \frac{V_b}{V}\left(\frac{m+2}{4\,(m+1)^2}\right) I(m) \qquad (18)$$

$$S_A(m) = \frac{A_b}{a}\left(1 + \frac{\lambda m}{1 + \lambda}\right)\left(\frac{m+2}{4\,(m+1)^2}\right) J(m) \qquad (19)$$

In the above equations A_b and V_b respectively denote the area and volume of the beam and λ is the width to depth ratio. The quantities $I(m)$ and $J(m)$ depend on the failure theory which is assumed to govern the fracture process. For the PIA analysis

$$I(m) = J(m) = 1 \qquad (20)$$

For the Batdorf-Heinisch failure theory it can be deduced from equations (4) and (5) that the quantities $I(m)$ and $J(m)$ are given by the integral expressions

$$I(m) = \frac{1}{2} \int_{-1}^{1} (f(z))^m dz \qquad (21)$$

$$J(m) = \frac{1}{2} \int_{-1}^{1} \{f(\cos(\pi(z+1)))\}^m \, dz \qquad (22)$$

$$f(z) = z^2 \left\{ A + B \sqrt{1 + C(1/z^2 - 1)} \right\} \qquad (23)$$

The above integrals can readily be evaluated numerically using, for example, the Gauss Legendre quadrature formulae [4].

For a uni-modal failure analysis, that is where failures are due to either volume defects or surface defects, equations (13) and (14) can be reparameterised and expressed as the standardised Weibull distribution

$$P_f(\sigma_b) = 1 - \exp\left\{ -\left(\frac{\sigma_b}{\sigma_o} \right)^m \right\} \qquad (24)$$

If σ_i, $i = 1,2,\cdots n$ denote the maximum bend stresses at fracture then it can be shown that the required Weibull parameters are given by the maximum likelihood equations [5]

$$\sigma_o = \left\{ \frac{1}{n} \sum_{i=1}^{n} (\sigma_i)^m \right\}^{\frac{1}{m}} \qquad (25)$$

where the Weibull modulus m is given by the solution of the equation

$$\frac{1}{m} = \frac{\displaystyle\sum_{i=1}^{n} \sigma_i^m \ln \sigma_i}{\displaystyle\sum_{i=1}^{n} (\sigma_i)^m} - \frac{\displaystyle\sum_{i=1}^{n} \ln \sigma_i}{n} \qquad (26)$$

For the uni-modal analysis only one set of parameter values for m and σ_o are derived from the fracture data. Once these estimates have been obtained the unit volume and area strengths follow from

$$\sigma_{ov} = \sigma_o \left(S_V(m) \right)^{\frac{1}{m}} \tag{27}$$

$$\sigma_{oa} = \sigma_o \left(S_A(m) \right)^{\frac{1}{m}} \tag{28}$$

ANALYSIS OF EXPERIMENTAL DATA

For the purpose of illustration the fracture data detailed in Table 2 was analysed using the material characterisation techniques presented in the previous sections. The Table 2 data has been derived from flexural tests on beams manufactured from the glass ceramic MACOR [6]. The beams were of dimensions 155 x 26 x 5 mm and were tested in accordance with the ASTM specification entitled "Flexural Testing of Glass". The distance between the central knife edges was 38 mm and in the analysis all failures were assumed to take place within that region. Within the central knife edges the bending moment is constant and hence in the testing of the beams it was assumed that they were fractured under pure bending conditions.

TABLE 2--<u>Maximum bend stress fracture data for glass ceramic (MACOR)</u>.

Max Bending Stress MPa	Max Bending Stress MPa	Max Bending Stress MPa
78.99	104.79	116.53
80.34	105.31	117.31
80.63	105.80	118.18
91.22	106.57	118.29
91.22	108.12	119.07
91.38	108.14	119.75
93.41	108.29	120.23
95.02	108.47	120.62
95.57	109.12	121.11
97.51	109.32	121.31
97.81	110.18	121.89
97.90	110.51	122.77
99.67	111.46	124.33
99.93	112.91	124.72
100.61	113.40	126.18
100.63	113.59	126.98
101.96	114.77	127.94
101.97	115.58	128.30
103.69	116.53	129.69

For the uni-modal analysis the maximum likelihood estimates of the Weibull parameters m and σ_o appearing in the reparameterised

Weibull equation (24) are respectively m = 10.5, σ_o = 114.47 MPa. The material failure strength characteristics σ_{ov} and σ_{oa} follow from equations (27) and (28) after evaluating the appropriate stress volume and stress area integral functions $S_V(m)$ and $S_A(m)$. The calculated values of σ_{ov} and σ_{oa} for the PIA and Batdorf-Heinisch failure theories are given in Table 3. A logarithmic failure probability plot corresponding to a uni-modal analysis of the fracture data is presented in Fig. 1. In that representation the probabilities $P_f^{(i)}$ assigned to the ranked fracture data $\sigma_b^{(i)}$, i = 1,2,$\cdots$n are the median ranked probabilities [7]

$$P_f^{(i)} = \frac{i - 0.3}{n + 0.4} \qquad (29)$$

TABLE 3--Failure characteristics derived from a
uni-modal volume and surface defect analysis.

Defect Parameter	PIA	Batdorf-Heinisch Failure Criteria			
		Maximum Stress		Maximum Energy	
		GC	PSC	GC	PSC
m_a	10.5	10.5	10.5	10.5	10.5
σ_{ov} (MPa)	190.3	145.6	147.2	150.7	155.3
m_V	10.5	10.5	10.5	10.5	10.5
σ_{oa} (MPa)	220.3	188.8	189.9	192.3	195.5

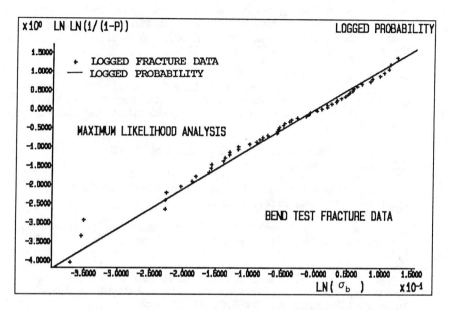

FIG. 1--Uni-modal logarithmic probability distribution.

CONCLUSIONS

A method has been presented for analysing strength test data to determine the volume and surface defect failure characteristics which are used in various probabilistic failure theories.   A uni-modal method of analysis has been described.

ACKNOWLEDGEMENT

This work has been carried out with the support of the Defence Research Agency, Ministry of Defence, U.K.

REFERENCES

[1]   Nemeth, N. N., Manderschied, J. M. and Gyekenyesi, J. P.,
       "Design of Ceramic Components with the NASA/CARES Computer
       Code," NASA Tech Memo 102369, 1990.

[2]   Batdorf, S. B. and Heinisch, H. L., "Weak Link Theory
       Reformulated for Arbitrary Fracture Criterion," Journal of the
       American Ceramics Society Incorporating Advanced Ceramic
       Materials, Vol. 61, 1978, pp 355-358.

[3]   Stanley, P., Fessler, H. and Sivill, A. D., "An Engineer's
       Approach to the Prediction of Failure Probability of Brittle
       Components," Basic Science Convention, Ceramics for Turbines and
       other High Temperature Engineering Applications, Cambridge, July
       1972.

[4]   Scheid, F., "Numerical Analysis," Schaums Outline Series, McGraw
       Hill Co., New York, 1968.

[5]   Hahn, G. J. and Shapiro, S. S., "Statistical Models in
       Engineering", John Wiley, New York, 1967.

[6]   Cooper, N. R., "Probabilistic Failure Prediction of Rocket Motor
       Components," Royal Military College of Science, Ph.D. Thesis,
       1988.

[7]   Johnson, L. G., "The Median Ranks of Sample Values in their
       Population with an Application to Certain Fatigue Studies,"
       Industrial Mathmatics, Vol 2, 1951, pp 1-9.

Prediction of the Behavior of Structural Components

John Cuccio[1], Alonso Peralta[2], Jeff Song[2], Peggy Brehm[3], Curtis Johnson[4], William Tucker[5], and Ho Fang[2]

PROBABILISTIC METHODS FOR CERAMIC COMPONENT DESIGN AND
IMPLICATIONS FOR STANDARDS

REFERENCE: Cuccio, J., Peralta, A., Song, J., Brehm, P., Johnson, C., Tucker, W., and Fang, H., **"Probabilistic Methods for Ceramic Component Design and Implications for Standards,"** Life Prediction Methodologies and Data for Ceramic Materials, ASTM STP 1201, C. R. Brinkman and S. F. Duffy, Eds., American Society for Testing and Materials, Philadelphia, 1994.

ABSTRACT: Probabilistic methods developed at Garrett Auxiliary Power Division of Allied-Signal Aerospace Company under the "Life Prediction Methodology for Ceramic Components of Advanced Heat Engines" program sponsored by The Department of Energy/Oak Ridge National Laboratory (DOE/ORNL) under contract No. 86X-SC674C (WBS Element 3.2.2.3) are presented. Statistical methods have been developed to estimate Weibull strength parameters and component reliability with confidence limits for structural ceramics. Estimates can be made using pooled strength data from specimens of multiple sizes and loading conditions, from multiple test temperatures, and from material with multiple strength distributions. Bootstrap and likelihood ratio techniques are used to calculate confidence intervals on parameters and reliability estimates from these complex pooled data sets. A large database was generated on one ceramic (NT154 silicon nitride) to verify the methods. These statistical methods guide the development of standards for more accurate parameter estimation, to define component reliability requirements with confidence limits, and to plan specimen tests for more efficient estimation of Weibull parameters and component reliability.

KEYWORDS: silicon nitride, strength, probabilistic, statistics, Weibull, specimen data, likelihood, confidence intervals

Probabilistic methods have become the standard for characterizing mechanical properties of structural ceramics and for predicting component reliability. Probabilistic methods are needed because of the brittle nature of ceramics and the large scatter in mechanical properties resulting from naturally-occurring processing flaws. Although the typical flaw sizes in structural ceramics are small (10 to 100 microns), failures initiate from these flaws due to the relatively low toughness (2 to 10 MPa m$^{1/2}$). This sensitivity to flaws

---

[1] Supervisor, [2] Senior Engineer, [3] Staff Engineer Special Projects, Garrett Auxiliary Power Division, Allied-Signal Aerospace Company, Phoenix, AZ 85010.

[4]Staff Scientist, and [5]Staff Statistician, General Electric Company, Corporate Research and Development, Schenectady, NY 12301.

significantly influences strength characteristics, including: 1) large
scatter in strength values; 2) size-dependent strength; 3) multiple
flaw types that produce multiple strength distributions; 4) dependence
of strength on multiaxial state of stress; and 5) anisotropic
(orientation-dependent) flaw distributions. In addition, strength
values for structural ceramics are typically dependent on temperature
and environmental conditions.

Probabilistic approaches are commonly used to model these strength
characteristics and are usually based on the two-parameter Weibull
distribution and weakest link theory [1]. Several expansions to the
Weibull distribution have been made [2-7] to model portions of the
mechanical characteristics listed above. Presented here is an approach
for combined modeling of these mechanical characteristics with
capabilities for calculating confidence intervals on estimates of
Weibull parameters and component reliability.

A summary of the experimental data and statistical derivations for
modeling ceramic strength is provided along with applications that
demonstrate these modeling capabilities and some interesting
observations. More detail reviews are available in program progress
reports [8].

DATABASE FOR DESIGN METHODS DEVELOPMENT

More than 1400 specimens of 14 geometries were fabricated and tested
to generate a statistically significant sampling of mechanical
properties from Norton/TRW Ceramics (NTC) NT154 silicon nitride. NT154
was selected for high strength and high temperature mechanical
capabilities, and NTCs experience making diverse sizes and shapes with
this material. All specimens were CIPped (cold isostatically pressed),
encapsulated HIPped (hot isostaticly pressed), and fully machined. The
fast fracture strength portion of this database is reviewed here.

Flexure and tensile specimens were used to generate fast fracture
data. Three sizes of flexure specimens were used; A and B in accordance
with ASTM Standard C1161, and a larger specimen denoted as size E. The
dimensions of these specimens and loading distances (height, width,
inner span, outer span) are; size A (1.5 x 2 x 10 x 20 mm), size B (3 x
4 x 20 x 40 mm) and size E (18 x 9 x 63.5 x 127 mm). The size E
specimen was tested on edge (i.e. the loading pins contacted the 9 mm
wide surfaces) to increase the ratio of stressed volume to stressed
area. The cylindrical tensile specimen has a 35 mm long and 6.3 mm
diameter gage section. Multiple types of specimens were selected to
generate surface and volume strength data over a wide range of specimen
sizes.

Considerable effort was made to generate and confirm consistent
material properties in all specimens. The large number of specimens and
variety of geometries required multiple processing batches at each
processing step. To minimize unwanted material inconsistencies
associated with multiple processing batches, the following precautions
were taken with assistance from NTC:

- A master lot of Si_3N_4 powder and sintering additives was used
- All specimens were machined from three billet sizes to minimize
  inconsistencies that could occur if too many billet sizes were used
- All of the ASTM A and B specimens (except spin disk cut-ups) were
  machined from square plates, and the size E and tensile specimens
  machined from cylindrical billets

- All test surfaces were machined to the same specifications and studies were performed to select machining vendors that could provide consistent properties
- In-process FPI, X-ray and visual inspection were performed but not used to screen specimens, to avoid inconsistent screening of different specimen geometries
- Small and large flexure specimens were sectioned from spin-disk billets to confirm material consistency in the thickest billet used.

HIP batches were identified as the most likely source of processing inconsistency if any were to occur according to NTC. A plan for allocating specimens for each type of test and test condition was defined to minimize and identify potential HIP batch inconsistencies. Tensile and E-size flexure specimens were processed in three groups, near the beginning, middle, and end of all processing dates to confirm consistency. A comparison of volume strength parameters for the three HIP dates are shown in Table 1. The deviation in parameters are not unexpected, as indicated by the 95 % confidence intervals of the parameter estimates. σ_o in Table 1 is the second Weibull parameter, a material property that can be thought of as the characteristic strength normalized to a unit size (i.e. a cubic millimeter) in uniform tension. The second Weibull parameter allows estimates for both size E flexure and tensile specimens to be directly compared. The methods used to calculate these parameters and confidence intervals are described in the next section.

TABLE 1--Comparison of volume-strength parameter estimates for size E flexure and tensile specimens that were HIPped on different dates. The median parameter estimates and 2.5% and 97.5% confidence intervals are listed.

SPECIMEN TYPE	HIP DATE	FAILURES Vol.	FAILURES Total	$\hat{\sigma}_o$*	Lower	Upper	$\hat{m}$	Lower	Upper
FLEX-E	6-90	3	26	2300	1240	15600	7.60	2.38	16.3
FLEX-E	10-90	7	48	2870	1670	12400	6.58	3.30	11.1
FLEX-E	3-91	5	26	1590	1240	3430	11.9	5.78	19.9
FLEX-E	all	15	100	2240	1631	4100	7.98	5.23	11.3
TENSILE	4-90	29	35	1700	1450	2130	8.46	6.36	10.9
TENSILE	10-90	10	22	2270	1650	4420	7.11	4.20	10.8
TENSILE	4-91	28	36	1670	1450	2110	9.42	7.01	12.1
TENSILE	all	67	93	1880	1660	2200	7.99	6.66	9.44

* units=$MPa \cdot \mathrm{mm}^{3/m}$

Average strengths from fast fracture tests are plotted in Figure 1. Fractography with optical microscopy was performed on every specimen. These data exhibit the expected strength characteristics listed above: large scatter in strength; size-dependent strength; temperature-dependent strength; and multiple strength distributions from surface and internal flaws. Final SEM fractography is underway to characterize origins that were not identified with optical microscopy. Also, anisotropic strength was observed for machined surfaces. Transverse and diagonal machined surface strengths were measured but are not reviewed in this paper.

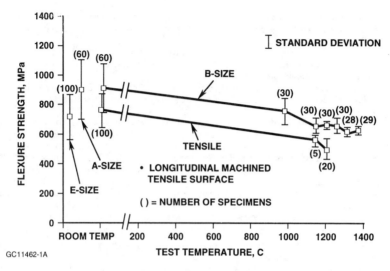

Fig. 1--Comparison of NT154 strengths from multiple specimen types.

ADVANCES IN PROBABILISTIC METHODS

Weibull [1] applied the weakest-link theory concept and developed the theory of brittle fracture, equation (1).

$$P_f = 1 - \exp - \int_V \left(\frac{\sigma}{\sigma_o} \right)^m dV \qquad (1)$$

Weibull showed that for brittle materials the observed strength depends on the specimen size and that the strength of large specimens or components can be predicted from small size specimens. Several investigators, notably Batdorf [4,5] and Evans [6] have derived new multiaxial approaches to replace Weibull's uniaxial and earlier multiaxial theories. The theory developed by Batdorf and Crose [4] assumes that the failure origins are from randomly oriented and randomly distributed microcracks; in contrast, Evans' theory assumes that the material has an elemental strength which characterizes the distribution of flaws in the material. These two different approaches have been proven to be the same for the same fracture criterion, numerically by Chao and Shetty [9] and analytically by Tucker and Johnson [10]. The authors chose to follow the Evans approach because it simplifies the formulation of the problem.

This section presents the methodology developed in the life prediction program to analyze data sets containing specimens of multiple sizes tested under several load conditions including different temperatures and multiaxial stresses. The methodology applies to the analysis of strength data from material with multiple competing (concurrent) flaw populations.

Pooled Data Analysis for a Single Failure Mode

For general cases where stress is a function of location in a stressed body, the integration in equation (1), with a generalization of the improvements by Evans, can be carried out and it results in

$$P_f = 1 - \exp\left\{ -IV\left(\frac{\sigma_{\max}}{\sigma_o}\right)^m \right\} \tag{2}$$

where I, the load factor, accounts for the gradient and multiaxiality of the stress field, and $\sigma_{\max}$ is the maximum stress in the body at the time of failure. The definition of the load factor I, for a volume failure mode, is given as [10]

$$I = \frac{1}{2\pi V} \int_V \int_0^{\frac{\pi}{2}} \int_0^{2\pi} \left(\frac{\sigma_e}{\sigma_{\max}}\right)^m \cos\phi \, d\psi \, d\phi \, dV \tag{3}$$

where I is a function of the stress distribution, σ_e is the effective stress field and it depends on the multiaxial fracture criterion selected, m is the Weibull modulus, and V is the total volume. IV has the physical meaning of effective volume. A parallel equation can be written for effective area. With the use of effective volume, the probability density function can be written as

$$f(\sigma_i) = m I_i V_i \sigma_o^{-1} \left(\frac{\sigma_i}{\sigma_o}\right)^{m-1} \exp\left\{ -I_i V_i \left(\frac{\sigma_i}{\sigma_o}\right)^m \right\} \tag{4}$$

Note that σ_i in the equation is the strength for each specimen i and $I_i V_i$ is the corresponding effective volume of that specimen. This probability density equation allows data from different types of specimens (e.g. bending and tensile) and different sizes of the same type of specimens to be combined for a single estimate of the Weibull strength parameters. An equation similar to equation (4) can be written for surface failure modes.

Weibull parameters in equation (4) for single failure mode data may be estimated with the maximum likelihood method [11]. The likelihood function of a sample is defined as the product of the values of the probability density function for the failure stress of each specimen, as shown by equation (5).

$$L(m, \sigma_o) = \prod_{i=1}^n f(\sigma_i, m, \sigma_o) \tag{5}$$

which yields the log likelihood function as

$$l = \sum_{i=1}^n \ln I_i + \sum_{i=1}^n \ln V_i + n\ln m + (m-1)\sum_{i=1}^n \ln \sigma_i - nm \ln \sigma_o - \sum_{i=1}^n I_i V_i \left(\frac{\sigma_i}{\sigma_o}\right)^m \tag{6}$$

where n is the total number of specimens. Unknown parameters m and σ_o can be determined when l reaches its maximum value. This is done by taking partial derivatives of (6) with respect to m and σ_o, setting them equal to zero and solving for $\hat{m}$ and $\hat{\sigma}_o$. It should be pointed out that since I is a function of m, and the m dependence of $I(m)$ term must be accounted for in the partial derivatives. This procedure yields two equations:

$$\frac{1}{\hat{m}} = \frac{1}{n}\sum_{i=1}^n \left\{ \left(\frac{\hat{I}_i'}{\hat{I}_i} + \ln \sigma_i\right) \left(\frac{n\hat{I}_i V_i \sigma_i^{\hat{m}}}{\sum_{i=1}^n \hat{I}_i V_i \sigma_i^{\hat{m}}}\right) \right\} \tag{7}$$

and

$$\hat{\sigma}_o = \left[\frac{1}{n} \sum_{i=1}^{n} I_i V_i \sigma_i^{\hat{m}} \right]^{\frac{1}{\hat{m}}} \tag{8}$$

where $\hat{m}$ and $\hat{\sigma}_o$ are the maximum likelihood estimates of Weibull modulus and the second Weibull parameter respectively, and $\hat{I}_i'$ is the first derivative of $\hat{I}_i$ with respect to $\hat{m}$. A closed-form solution for $\hat{m}$ from equation (7) is not possible; an iteration process must be used.

Using effective size in Weibull strength calculation permits pooling strength data from different specimen types and sizes for strength parameter estimation. The extension of these methods to competing failure mode strength data is reviewed in the following section.

Pooled Data Analysis for Multiple failure modes

In structural ceramics, more than one flaw population is usually present. These flaw populations are generally competing, i.e. these multiple flaw populations exist in each specimen and compete to cause failure. The maximum likelihood method for analysis of competing strength distributions requires that fracture origins for each specimen be identified. Although parameter estimates can be obtained without fractography it is very inefficient and requires significantly larger sample sizes.

Strength distributions associated with the different flaw populations can each be described with independent sets of Weibull parameters. In order to determine the best estimates of Weibull parameters for a given failure mode, it is necessary to perform maximum likelihood analysis while censoring the strength data points associated with other competing failure modes.

The likelihood function for censored data contains the product of probability density functions of the non-censored data and the product of the cumulative survival probability functions of the censored data, as shown below:

$$L = \prod_{i=1}^{r} f(\sigma_i) \prod_{j=r+1}^{n} S(\sigma_j) \tag{9}$$

where r is the number of specimens failed by the flaw population under consideration, and n is the total number of specimens. The number of censored data points is the difference between n and r. $S(\sigma_j)$ is the probability of the flaw type under consideration surviving a stress level of σ_j.

Following the same procedure used for maximum likelihood analysis of single failure mode data, two equations result:

$$\frac{1}{\hat{m}} = \frac{1}{\sum_{k=1}^{n} \hat{I}_k V_k \sigma_k^{\hat{m}}} \sum_{k=1}^{n} V_k \sigma_k^{\hat{m}} \left[\hat{I}_k' + \hat{I}_k \ln(\sigma_k) \right] - \frac{1}{r} \sum_{i=1}^{r} \left[\frac{\hat{I}_i'}{\hat{I}_i} + \ln(\sigma_i) \right] \tag{10}$$

$$\hat{\sigma}_o = \left[\frac{1}{r} \sum_{i=1}^{n} I_i V_i \sigma_i^{\hat{m}} \right]^{\frac{1}{\hat{m}}} \tag{11}$$

Similar to equation (7), equation (10) requires iteration to solve for $\hat{m}$. Our experience to date shows that the presence of censored data does not significantly impact numerical efficiency.

The GE, SiC data [11] in Table 2 is a useful data set to demonstrate multimodal strength analysis capability for pooled data.

TABLE 2-- Comparison of parameter estimates for different sizes of SiC specimens and estimates from the pooled data

	Flexure Specimen Geometry	Observa-tions	$\hat{m}$	$\hat{\sigma}_o$ MPa $mm^{a/m}$
Surface a=2	3-pt Size A	11	15.2	443
	4-pt Size A	16	9.5	460
	3-pt Size B	12	12.7	459
	4-pt Size B	35	14.2	440
	3-pt Size C	9	20.6	412
	4-pt Size C	12	13.1	474
	Combination	95	14.3	444
Edge a=1	3-pt Size A	6	12.2	480
	4-pt Size A	1	7.3	732
	3-pt Size B	4	10.4	505
	4-pt Size B	12	14.9	444
	3-pt Size C	6	11.3	459
	4-pt Size C	4	12.4	473
	Combination	33	11.5	488
Volume a=3	3-pt Size A	1	34.5	403
	4-pt Size A	0	-	-
	3-pt Size B	2	12.8	442
	4-pt Size B	1	9.5	611
	3-pt Size C	3	17.5	411
	4-pt Size C	2	59.3	337
	Combination	9	18.9	411

All specimen geometries were in accordance with ASTM Standard C1161. The dimensions of size A and B specimens were previously noted, dimensions for Size C are; 6 mm (height), 8 mm (width), 40 mm (inner span), and 80 mm (outer span). The last column in Table 2 is the second Weibull parameter which is a material property. Therefore, we expect estimates of σ_o to be nearly equal for different size specimens when sample sizes are large and the underlying weakest link size scaling assumptions apply. The parameters in the "combination" rows were from pooled data analyses. The data from all six specimen sizes were combined into one data set and parameters were calculated using equations (10) and (11). An encouraging outcome from this example is the ability of this censored data analyses method to provide parameter estimates for data sets with only one non-censored (failed) data point. Parameter estimates were obtained for three groups of data with only one non-censored data point, indicating that some information is extracted from the censored data.

Confidence Intervals for Component Reliability

With a limited size test matrix, one can never exactly predict component reliability or determine the parameters which define the strength distributions. The uncertainty associated with the estimators

can be quantified by the size of the confidence bounds. Two techniques
for calculating confidence intervals are used; bootstrap [11,12] and
likelihood ratio [11,13]. These methods were extended to analyses of
specimens with multiple sizes and loading conditions by Johnson and
Tucker [3]. Provided here are further extensions of the bootstrap and
likelihood ratio techniques to specimens and components with competing
failure modes (i.e. censored data analyses), and multiple test-
temperature data.

Bootstrap Techniques for Confidence Bounds--The parametric bootstrap
techniques is a simulation method (Monte Carlo simulation) applied here
to quantify the uncertainty in component reliability estimates from data
with specimens of multiple sizes and loading conditions, and for
material with competing strength distributions. This technique uses the
parameter estimates $(\hat{m}_j, \hat{\sigma}_{oj})$, of failure mode j, obtained from the
maximum likelihood analysis of an original data set to generate a large
number of simulated data sets. Each simulated data set has the same
number of strength observations associated with each specimen type and
size as the original data set. In each simulate group of specimens the
proportion of failures associated with each competing failure mode may
vary somewhat from the original data.
   A large number of simulated data sets (typically 1000) are generated
from the same set of Weibull parameters. Maximum likelihood analysis is
performed for each simulated data (i) set to determine the estimators
$\tilde{m}_{ji}, \tilde{\sigma}_{o,ji}$ for each flaw distribution (j) in the original data set.
   The combined failure probability of all the competing flaw
populations can be calculated by

$$P_f(z) = 1 - \exp\left\{-\sum_{j=1}^{k} I_j V_j \left(\frac{z}{\sigma_{o,j}}\right)^{m_j}\right\} \qquad (12)$$

where z is the given design stress level, here k is the total number of
competing flaw populations. The combined failure probability is
calculated for each set of $\tilde{m}_j, \tilde{\sigma}_{o,j}$ values resulting in a set of combined
failure probabilities corresponding to the number of simulated data
sets. These probability values are sorted into ascending order. The
confidence bounds corresponding to the 95 percent confidence interval
are obtained by picking the probability values at the 25th and the 975th
ordered positions.
   The computation required for the bootstrap technique can be intensive
for large problems and it has difficulty with strength distributions
that have only a few occurrences in the original data set. These
failure modes with few observations typically result in zero occurrences
in a portion of the simulated sets, which prohibits confidence interval
calculation using the procedure described here. However, the bootstrap
method can be adapted to complex problems, such as confidence interval
calculations for data sets with competing failure modes, multiple
specimen sizes and multiple temperature strength data as demonstrated in
a later section.

Likelihood Ratio Techniques for Confidence Intervals--The likelihood
ratio method is well-established as a basis for computing confidence
limits on predictions of failure probability due to a single failure
mode. The following derivation extends the method to permit computation
of confidence limits on combined failure probability in the presence of
multiple competing flaw populations.

An example with three competing failure modes was chosen for the derivation; the technique extends to any number of competing failure modes.

The log likelihood for such a data set is given by:

$$l = \sum_{i=1}^{n}\left[\ln\left\{f_1(\sigma_i)^{\delta_{1i}} S_1(\sigma_i)^{\delta_{2i}} S_1(\sigma_i)^{\delta_{3i}}\right\} + \ln\left\{S_2(\sigma_i)^{\delta_{1i}} f_2(\sigma_i)^{\delta_{2i}} S_2(\sigma_i)^{\delta_{3i}}\right\}\right.$$
$$\left. + \ln\left\{S_3(\sigma_i)^{\delta_{1i}} S_3(\sigma_i)^{\delta_{2i}} f_3(\sigma_i)^{\delta_{3i}}\right\}\right]$$

(13)

where $f_j(\sigma_i)$ is the density function for strength σ_i for failure mode j, and $S_j(\sigma_i)$ is the survival probability of jth failure mode to exceed strength σ_i. $\delta_{ij} = 1$ if the i specimen failed from mode j; otherwise, $\delta_{ij} = 0$.

The likelihood ratio method is based on the direct use of the likelihood ratio statistic

$$W(\beta) = 2\left[\hat{l} - \hat{l}_\beta\right]$$

(14)

where $\hat{l}$ is the value of the log likelihood function evaluated at the joint maximum likelihood estimate of all parameters, β is any parameter (i.e. m or σ_o), and $\hat{l}_\beta$ is the log likelihood evaluated at a fixed value of parameter β and the maximum likelihood estimates of all other parameters conditional on the given value of β.

$W(\beta)$ has, approximately, a chi-squared distribution with one degree of freedom. This yields a $1 - \alpha$ confidence region on parameter β:

$$\left\{\beta : W(\beta) \leq \chi^2_{1,\alpha}\right\}$$

(15)

Obtaining confidence limits on a quantity such as reliability requires substitution for and elimination of one of the distribution parameters, e.g. σ_{o1}, (1 indicates the first failure mode) in the following manner:

For a design stress z, the reliability is given by

$$R(z) = S_1(z)S_2(z)S_3(z)$$

(16)

where

$$S_j = \exp\left(-I_{jc}V_{jc}\left(\frac{z}{\sigma_{oj}}\right)^{m_j}\right)$$

and the subscript j refers to a particular failure mode, and subscript c refers to a reference component (perhaps one of the specimen types).

Defining $\gamma = \ln\left(\frac{1}{R(z)}\right)$ and $\gamma_{jz} = \ln\left(\frac{1}{R_j(z)}\right) = I_{jc}V_{jc}\left(\frac{z}{\sigma_{0j}}\right)^{m_j}$ and applying a bit of algebra to (16) yields:

$$\gamma = \ln\left(\frac{1}{R(z)}\right) = \gamma_{1z} + \gamma_{2z} + \gamma_{3z}$$

The probability density function can now be written as:

$$f_j(z) = \frac{I_{ji}V_{ji}}{I_{jc}V_{jc}}\gamma_{jz}m_j\frac{\sigma_i^{m_j-1}}{z^{m_j}}\exp-\left[\frac{I_{ji}V_{ji}}{I_{jc}V_{jc}}\gamma_{jz}\left(\frac{\sigma_i}{z}\right)^{m_j}\right] \tag{17}$$

The resulting likelihood statistic from (14) and the substitution of (17) is:

$$W(R(z)) = 2\left[l(\hat{m}_1,\hat{m}_2,\hat{m}_3,\hat{R}(z),\hat{\sigma}_{o2},\hat{\sigma}_{o3}) - l(\hat{m}_{1,R(z)},\hat{m}_{2,R(z)},\hat{m}_{3,R(z)},R(z),\hat{\sigma}_{o2,R(z)},\hat{\sigma}_{o3,R(z)})\right] \tag{18}$$

Quantities subscripted with $R(z)$ are values conditional on the current value of $R(z)$.

Determining the right-hand log likelihood in (18) is equivalent to a constrained optimization problem:

Maximize $\hat{l}_{R(z)}$

such that $\ln\left(\dfrac{1}{R(z)}\right) = \gamma_{1z} + \gamma_{2z} + \gamma_{3z}$

as can be shown with some manipulation.

Using Lagrange multipliers changes the problem into the format of an unconstrained optimization problem with the additional parameter λ:

Maximize: $$L = \hat{l}_{R(z)} + \lambda\left(\gamma - \gamma_{1z} + \gamma_{2z} + \gamma_{3z}\right) \tag{19}$$

Iterating on estimates of $R(z)$ (or γ) eventually satisfies the equality in (15). Equation (19) must be maximized at each step. An efficient method of maximizing (19) entails setting the partial derivatives of L equal to 0 and solving for the requisite m_j's for a given $\lambda \neq 0$. This works well for $\lambda > 0$ and handily gives an upper bound on reliability. The lower bound is of much greater interest, however, and this corresponds to $\lambda < 0$, which tends to give numerical difficulties. The most promising approach attempted to date uses a quasi-Newton search algorithm in the space defined by the m_j's. This method was successfully applied to a set of NT154 strength data, as shown in Figure 2.

The data set analyzed for Figure 2 was particularly challenging. It combines tensile and flexure specimen strength data and has seven failure modes (listed in the figure). These data have both competing and non-competing failure modes. For example one set of flexure specimens has three competing failure modes; surface (SURFACE(F)), volume (VOLUME), and chamfer (CORNER) However, the three surface distributions; as-processed (AP-SURFACE), machined tensile specimen (SURFACE(T)), and machined flexure specimen (SURFACE(F)) do not occur in the same specimens and are not competing. This example demonstrates the capability of both bootstrap and likelihood ratio technique for calculating similar confidence intervals for highly censored data sets.

Additional capabilities of these methods for calculating confidence intervals for reliability predictions are demonstrated in Figures 3 and 4. The effect of size scaling on confidence intervals is shown in Figure 3. This is a Weibull strength plot of the NT154 tensile specimen volume failures. The top set of lines are estimates from the maximum likelihood (median line) and likelihood ratio techniques (95% confidence bands). The lower set of lines are failure probabilities predicted for an effective volume of one cubic millimeter. The confidence intervals translate vertically when predictions are made for a different size specimen or component. This is expected, confidence intervals should be narrowest in the region where failure stresses are sampled. As a result, the highest confidence for predicting the strength of future tensile specimens is at approximately 50% probability of failure.

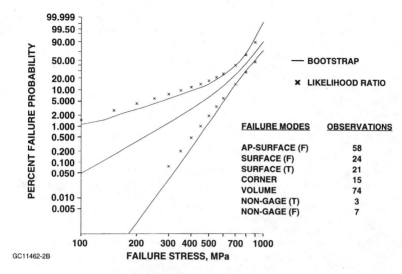

GC11462-2B

Fig. 2--Comparison of bootstrap and likelihood ratio confidence
intervals for combined failure probability in the presence
of multiple flaw populations.

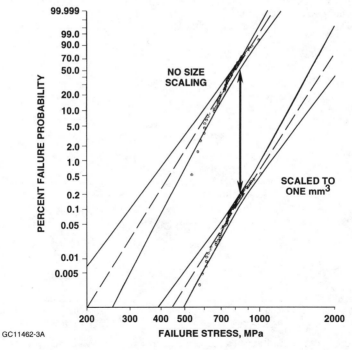

GC11462-3A

Fig. 3--Confidence intervals translate vertically with size scaling
as demonstrated by these plots (with and without size
scaling) of the tensile volume strength data.

However, prediction of strengths for a cubic millimeter effective volume
has the least uncertainty around .2% failure probability. And for
prediction of larger volumes there is less uncertainty for failure
probabilities above 50%. Failure probabilities of interest for
component design are typically 1% or less.

More realistic conditions for component reliability predictions
include multiple competing flaw distributions with separate size scaling
for surface, volume, and edge length. An example with these complications is shown in Figure 4. This figure shows the data and failure
probability lines for NT154, size-E flexure strength tests. The
straight dashed lines are the failure probabilities for the three
independent failure modes and the solid line is the combined failure
probability. Figure 4(B) shows a reliability prediction from the same
data. The independent and combined failure probability lines have all
shifted as a result of scaling to a component with a different
(hypothetical) size. The lower confidence bound (relative to strength)
is also plotted. The specimen data for all failure modes can no longer
be plotted in a combined manner because the proportion of failures from
each failure mode is different for this component than the specimen used
to obtain strength data. However, to get a physical sense for the
location of the data relative to the predicted lines a reranking technique [14] for censored data was used to plot the data separately for
each failure mode. This plot shows the highest confidence occurs near
the center of the range of specimen failure stresses, similar to what
was shown for individual failure modes in Figure 3. Figure 4 demonstrates the capability for calculating confidence intervals on component
reliability in the presence of multiple competing failure modes.

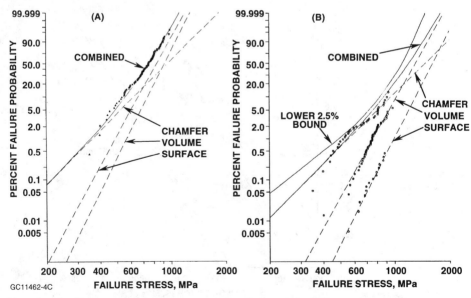

Fig. 4--Confidence intervals on predictions with size scaling and compet
        ing failure modes. Plot A is the non-scaled Size E data. Plot B
        Shows the reliability line and lower confidence bound calculated
        from size E data with scaling for length, area, and volume.

Multiple Temperature Capability

Methods for modeling temperature-dependent strength are important for designing ceramic components for high-temperature engine applications. The method proposed here, based on maximum likelihood statistics, allows Size E data with scaling for length, area, and volume strength data from multiple temperatures to be combined for a single estimate of temperature-dependent strength parameters and for reliability predictions with confidence intervals for high-temperature components.

The method for pooling data from multiple temperatures is based on an assumption that the Weibull modulus, m, for a given strength distribution is independent of temperature under fast fracture conditions and thus only the second Weibull parameter is a function of temperature. This assumption is unconventional but is supported in two ways as follows: the first support is from analysis of NT154 fast fracture data at varying temperatures as measured by the University of Dayton Research Institute (UDRI) and (AP). Weibull moduli for each failure mode are plotted versus temperature in Figure 5. The variation of m with respect to temperature is well within the 95% confidence bands for each of the four failure modes analyzed, and there is no obvious trend of m to increase or decrease with temperature.

The second supporting argument for the assumption of temperature independence of m results from consideration of strength controlling flaw distributions in the absence of slow crack growth during strength testing, and in a material that exhibits linear elastic fracture (no R-curve behavior, etc.). In this case, the m value of a given distribution and its variability are not expected to change by simply increasing

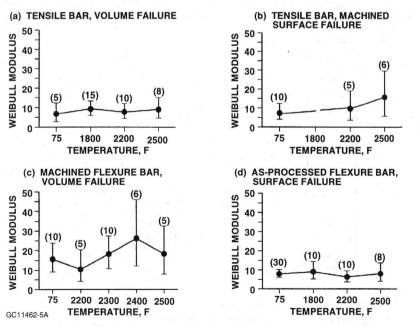

Fig. 5- Weibull moduli of NT154 show no obvious temperature dependence. The numbers of data points at each temperature are shown in parentheses.

the test temperature; therefore, m should not be temperature dependent.

Pooling fracture data from multiple test temperatures is greatly facilitated by this assumption of temperature independence of m.   The above arguments support, but do not prove, the validity of the assumption.   There are numerous examples of data sets in the literature that suggest a temperature dependence of m.   In light of the above discussion, it is proposed that one or more of the following factors are the cause of such observations: slow crack growth during loading; lack of proper censoring of multiple strength distributions; creation of new flaws from oxidation, etc.; temperature dependent R-curve behavior; and/or insufficient sample size.

Based on the assumption of temperature independence of m_j, then only modeling the variability of $\hat{\sigma}_{oj}$ with temperature is necessary. This is accomplished with an additional variable C_{jq} which is define as;

$$C_{jq} = \frac{\sigma_{ojq}}{\sigma_{ojr}} \qquad (20)$$

Where σ_{ojq} is the second Weibull parameter for the strength distribution j at temperature q, and σ_{ojr} is the second Weibull parameter at a reference temperature (room temperature was used).   This temperature scaling variable C_{jq} can be added to the maximum likelihood equations (10) and (11).   The resulting equations allow strength data with multiple competing strength distributions from multiple temperatures to be pooled for estimation of C_{jq} along with $\hat{m}_j$ and $\hat{\sigma}_{ojq}$.

An example of combined temperature strength data analysis is shown in Figure 6.   These test data of NT154 were generated in UDRI and AP test

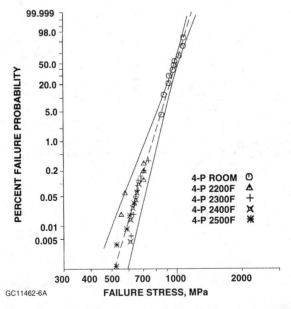

GC11462-6A

Fig. 6-- Multiple temperature data scaled to room temperature to determine the Weibull modulus for all of the data.

laboratories. The rank positions of the elevated temperature strength
data were adjusted to plot all data with respect to room temperature.
For example, consider the median strength specimen from a set of
elevated temperature strength data. If the median strength at that
temperature were equivalent to the 5% strength at room temperature then
the median strength specimen would be plotted at 5% failure probability
on the room temperature Weibull plot. This approach was used to plot all
of the elevated temperature data in Figure 6. The temperature scaling
variable C_{jq} was incorporated in the bootstrap method to calculate the
confidence intervals plotted in Figure 6. C_{jq} was allowed to change for
each simulation. Since the uncertainty in $\hat{C}_{jq}$ is dependent on the
number of specimens tested at a given temperature both the width and
location of the confidence intervals change when the failure
probabilities are plotted for different temperatures.
     The methods for analysis of multiple temperature strength data
combined with the size scaling and censored data analysis techniques for
competing strength distributions also applies to component predictions.
These methods provide prediction techniques that allow data from
specimens of multiple sizes, loading conditions, tested at several
temperatures and with competing strength distributions to be combined to
predict the strength of components with arbitrary size, temperature, and
stress distributions.

INDUSTRY STANDARDS FOR CHARACTERIZING CERAMIC STRENGTHS

     The statistical methods described in this paper can provide accurate
estimates of ceramic strength parameters and component reliability, and
guide planning of specimen tests to generate data for these estimates.
Industry standards in each of these areas would offer more consistent
and accurate strength characterization to help ceramic material
suppliers and component designers to develop materials and component
designs that are adequate for production applications.
     Strength parameters are estimated to determine if a material has
adequate strength for an application, or to characterize the properties
of a new material. To adequately describe the strength of a ceramic,
the Weibull parameters should be presented for each independent strength
distribution. Ceramics generally have multiple strength distributions,
and to ignore this and treat them as having only one strength distribution will typically result in an erroneous assessment of strength
properties. Proper treatment of multiple strength distributions
requires fractography and censored data analysis techniques. In
addition, the characteristic strength should be normalized to a unit
size (e.g. mm^2 or mm^3); which is referred to in this paper as the second
Weibull parameter. The second Weibull parameter is a material property
which can be compared for different materials without knowing the size
and loading conditions of the specimens used for strength testing.
Parameter estimates should be presented with confidence intervals to
quantify sampling uncertainty from the number of specimens tested, size
of the specimens tested, and number of failures obtained for each
strength distribution. The techniques described in this paper allow
confidence intervals to be calculated for data sets with multiple
strength distributions and other complications.
Ceramic components are typically designed to reliability values without
confidence levels because methods for calculating confidence intervals
on predictions with size scaling and multiple distributions are not

readily available. Simple reliability prediction does not indicate any
of the uncertainty due to size of the database (i.e. point estimation of
reliability does not indicate if a prediction comes from a database of 5
or 500 specimens). The confidence interval calculation techniques
described here account for both the number and size of specimens tested
as well as the number of failures sampled from each strength
distribution. Component design with confidence limits is necessary if
we are to have sufficient information to manage the risks associated
with designing production-quality ceramic components.

Progress can also be made in reducing the cost and increasing the
quality of ceramic strength data. More specimens are required to
characterize the strength of ceramics than metals, due to the larger
scatter in strength of ceramics. The methods described in this paper
can be used to perform tradeoff studies to define the size and number of
specimens that will provide accurate parameter estimates or component
predictions (based on the width of the confidence intervals) for a
specified investment. These tradeoffs can be performed with slight
variations of the methods described here which will allow confidence
intervals to be predicted from pooled specimen data with multiple sizes
and multiple test temperatures. Since the optimum data set is
dependent on the application (e.g. volume properties from large
specimens are more important for a large component with a highly
stressed volume than for a small component with stresses concentrated
near the surface), it would be more useful to standardize an approach
for defining a database than to try to define a database that meets the
needs of all applications.

The methods suggested here are intended to help the ceramics
community agree on how better to characterize and represent ceramic
strength properties. This is important if ceramic component manufactur-
ers are to supply materials with adequate properties for production
applications, and if designers are to ensure that ceramic components
will meet reliability objectives by design, and not by costly and time
consuming trial and error. Progress is being made in the American
Society for Testing and Materials, Advanced Ceramics Committee C28, to
develop some of these techniques into standards. The efforts need
ongoing support to ensure that ceramic properties and component design
continue to improve to meet the needs for production applications of
ceramics.

SUMMARY

A Weibull based statistical model for estimating strength parameters
and component reliability with confidence intervals has been presented.
This model provides capabilities for:

- censored analysis of competing strength distributions
- analysis of data from specimens with multiple sizes
- analysis of data from specimens with multiple loading conditions
- analysis of data from multiple temperature tests
- calculation of confidence intervals on parameter estimates
- calculation of confidence intervals on reliability estimates

A combined probability density function and the associated likelihood
function were presented. Resulting analysis methods apply to pooled
data sets from specimens of multiple sizes, types, and test
temperatures.

Progress has also been made in generating an extensive database on
NT154 for methods confirmation. The fast fracture portion of that

database was used along with other NT154 and SiC data to demonstrate the methodologies reviewed in this paper. More thorough investigation of the methods will be completed after the NT154 testing and fractography are complete, including room and elevated temperature testing of the spin disk shown in Figure 7.

Methods proposed in this paper have the potential for providing more accurate estimates of strength parameters, more accurate estimates of component reliability, and guidance for planning specimen tests to obtain accurate estimates of strength parameters and component reliability. Industry standards in each of these areas would provide more consistent and accurate strength characterization to help material suppliers and component designers to develop materials and component designs that are adequate for production applications.

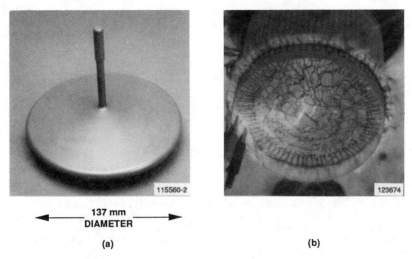

137 mm
DIAMETER

(a)   (b)

Fig. 7-- a) NT154 spin disk for methods confirmation. b) High
speed photo of a spin disk soon after burst.

ACKNOWLEDGMENTS

The support and latitude provided by Charles Brinkman (Program Monitor) and Dr. Ray Johnson, (Program Manager) in the "Life Prediction Methodology for Ceramic Components of Advanced Heat Engines" program sponsored by Oak Ridge National Laboratories and Department of Energy has greatly contributed to the progress being made in this program.

REFERENCES

[1]   Weibull, W., "A Statistical Theory of the Strength of Materials," Royal Swedish Academy of Engineering Sciences Proceedings, Volume 151, pp 1-45, 1939

[2]   "Ceramic Gas Turbine Engine Demonstration," Garrett Turbine Engine Company, ARPA/NAVAIR, N00024-76-5352, 1980.

[3]   Johnson, C. A., and Tucker, W. T., "Advanced Statistical Concepts
      of Fracture in Brittle Materials," Engineered Materials Handbook
      Volume 4: Ceramics and Glasses, ASM, pp 709-715, 1992.

[4]   Batdorf, S. B. , and Crose, J. G., "A Statistical Theory for the
      Fracture of Brittle Structures Subjected to Nonuniform Polyaxial
      Stresses," Journal of Applied Mechanics, 41, pp 459-464, 1974.

[5]   Batdorf, S. B. , and, Heinisch, H. L., "Weakest Link Theory
      Reformulated for Arbitrary Fracture Criterion," Journal of the
      American Ceramic Society, Vol 61 (No. 7-8), 1978, pp 355-358,
      1978.

[6]   Evans, A. G., "A General Approach for the Statistical Analysis of
      Multiaxial Fracture," Journal of the American Ceramic Society, 61,
      7-8, pp 302-308, 1978.

[7]   Nemeth, N. N., Manderscheid, J. M., and Gyekenyesie, J. P., "
      Ceramic Analysis and Reliability Evaluation of Structures (CARES),
      NASA Technical Paper 2916, 1990.

[8]   Johnson, D. R., "Ceramic Technology Project Semiannual Progress
      Report for April 1991 through September 1991", ORNL/TM-11984, pp
      320-344, 1992.

[9]   Chao L. and Shetty D. K., "Equivalence of Physically Based
      Statistical Fracture Theories for Reliability Analysis of Ceramics
      in Multiaxial Loading," J. AM. Ceram. Soc., 73, 7, 1917-1921, 1990

[10] Tucker, W. T., and Johnson, C. A., "The multiaxial Equivalence of
      Stressed Volume", Life Prediction Methodology and Data for Ceramic
      Materials, ASTM STP 1201, C. R. Brinkman, and S. F. Duffy, Eds.,
      American Society for Testing and Materials, Philadelphia, 1993.

[11] Johnson, C. A., and Tucker, W. T., "Weibull Estimators for Pooled
      Fracture Data", Life Prediction Methodology and Data for Ceramic
      Materials, ASTM STP 1201, C. R. Brinkman, and S. F. Duffy, Eds.,
      American Society for Testing and Materials, Philadelphia, 1993.

[12] Efron, B. and Tibshiriami, R., "Bootstrap Methods for Standard
      Errors, Confidence Intervals, and Other Measures of Statistical
      Accuracy," Statistical Science, Vol. 1, pp. 54-57, 1986.

[13] Cox, D. R., and Oakes, D., "Analysis of Survival Data", Chapman
      and Hall, Chapt. 3.3, 1984.

[14] Johnson, L. G., "The Statistical Treatment of Fatigue
      Experiments," Research Laboratories, General Motors, pp. 44-50,
      1959.

Osama M. Jadaan[1]

LIFITIME PREDICTION FOR CERAMIC TUBULAR COMPONENTS

REFERENCE: Jadaan, O. M., "Lifetime Prediction for Ceramic Tubular Components," Life Prediction Methodologies and Data for Ceramic Materials, ASTM STP 1201, C. R. Brinkman and S. F. Duffy, Eds., American Society for Testing and Materials, Philadelphia, 1994.

ABSTRACT: The main objective of this research is to develop experimental and analytical methodologies to predict the lifetimes for internally pressurized SiC tubes from the lifetimes of simple specimens subjected to similar delayed failure modes. In general, two different mechanisms are responsible for delayed failure behavior in ceramics, depending on the material, microstructure, size of inherent flaws, and the level af applied stress. These delayed failure mechanisms are slow crack growth (SCG) and creep rupture. In this paper, a methodology to predict the lifetimes for sintered alpha silicon carbide (SASC) tubes, expected to fail due to SCG mechanism, will be shown. This methodology involved experimental determination of the SCG parameters for the SASC material and the scaling analysis to project the stress rupture data for small specimens (O-rings and compressed C-rings) to large tubular components. Also included in this paper is a methodology to predict the lifetimes for internally pressurized reaction bonded silicon carbide (SCRB210) tubes, for which delayed failure behavior is expected to be controlled by creep rupture mechanism. Finite element analysis (FEM) in association with the Monkman-Grant creep rupture criterion, were used to predict the lifetimes for the SCRB210 tubes. The relationship between the two delayed failure mechanisms, specimen size and applied stress level will also be discussed.

KEYWORDS: Silicon carbide, slow crack growth, creep, Weibull statistics.

Components with tubular geometry are widely used in structural applications. Circular cylinders are perhaps the most used in heat exchanger, radiant tube, power generator and other thermal applications. Tubular components are also used in applications such as space-radiator tubes, vacuum furnace containment cylinders, rocket exhausts, and seal faces in automotive water pumps (rings). The circular cylinder constitutes the basic element in boilers, steam or gas turbines and gas compressors to mention a few. Thus the tubular geometry is an essential element in a wide variety of industrial and structural applications.

Currently, ceramic materials are being contemplated for thermal and structural applications because, compared to metals, they can withstand high operating temperatures, thereby providing greater thermal efficiency in addition to improved oxidation and corrosion resistance. Furthermore, ceramics in general possess low coefficient of thermal

---

[1] Assistant Professor, College of Engineering, University of Wisconsin-Platteville, Platteville, WI 53818.

expansion and high thermal conductivity, henceforth being resistant to thermal shock. Ceramics are also ideal for high temperature applications because they maintain relatively high strength at such temperatures. Because of these desirable properties of ceramic materials at high temperatures and because of the widespread use of the tubular geometry in thermal and thermomechanical applications, interest in studying the feasibility of utilizing ceramic tubular components in actual structural applications began.

However, ceramic materials exhibit time-dependent failure behavior at high temperatures. In general, two different delayed failure mechanisms are responsible for this behavior, depending on the level of applied stress , microstructure, material and size of preexisting flaws. These mechanisms are slow crack growth and creep rupture.

At high stresses or in the presence of large inherent flaws, delayed failure is controlled by the SCG mechanism, while at low stresses delayed failure is caused by the creep rupture mechanism [1,2]. Failure due to creep rupture mechanism occurs in the absence of crack propagation, where the typical microstructural damage is not limited to the vicinity af a propagating crack [3]. On the other hand, a SCG-induced delayed failure initiates at the most critical flaw, which grows to a critical size, after which catastrophic failure occurs [4].In terms of longevity of rupture life for a loaded ceramic component, substantially longer lifetimes are expected if failure is controlled by creep rupture rather than SCG.

The objective of this paper is to provide analytical and experimental methodologies to predict the lifetimes of thermomechanically loaded ceramic tubular components, from reliability analysis conducted on simple specimens subjected to similar delayed failure modes. The first portion of this article will emphasize the lifetime prediction methodology for ceramic tubular components exhibiting time dependent failure mode controlled by SCG. Dynamic fatigue data obtained from testing O-ring and C-ring specimens tested in compression at 1200 ° and 1300 °C were used to calculate the SCG parameters for the SASC material. These SCG parameters were then used to predict the lifetimes of internally heated and pressurized SASC tubes. The effect of a threshold stress intensity factor, below which no SCG occurs, was incorporated into this analysis. The O-rings and C-rings in compression were selected because of their simple tubular configuration and ease of machining and loading. Another reason these two specimen configurations were selected is because the O-ring specimens fail due to flaws associated with the inner surface of the tube while the compressed C-ring specimens fail due to flaws associated with the outer surface of the tube, thus permitting comparison of SCG behavior at both surfaces of the tube.

In the second portion of this paper, an analytical methodology is introduced to predict the lifetimes for internally heated and pressurized SCRB210 tubes which initially were expected to fail due to creep rupture mode. The Monkman-Grant creep rupture criterion was utilized to predict the lifetimes for the tubes by using creep rupture data obtained from testing coupon tensile specimens. Finite element analysis (FEM) was used to model the stress and creep rate distributions for the SCRB210 tubes. To test the validity of this analysis, long SCRB210 tubes were tested at 1300 °C by internally pressurizing them to a subcatastrophic stress level and measuring the time to failure. The measured and predicted lifetimes for the tested tubes, based on creep analysis, were compared and conclusions were drawn. Experimental and analytical evidences indicated that the SCRB210 tubes did not fail due to creep rupture, but rather due to crack growth.

I) LIFETIME PREDICTION FOR CERAMIC TUBULAR COMPONENTS THAT FAIL DUE TO SCG MECHANISM

1) Experimental Procedure:

Sintered Alpha Silicon Carbide (SASC)[1] was selected for this part of the study because it does not creep below 1500 °C and ,therefore, its delayed failure behavior, for that temperature range, is controlled by SCG [5,6,7,8]. This material features a nearly uniform SiC grain structure ranging in size between 2-5 μm and contains approximately 3% porosity (see figure 1). The SASC long tubes (1.83 m long with mean outside diameter of 43.8 mm and mean wall thickness of 4.8 mm) from which the O-ring and C-ring specimens were cut, were formed by an extrusion process using water based binder system.

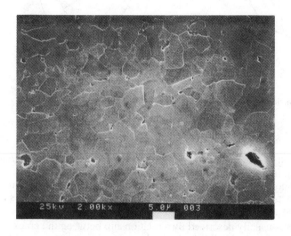

Figure 1. SEM micrograph showing the microstructure for the SASC material.

The O-ring and C-ring specimens (see figure 2) were cut in widths of 9.5 mm from the long SASC tubes. Side surfaces of the O-ring and C-ring specimens were ground to within 0.08 - 0.13 mm oversize and then lapped using 15 μm sized diamond lapping medium to the final dimensions . A notch size of 9.5 mm was cut from a typical O-ring specimen to form the C-ring configuration.

Dynamic fatigue tests (strength vs. stressing rate) were conducted on SASC O-rings and C-rings in diametral compression at 1200 ° and 1300 °C. These specimens were oxidized prior to testing for 24 hours at 1200 °C. The dynamic fatigue tests were conducted in an Instron machine[2] at cross head speeds of 0.0005, 0.005, 0.05 and 0.5 cm/min. These cross head speeds correspond to maximum stressing rates of 0.35, 3.5, 35 and 350 MPa/sec for the O-ring specimens and 0.1, 1.0, 10 and 100 MPa/sec for the compressed C-ring specimens. At least seven specimens were tested at each stressing rate. Fractographic examination of each tested specimen was conducted in order to asses the source of failure and to measure the critical flaw size.

---

[1] SASC, Carborandum Company, Niagara Falls, New York.
[2] Instron model 4202, Instron Corp., Canton, Mass.

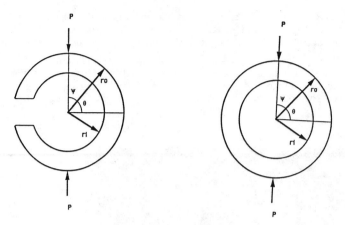

Figure 2. Schematic representation of O-ring and compressed C-ring specimens.

2) Analysis

2.1)Calculation of SCG parameters-- The strength of ceramic materials is controlled by inherent or pre-existing flaws. When a ceramic material is subjected to mechanical and/or thermal stresses, the inherent flaws can grow to a critical size after which catastrophic failure will occur. This phenomenon for which growth of flaws occurs at stress intensities below the critical stress intensity, K_{IC}, is known as slow crack growth .

Slow crack growth is usually described by a relationship between the crack tip velocity, v, and the stress intensity factor, K_I. This relationship is used to evaluate the SCG parameters necessary for lifetime prediction. The most common relationship used to model SCG behavior is described through the following equation [4,9,10]:

$$v = A\,K_I^N ,\tag{1}$$

where A and N are SCG parameters which depend on the material, environment, and temperature.

The SCG parameters, A and N, can be determined by correlating the strength of the material to the applied stressing rate, $\dot{\sigma}$, for a given temperature and environment as follows [11]:

$$\sigma_{df} = \left\{ \frac{2\,(N+1)}{A\,Y^N\,(N-2)}\,(c_i^{(2-N)/2} - c_f^{(2-N/2)})\,\dot{\sigma} \right\}^{\frac{1}{N+1}}\tag{2}$$

where σ_{df} is the fracture stress at a given stressing rate, c_f is the final flaw size before failure, c_i is the initial flaw size, and Y is the crack boundary correction factor (Y=1.24) [12].

Equation 2 describes a relationship between fracture stress and stressing rate for the simple case when the threshold stress intensity factor, K_{Ith}, vanishes.Some ceramics including the SASC material exhibit a significant threshold stress intensity factor [13,14,15]. In such cases, equation 2 is not valid for describing the dynamic fatigue relationship. Based on analysis introduced by Evans [16], a more general description for the dynamic fatigue curve that incorporates the effect of K_{Ith} can be derived [17],

$$\sigma_{df} = \frac{K_{Ith}}{Y\sqrt{c_i}}\{1 + \frac{2Y(N+1)c_i^{\frac{N+1}{2}}}{A(N-2)K_{Ith}^{N+1}}[c_i^{\frac{2-N}{2}} - (\frac{Y\sigma_{df}}{K_{IC}})^{N-2}]\dot{\sigma}\}^{\frac{1}{N+1}} \quad . \tag{3}$$

If the parameters σ_{df}, K_{Ith}, K_{IC}, c_i, Y, and $\dot{\sigma}$ are known then the SCG parameters could be estimated from equation 3, as was done in this study.

Yavuz [15] have demonstrated that if the threshold stress intensity is equal to or greater that 60% of the fracture toughness value, a significant change in the dynamic fatigue curve will result. Therefore, for materials that exhibit a high threshold stress intensity to fracture toughness ratio, the SCG parameters obtained from least square fit of equation 2 will be in error when compared to experimental results from crack growth measurements. However, by using equation 3 which takes into account the influence of the threshold stress intensity factor, the crack growth parameters can be estimated more accurately.

Nonlinear regression analysis was used to calculate the SCG parameters A and N by fitting the dynamic fatigue data to equation 3. The DUD nonlinear regression procedure available with the SAS statistical program [18] was utilized for this purpose. Although the nonlinear regression procedure of the SAS program utilizes an iterative solution to yield optimized sets of SCG parameters (minimize sum of square errors), these parameters may not correspond to the actual material behavior. Therefore, a criterion is necessary to determine the appropriate SCG parameters which correctly describe the behavior of the material. This criterion was obtained from a theoretical derivation by Chuang [10,19] in which he gave a relationship to calculate the crack threshold velocity, v_{th}. In his derivation, Chuang assumed that the crack propagates along the grain boundary by a coupled process of surface and grain boundary self diffusion, and that steady state conditions prevail. This relationship is given by:

$$v_{th} = 8.13 \frac{D_s^4 \Omega^{7/3}}{KT\gamma_s^2}\{\frac{E}{(1-\nu^2)D_b\delta_b}\}^3 \quad , \tag{4}$$

where Ω is the atomic volume, $D_b\delta_b$ is the grain boundary diffusivity, D_s is the surface diffusivity, γ_s is the surface energy, E is Young's modulus, ν is Poisson's ratio, K is Boltzman's constant (1.38×10^{-23} J/°K), and T is the absolute temperature.

Using equation 4 and substituting values for the parameters corresponding to the SASC material (obtained from references 10 and 14), the crack threshold velocity was calculated to be 1.0×10^{-9} and 1.1×10^{-9} m/sec at 1200° and 1300 °C, respectively. This result agrees ,to a certain degree, with slow crack growth velocity measurements conducted by Yavuz [15] for aluminum doped SASC material. He measured v_{th} to be 6×10^{-9} and 2.3×10^{-8} m/sec at 1200° and 1300 °C, respectively.Thus, at this temperature range, v_{th} can be considered to be 1.0×10^{-9} m/sec. Substituting this value in equation 1, the following criterion results:

$$v_{th} = 1.0 \times 10^{-9} = A K_{Ith}^N \quad . \tag{5}$$

Equations, 3 and 5, can now be used to extract the appropriate SCG parameters.

In order to utilize the dynamic fatigue relationship, expressions for the stressing rates of O-ring and compressed C-ring specimens are needed. For the O-ring specimen (fig. 2), $\dot{\sigma}$ is given by [20]:

$$\dot{\sigma} = \frac{E I \left[0.637 \frac{r_a x}{I} - \cos(\theta) \left(\frac{1}{A} + \frac{r_a x}{I} \right) \right]}{0.298 \, r_a^3} \, \dot{y} \, ,$$ (6)

where,   E = Young's modulus
         A = $b(r_o - r_i) = b t$, b = width, t = wall thickness
         I = $1/12 (b/t^3)$ $r_a = (r_i + r_o)/2$
         x = $r_a - r$ $\dot{y}$ = ram cross head speed.

and for the C-ring specimen tested in compression, $\dot{\sigma}$ is given by [20]:

$$\dot{\sigma} = \frac{2 E I}{\pi \, r_a^3 \, b \, t} \left[\frac{R (r_o - r_a)}{r_o (R - r_a)} \cos(\theta) \right] \dot{y}$$ (7a)

where, $R = t/\ln (r_o/r_i)$. (7b)

2.2) Lifetime prediction--The time to failure for a component under constant applied stress, σ_a, can be calculated from the following equation [11]:

$$t_f = B \, \sigma_{max}^{N-2} \, \sigma_a^{-N}$$ (8a)

where σ_{max} is the inert strength where no slow crack growth occurs before failure and B is calculated from the following equation:

$$B = \frac{2}{A Y^2 (N-2) K_{IC}^{N-2}} .$$ (8b)

In the case where the material being investigated exhibits a substantial threshold stress intensity factor, K_{Ith}, the time to failure can still be predicted from equations 8. The effect of a substantial threshold stress intensity factor will influence lifetime prediction in the sense that the SCG parameters will significantly differ from the case when K_{Ith} is not considered.

Another approach to predicting the lifetime of ceramic components involves the projection of stress rupture data for simple specimens to predict the stress rupture distribution for large components. This objective can be accomplished by substituting the following equation ,

$$\sigma_{max} = \sigma_{0v} (KV)^{-1/m} [\ln (\frac{1}{1-F})]^{1/m} \, ,$$ (9)

where σ_{0v} is the Weibull scale parameter based on volume analysis, KV is the effective volume, m is the Weibull modulus, and F is the failure probability, into equation 8a, resulting in the following expression:

$$t_f = \frac{2 \, \sigma_a^{-N}}{A Y^2 (N-2) K_{IC}^{N-2}} \sigma_{0v}^{N-2} (KV)^{\frac{2-N}{m}} [\ln(\frac{1}{1-F})]^{\frac{N-2}{m}} .$$ (10)

For a given failure probability, F, and assuming the SCG and Weibull parameters to be material properties that do not depend on the shape, size and loading configuration of a given specimen, the following relationship can be derived:

$$\frac{t_{f1}}{t_{f2}} = (\frac{Y_2}{Y_1})^2 \; (\frac{\sigma_{a2}}{\sigma_{a1}})^N \; (\frac{K_2 V_2}{K_1 V_1})^{\frac{N-2}{m}} \; . \tag{11}$$

Subscripts 1 and 2 refer to the two specimen configurations in question. The crack border correction factor, Y, is included in equation 11 because this factor depends on the specimen and loading configuration. Equation 11 can be used to generate an entire stress rupture curve for a given specimen configuration from one stress rupture data point of another specimen configuration. Note that equation 11 reduces to the familiar stress rupture equation, $\sigma_{a1}/\sigma_{a2} = (t_{f2}/t_{f1})^{1/N}$, when the two specimen configurations to be compared are identical.

Equation 11 can also be written in the following form:

$$\frac{t_{f1}}{t_{f2}} = (\frac{Y_2}{Y_1})^2 \; (\frac{\sigma_{a2}}{\sigma_{a1}})^N \; (\frac{\sigma_2}{\sigma_1})^{2-N} \; , \tag{12}$$

where σ_1 and σ_2 are fracture strengths corresponding to equal failure probabilities for the two specimen configurations.

Equation 11 assumes that failure is predominantly associated with volume flaws. When failure is predominantly associated with surface flaws , then the terms σ_{0V} and KV should be replaced with the Weibull scale parameter based on area analysis ,σ_{0A}, and the effective area, KA, respectively. The effective volume (KV) and effective area (KA) expressions for the O-ring, C-ring and internally pressurized tube specimens can be obtained from references 20 and 21.

3) Results and Discussion

3.1) Dynamic fatigue-- Figure 3 shows the fracture stress vs. stressing rate for the SASC C-ring specimens tested in compression at 1200 ° and 1300 °C. Similarly, Figure 4 shows the fracture stress vs. stressing rate for the SASC O-ring specimens tested at 1200 ° and 1300 °C. As expected, for both specimen configurations at 1200 ° and 1300 °C, the strength decreased as the stressing rate decreased. Figure 3 indicates that the strength for the C-ring specimens tested in compression decreased at 1200 ° and 1300 °C, as the stressing rate decreased from 100 MPa/sec to 1 MPa/sec. However, at a stressing rate of 0.1 MPa/sec the strength increased slightly. The O-ring specimen exhibited similar behavior. Figure 4 indicates that the strength decreased at both 1200 ° and 1300 °C, as the stressing rate decreased from 350 MPa/sec to 3.5 MPa/sec . However, at a stressing rate of 0.35 MPa/sec, a strength improvement was again detected.

The strength improvement at the lowest loading rate is most likely due to a flaw-blunting mechanism, which has been shown by other investigators [14,22], to operate below a certain loading rate. One mechanism of flaw blunting is the occurrence of microscopic creep regions just ahead of the crack tip, where the stress intensity is greatest. Due to flow in this region, stress intensity relaxation occurs at the crack tip, which in turn results in strength improvement. At such low stressing rate levels, two competing mechanisms are involved in the delayed failure behavior. The first is the slow crack growth mechanism, which is responsible for increasing the size of the failure-initiating flaw and, thus, lowering the strength. The second mechanism is responsible for the development of microscopic creep zones ahead of the crack tip causing stress relaxation.

At the slowest stressing rate, the second mechanism (microscopic creep) appears to be dominant over the first (slow crack growth), and thus we observe the slight strength improvement. Therefore, in order to correctly obtain the slow crack growth parameters from the dynamic fatigue data, it is necessary to perform the analysis using the data obtained at the three highest stressing rates only. In doing so, slow crack growth is the dominant mechanism responsible for delayed failure. The lifetime prediction for the SASC tubular components will not be compromised because slow crack growth in this material takes place only at stress intensities above the threshold stress intensity level, K_{Ith}, where microscopic creep zones do not develop [19].

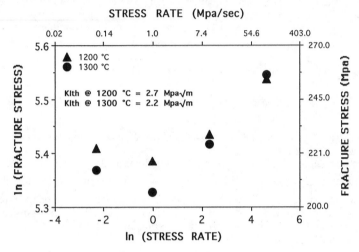

Figure 3. Strength vs. stressing rate for oxidized SASC C-ring specimens tested in compression at 1200 ° and 1300 °C.

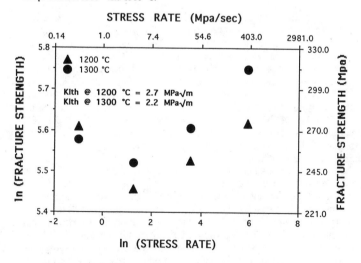

Figure 4. Strength vs. stressing rate for oxidized SASC O-ring specimens tested at 1200 ° and 1300 °C.

Nonlinear regression analysis, in association with equations 3 and 5 were utilized to extract the SCG parameters from the dynamic fatigue data. In order to perform this analysis, the initial critical flaw sizes must first be measured from the fracture surfaces of the broken specimens.Table 1 lists the strength, initial flaw size, and fracture toughness of oxidized SASC O-rings and C-rings as a function of stressing rate at 1200° and 1300°C. The dynamic strengths listed in this table were calculated at the point of fracture.At least seven specimens were tested at each stressing rate. The threshold stress intensity values for the SASC material at 1200 °C and 1300 °C were obtained from reference 14 and were adjusted to the K_{IC} value found in this study by assuming that the ratio of the threshold stress intensity factor to critical stress intensity factor is constant.

TABLE--1: Strength, initial flaw size, and apparent fracture toughness of SASC material as a function of stress rate for the O-ring and compressed C-ring specimens at 1200°C and 1300°C

Specimen Config.	Temp. (°C)	σ (MPa/s)	σ_{df} (MPa)	C_i (μm)	K_{IC} (MPa $\sqrt{}$m)
O-ring	1200	350	275.7 (36)	197	5.0
		35	251.2 (43)	217	4.4
		3.5	234.2 (27)	193	4.0
		0.35	273.3 (30)	163	4.5
	1300	350	313.5 (28)	138	4.4
		35	271.7 (24)	138	3.9
		3.5	249.8 (42)	141	3.9
		0.35	264.9 (28)	151	3.9
C-ring in Compression	1200	100	253.9 (16)	250	4.9
		10	229.3 (30)	235	4.5
		1.0	218.2 (39)	230	4.2
		0.1	223.6 (40)	184	4.6
	1300	100	256.0 (25)	240	4.7
		10	225.4 (28)	220	4.0
		1.0	205.4 (42)	190	3.9
		0.1	214.7 (23)	140	---

(Values in parenthesis correspond to standard deviations.)

Table 2 lists the slow crack growth parameters for oxidized SASC compressed C-ring and O-ring specimens at 1200 ° and 1300 °C. The slow crack growth exponents, N, are in remarkable agreement with those obtained from stress rupture studies published in the open literature. Table 3 contains a summary of slow crack growth exponents gathered from different publications for the SASC material. At 1200 °C, Govila [8] calculated N to be 27, while Quinn and Katz [23] found N to be 25 for the 1980 SASC material. These compare to the current findings of 19 and 23 for the compressed C-ring and O-ring specimens, respectively. At 1300 °C, Govila [8,24] calculated N to be 16 and 21 from stress rupture tests conducted on flexure bars and tensile specimens, respectively. These compare to the current findings of 17 and 23 for the compressed C-ring and O-ring specimens, respectively.

TABLE--2: Slow crack growth parameters for oxidized SASC compressed C-ring and O-ring specimens at 1200° and 1300°C.

TEMPERATURE

CONFIGURATION	1200°C		1300°C	
	A	N	A	N
C-ring in compression	1.0 E-17	18.7(7.0)*	1.0 E-15	17.0(7.9)
O-ring	1.0 E-19	23.2(9.7)	1.0 E-17	22.7(5.5)

(Values between parenthesis correspond to standard deviations.)

TABLE--3: Summary of slow crack growth exponents for sintered alpha silicon carbide.

Temperature (°C)	N	Test	Comments	Reference
1000	41	dynamic fatigue ($K_{I th}=0$)	preoxidized flexure bars	[22]
1300 1400	16 13	stress rupture	flexure bars	[8,24]
1200 1300	27 21	stress rupture	tension	[8]
1200	47	dynamic fatigue ($K_{I th}=0$)	flexure bars	[34]
1200	41	stress rupture	flexure bars (78 material)	[23,25]
1200	25		flexure bars (80 material)	
1200 1400	49 24	dynamic fatigue ($K_{I th}=0$)	preoxidized, flexure bars	[14]
1200	62	dynamic fatigue ($K_{I th}=0$)	preoxidized, flexure bars	[36]
1200	28	dynamic fatigue ($K_{I th}=2.25$ MPa$\sqrt{m}$)	flexure bars	[15]

3.2) Lifetime prediction for SASC tubular components--The two lifetime prediction
methods described previously will be used in this section to predict the lifetime for SASC
tubular components subjected to sustained thermomechanical loading (internally heated and
pressurized).Figure 5 displays the predicted lifetimes for SASC long tubes subjected to
sustained internal pressure at 1300 °C. The SCG parameters used to predict the lifetimes
for these tubes were extracted from dynamic fatigue data of compressed C-ring specimens
tested at 1300 °C (see table 2). The open triangular legend in figure 5 corresponds to the
stress rupture distribution for the SASC tubes as predicted from the fracture mechanics
approach (equation 8). The average fast fracture strength for the SASC tubes was
experimentally measured in a previous study to be 110 ± 7 MPa [20,25]. Therefore,
lifetime prediction for SASC tubes was calculated at applied stress levels ranging between
70 and 100 MPa. The lower applied stress level, 70 MPa, corresponds to 65 % of the
average fast fracture strength of the SASC tubes which is near the threshold level for SCG.
At this stress level, and according to the fracture mechanics approach for lifetime
prediction, the tube is expected to survive for 3.8 hours while at an applied stress of 100
MPa the tube is expected to fail in 0.01 hours.

Figure 5 also shows the predicted stress rupture distributions for internally pressurized
SASC tubes at 1300 °C (using equations 11 or 12) as predicted from the stress rupture
distributions for tensile and flexural specimens. The stress rupture data for the tensile and
flexural SASC specimens were obtained from reference 8. For this approach and using the
tensile stress rupture data, the SASC tubes were expected to fail as early as 63 hours and as
long as 13840 hours at an applied stress of 70 MPa. At an applied stress of 100 MPa, the
tubes are expected to fail in times ranging between 0.2 and 32 hours. Using the flexural
stress rupture data, the SASC tubes are expected to survive for as little as 2196 hours and
as long as 93570 hours at an applied stress of 70 MPa. At an applied stress of 100 MPa,
the tubes are predicted to survive between 5 and 218 hours. The large scatter in the
predicted lifetimes of the SASC tubes is due to the large scatter in the measured lifetimes

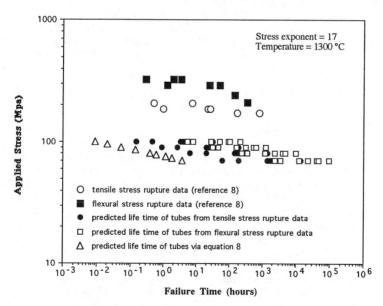

Figure 5. Predicted lifetimes of SASC long tubes at 1300 °C. N=17 was calculated from
dynamic fatigue data of compressed C-ring specimens tested at 1300 °C. The stress rupture
data for the tensile and flexural specimens were obtained from reference 8.

of the tensile and flexural specimens. Predicted lifetime curves for SASC tubular components using SCG parameters obtained from dynamic fatigue testing of O-ring specimens can be constructed in a similar manner.

Scaling analysis to predict the lifetimes for the SASC tubular components was based on volumetric considerations. The SASC material contains 3% porosity that is distributed uniformly throughout the material. These porosities constitute potential failure initiation sites. Fractographic analysis performed on the fracture surfaces for the C-ring and O-ring specimens verified that failures originated at surface and subsurface defects. Govila [8] found that failures in the stress rupture tested tensile and flexural specimens also initiated at subsurface and surface porosity locations and at machining damage sites . Since failure in the SASC specimens initiated at internal and surface defects, volume based analysis was used to construct the predicted lifetime distributions in figure 5. Fractography should be used to determine whether lifetime prediction should be based on volumetric ar area analyses techniques.

From the above discussion, it is observed that the fracture mechanics approach yielded the most conservative lifetime prediction for the SASC tubes. In the second approach, the flexural stress rupture data yielded very long predicted lifetimes for the SASC tubes. In the middle, fell the predicted lifetimes for the SASC long tubes as predicted from the tensile stress rupture data. The predicted lifetimes for the SASC tubes based on the flexural stress rupture data should be treated with caution, since the stress distribution for this specimen configuration is significantly different from that for the thermomechanically loaded tubes. Alternatively, the uniform stress distribution for the tensile specimen simulates closely that for the internally pressurized tube. Thus, the stress rupture data for the tensile specimen configuration is expected to better predict the delayed failure behavior for the SASC tubes. Furthermore, the flexural specimens were ground prior to testing, thus altering the inherent flaw population at the surfaces of these specimens. Since failure in flexural specimens is predominantly surface related, then the flexure stress rupture data does not represent realistically the delayed failure behavior for the tubes.

Extensive effort was directed towards obtaining experimentally measured stress rupture data for the SASC long tubes at 1300 °C. However, most of these tubes failed during pressurization due to their out of round nature and the difficulty of pressure sealing them. Therefore, no reliable stress rupture data for the SASC long tubes was gathered.

II) LIFETIME PREDICTION FOR CERAMIC TUBULAR COMPONENTS THAT FAIL DUE TO CREEP RUPTURE MECHANISM:

1) Material and Experimental Procedure

1.1) Material-- Reaction-bonded silicon carbide material (SCRB210)[3] was used for this part of the study because, based on tests conducted on coupon specimens, this material displays delayed failure behavior controlled by creep rupture mechanism. The SCRB210 material has a bimodal SiC grain size distribution were small grain sizes range from 2 to 5 μm and large grain sizes range between 25 and 75 μm (figure 6). This material contains 19 vol% free silicon. At high temperatures, the silicon phase becomes viscous, thus contributing to the creep behavior for the SCRB210 material.

The SCRB210 tubes were manufactured using slip cast process. These tubes had uniform circular cross sections and their inner surfaces were decorated with silicon nodules remnant from fabrication. From previous investigations, the strength and fracture toughness at 1300 °C for diametrically compressed O-ring specimens (outside diameter =

---

3 SCRB210, Coors Ceramics, Golden, Colorado.

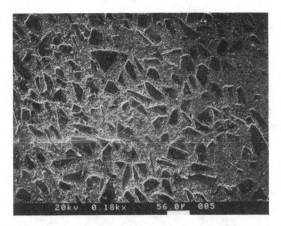

Figure 6. SEM micrograph showing the microstructure for the SCRB210 material.

44 mm, wall thickness = 4.8 mm, and width = 9.5 mm) were determined to be 301± 49 MPa and 4.1 MPa√m, respectively [20,26]. Long SCRB210 tubes (same diametral dimensions as the O-ring specimens but with lengths of 660 mm) were also tested by rapidly pressurizing them to failure at 1200 °C and were found to have an average strength of 78 ± 17 MPa. SCRB210 O-rings were also fractured at 1200 °C and their average strength was found to be equal to that at 1300 °C, indicating that the average strength for the SCRB210 material does not vary between 1200 ° and 1300 °C.

1.2) Sample preparation-- The creep rupture data for the SCRB210 material were obtained by testing small tensile specimens at temperatures ranging between 1250 ° and 1350 °C [29].These tests were performed by Wiederhorn et al. at NIST as part of an overall research program to devise test methodologies to determine the reliability for ceramic tubular components. Figure 7 shows a schematic representation of the specimen geometry and dimensions for the uniaxial tensile specimen.

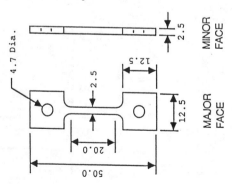

Figure 7. Schematic representation for the tensile creep test specimen.

The SCRB210 tubes were cut to lengths of 66 cm from the 183 cm long tubes supplied by the manufacturer. Machining was required on the inside surfaces of the open ends of the tubes in order to remove the silicon nodules. Smooth and round surfaces were needed to accommodate the end pressure seals. Details regarding machining and preparation of tubes are reported in references 20,25,26.

1.3) Experimental apparatus and delayed failure testing of long tubes-- The long tube components were tested in a specially designed high-temperature internal pressure tube burst apparatus. This facility consisted mainly of four components: 1) the pressure seals, 2) the end plates, 3) the water cooled end caps, and 4) the test chamber. A schematic representation of the tube burst test apparatus is shown in figure 8.

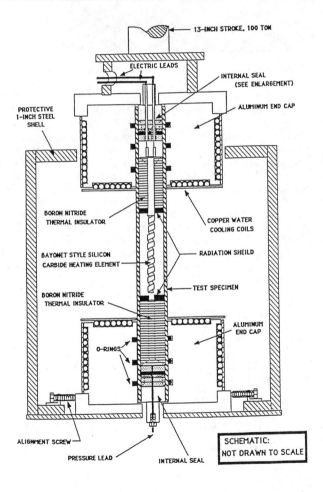

Figure 8. Schematic representation for the tube burst test apparatus.

In testing a long tube at high temperature, the long tube was positioned vertically in the test chamber. The pressure seal which was inserted into the top end of the tube is equipped with four electrical lead wires which run through the shaft of the seal. Two electrical lead wires were used to connect to and suspend crusilite DS[4] bayonet style heating element inside the tube. The two other leads were connected to a type B thermocouple which was used to measure the temperature at the inside surface of the tube. At the outside surface of the tube, additional thermocouples were placed along the length and circumference of the tube in order to monitor the axial and circumferential temperature distributions. All four electrical lead wires were connected to a programmable temperature controller[5], which was used to heat the heating element and monitor the temperature at the inside surface of the tube.

The lower pressure seal which was inserted into the bottom end of the tube contains a 1.6 mm diameter hole through the center of the shaft to allow gas inlet to the tube. An air driven gas booster[6] capable of pressurizing up to 69 MPa was used for this purpose. A pressure gage[7] equipped with a stop needle was also connected to the pressure line going from the pressure booster-pressurized gas tank assembly to the tube. Because of the method the tubes were mounted (tube is free floating on a rubber O-ring seal which is part of the pressure seal assembly), they were not restrained axially. This means that a negligibly small axial stress state exists along the length of tube. A series of relays, a solenoid, a timer, and a high-low pressure gage were also connected to the facility to measure the time to failure and to shut down the system upon failure of the tube. Details regarding the design and operation of the tube burst test apparatus are presented in references 20,25,26.

2) Creep Analysis for Internally Heated and Pressurized Tubes

In order to predict the lifetime for internally pressurized ceramic tubular components subjected to creep rupture loading, the stress and strain rate distributions as a function of time must be known. Combining this information with the Monkman-Grant creep rupture criterion [27], which was found to describe well the creep rupture behavior for the SCRB210 material, the lifetime for a given loaded tube can be predicted. The Monkman-Grant rupture criterion suggests that an empirical relationship exists between the rupture life, t_f, and steady state creep rate, ε. This relationship has the following form:

$$\dot{\varepsilon}^m t_f = C \tag{13}$$

where m and C are material constants. Therefore, equation 13 can be used to predict the lifetime for an internally pressurized tube if the creep rupture constants for the material are known and if the creep rate at failure can be calculated.

When a tubular component is internally pressurized at high temperatures, the tangential stress distribution across the wall of the tube will be initially elastic and can be described by the elastic thick-wall tube theory. This equation indicates that the maximum stress occurs at the inner surface of the tube (for small radial temperature gradients). However, if the internal pressure is sustained long enough for steady state creep to occur (described by the Dorn equation, $\varepsilon = D\sigma^n$, where D is a constant which includes the temperature, grain size,

---

[4] Kanthal Ltd., Perth, Scotland

[5] CM furnaces, Bloomfield, New Jersey

[6] Haskell Inc., Burbank, CA.

[7] Heise Company, Newton, CT

diffusion and elastic properties for the given material, σ is the applied stress, and n is the creep exponent), then tangential stress redistribution will occur.

Finite element analysis for a typical internally pressurized thick wall tube subjected to creep rupture loading was performed to describe the stress and creep rate distributions through the wall thickness for the tube. The ANSYS finite element code [28] was selected to perform this analysis. Figure 9 shows the tangential stress distributions along the wall thickness for an internally pressurized tube, for both elastic loading (at t = 0) and stationary steady-state creep conditions. The creep parameters for the SCRB210 material at 1300°C ($D=1.6 \times 10^{-20}$ and $n=7.3$) [29] and an internal pressure of 8.8 MPa, corresponding to an actual test conducted in this study, were used to perform the analysis and construct figure 9. Figure 9 also displays the stationary creep rate distribution across the tube's wall.

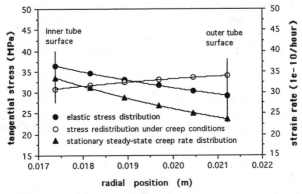

Figure 9. Tangential stress and creep rate distributions vs. radial position for a thick-wall tube under elastic and stationary steady-state creep conditions.

The effect of a radial temperature gradient, ΔT, on the creep rate distribution was also studied. Figure 10 shows the creep rate distribution at the inner surface of the tube described above as a function of time. This figure displays the creep rate distributions for the cases when no radial temperature gradient exists (ΔT=0) and when ΔT=6.6 °C (average experimental measurements of radial temperature gradients for tested tubes). As can be seen from the figure, the creep rate decreases as time elapses approaching the stationary creep rate value. This figure also indicates that the effect of ΔT diminishes as time elapses. Furthermore, figure 10 shows that the existence of a radial temperature gradient reduces the creep rate at the inner surface of the tube, thereby improving the tube's long term reliability. This is because ΔT imposes a beneficial compressive stress field at the inner portion of the tube. Therefore, since ignoring the radial temperature gradient would result in more conservative analysis, ΔT was ignored in all subsequent analyses.

3) Results and Discussion

Long term testing of internally pressurized SCRB210 tubes was conducted at 1250 ° and 1300 °C in order to experimentally verify the analytical procedure described in the previous section. Two tubes were tested under sustained internal pressure and at the temperatures mentioned above.

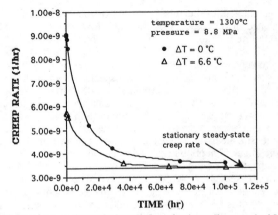

Figure 10. Creep rate vs. time at the inner surface of an internally pressurized SCRB210 tube tested at 1300 °C.

The first tube was pressurized to 8.8 MPa at 1300 °C. This pressure level induced a maximum tangential stress of 40 MPa and critical strain rate of $2.8 \times 10^{-12} \, s^{-1}$ at the inner surface of the tube. These stress and creep rate levels were calculated at the time of failure which was 3.8 hours in this case. The creep parameters used to calculate the stress and creep rate were 1.6×10^{-20} and 7.3, respectively.

Figure 11 shows the polynomial Monkman-Grant representation of creep rupture data obtained from testing tensile SCRB210 specimens at 1250 °, 1300 °, and 1350 °C [29]. From this plot and for a creep rate of $2.8 \times 10^{-12} \, s^{-1}$, the tube was expected to survive for 120,000 hours. Needless to say, this prediction is in poor agreement with the experimentally measured lifetime of 3.8 hours.

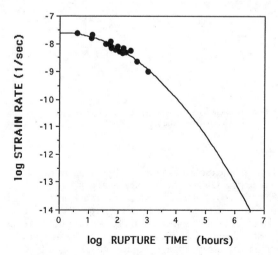

Figure 11. Polynomial Monkman-Grant representation of creep rupture data obtained by testing tensile SCRB210 specimens at 1200 °, 1250 ° and 1300 °C.

The long lifetime predicted for this tube is understandable because of the low applied stress level which resulted in very slow creep rate. This low stress level (40 MPa) corresponds to 1.7 % failure probability in fast fracture mode for the long tubes (using Weibull interpolation). The reason such a low stress level was selected, is because the tubular components are very weak (average strength = 78.0 ± 17 MPa at 1200 °C) when compared to tensile and other simple specimens. Indeed, the average fast fracture strength for SCRB210 long tube components is one third the average strength for the tensile specimens used to collect the creep rupture data shown in figure 11 [30].

Figure 12 shows the Weibull strength distribution for SCRB210 long tube components tested at 1200 °C. The area between the two vertical lines in this figure indicates the stress range over which the tensile creep data were collected. It is clear from this figure that the tensile creep data were obtained at applied stress levels which would have resulted in immediate failure for the tubular components. For this reason, the tubes could not be subjected to such high stress levels. Note that the creep rate used to predict the creep rupture life for the tube was out of range of the collected creep rupture data shown in figure 11. At this low creep rate, it is not even known whether creep damage is operative or not, depending on the magnitude of the threshold stress for creep damage in the SCRB210 material.

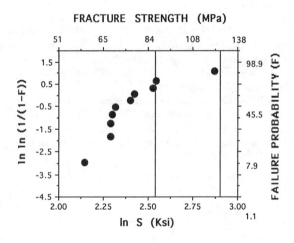

Figure 12. Weibull distribution for the SCRB210 tubes at 1200 °C. The region between the two vertical lines indicate the stress range over which the tensile creep data were collected.

In an effort to determine the mode of failure, all fragments were collected and used to reconstruct the burst tube (figure 13). The primary source of failure can be located by tracing the path of the crack growth and from fracture surface analysis. Unfortunately, the tube was heavily damaged at the region were failure is suspected to have initiated and fractographic examination of this section could not be performed. However, since the tube is axisymetrically stressed, then the fracture surface of any fragment of the tube, within the hot zone, can be examined for creep damage. Figure 14 shows an SEM micrograph of the fracture surface of a fragment chosen from the hottest section of the tube. As can be seen from this figure, no creep damage was detected, but rather the fracture surface is indicative of a fast fracture mode.

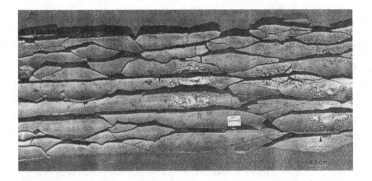

Figure 13. Reconstructed SCRB210 tube tested under creep rupture conditions at 1300 °C.

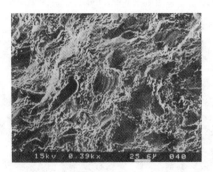

Figure 14. SEM micrograph showing the fracture surface of a fragment of an SCRB210
tube tested under creep rupture conditions at 1300 °C.

The second tube was pressurized to 7.2 MPa at 1250 °C. This pressure level induced a
maximum stress level of 32.4 MPa and maximum creep rate of $1.7 \times 10^{-13} \, s^{-1}$ at the inner
surface of the tube. The creep parameters used to calculate the stress and creep rate for the
tube were 3.1×10^{-22} and 8.0, respectively [29]. This pressure level was maintained for
11.3 hours at which time the pressure was raised to 9.6 MPa. Two minutes into this stress
step the tube burst. At pressure of 9.6 MPa, the maximum stress was calculated to be 43.5
MPa. This stress level corresponds to 2.8 % failure probability in fast fracture mode for the
long tube components.

Again, the Monkman-Grant creep rupture criterion predicted significantly longer
lifetime for the tube than the observed lifetime, indicating that creep was not the primary
mode of delayed failure. This tube contained large defects remnant from processing. Figure
15a shows an SEM micrograph of a surface connected pore which could be the result of
incomplete infiltration of silicon around the silicon carbide grains. Figure 15b shows an

SEM micrograph of a large silicon lake which existed on the fracture surface of another fragment of this ruptured tube[8] . For such large defects, even low stress levels can induce high stress intensities which could cause crack growth to occur. Flaw sizes up to 2.5 mm were encountered in some of the SCRB210 long tubes which have wall thicknesses of 4.7 mm. Using the Newman and Raju [31] approach to calculate the stress intensity factor for an internal surface flaw of size 2.5 mm in an internally pressurized tube (pressure = 40 MPa for the first tube tested), the stress intensity was calculated to be 3.8 MPa $\sqrt{m}$. This stress intensity value is 93% of K_{IC} for the SCRB210 material, 4.1MPa $\sqrt{m}$. At such high stress intensity levels crack growth could occur resulting in delayed failure of the SCRB210 tubes. The second tube was initially stressed at 32.4 MPa which resulted in a stress intensity factor of 3.1 MPa $\sqrt{m}$. At this stress intensity level the tube survived for 11.4 hours. However, when the stress level was raised to 43.4 MPa yielding a stress intensity factor of 4.1 MPa $\sqrt{m}$, failure resulted almost immediately. The following scenario can be proposed to explain the above results. A large critical flaw (up to 2.5 mm) initially existed in the tube. However, the size of this flaw was not large enough to cause immediate failure at the lower stress level. On the other hand the applied stress intensity was large enough to cause the crack to grow during the 11.4-hour hold period. During that time the crack grew to a critical size such that when the stress level was raised, failure resulted immediately.

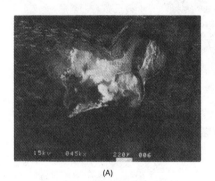
(A)

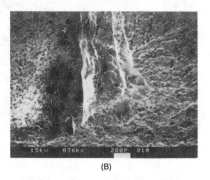
(B)

Figure 15. SEM micrographs of fracture surfaces of fragments of an SCRB210 tube tested under creep rupture conditions at 1250 °C and showing (A) large surface connected pore, and (B) silicon lake.

Quinn [32] conducted stress rupture and stepped temperature stress rupture (STSR) tests on a grade of siliconized silicon carbide (NC435) which contained 18 to 35 v% silicon and fine unimodal silicon carbide grains. His results indicated that this material exhibited delayed failure behavior at 1200 °C and that the fracture surfaces of the failed flexural specimens showed distinct features associated with slow crack growth. In another study, Quinn [33] studied the stress rupture behavior of another type of siliconized silicon carbide material. This material contained 20 v% free silicon and fine silicon carbide grains. He concluded that this material was very resistant to creep failure. For the few specimens which failed in time-dependent fashion, he could not identify the mechanism of

---

[8] The manufacturer has made many improvements to minimize these types of defects in subsequent productions of tubes.

delayed failure because of the unclear markings on the fracture surface. However, he ruled out creep fracture due to the small measured strains at failure.

From the results and discussion presented above, it is apparent that the size of the processing defects that exist in the SCRB210 tubes is the main variable controlling the design criteria. Creep is not the limiting design criterion for the SCRB210 tubular components since the stress levels which can be applied to these components, without causing immediate failure, are small and therefore will not induce appreciable creep rates. Instead, another delayed failure mechanism, crack growth, is responsible for the time - dependent failure of the SCRB210 tubes. Such crack growth could not have occurred at such low stress levels had the defects been smaller in size.

The relationship between the delayed failure mode (SCG or creep rupture) and component size (directly related to inherent flaw sizes) is shown schematically in figure 16. Note that the stress rupture curve, when delayed failure is controlled by SCG, shifts to lower applied stress levels as the component sizes / defect sizes increase as was shown previously in section I when the stress rupture curve for the SASC tensile specimens was projected to predict the stress rupture curve for the SASC tubes.However, the creep rupture curve is unique for all component sizes. Therefore, for a given applied stress level (horizontal line), the delayed failure mode would change from SCG for the tubular components to creep rupture for the tensile specimens. Unfortunately, at the present time, SCG parameters for the SCRB210 material are not available, thus precluding further analysis to predict the lifetimes for the tubes based on SCG delayed failure analysis.

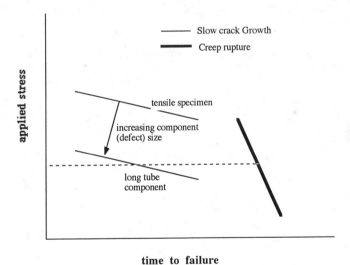

time to failure

Figure 16. Schematic representation showing time to failure vs. applied stress as a function of delayed failure mode and component size (inherent flaw size distribution).

CONCLUSIONS

Methodologies to predict the lifetimes for ceramic tubular components exhibiting delayed failure behavior due to SCG and creep rupture mechanisms were introduced. Dynamic fatigue testing of O-ring and compressed C-ring specimens was used to obtain the SCG parameters for the SASC material, expected to fail due to the SCG delayed failure mechanism. The effect of threshold stress intensity factor, below which no SCG occurs, was incorporated into this analysis. The calculated SCG parameters were in agreement with published data obtained from stress rupture testing of SASC flexure and tensile specimens. An analytical basis for projecting stress rupture curves constructed by testing coupon specimens to predict stress rupture curves (lifetimes) for actual structural components was introduced. FEM was used in association with the Monkman-Grant creep rupture criterion to predict the lifetimes of internally pressurized SCRB210 tubes, which initially were expected to fail due to creep rupture mechanism. Measured lifetimes for SCRB210 tubes subjected to sustained internal pressure were significantly shorter than predicted lifetimes when the analysis was based on delayed failure controlled by creep rupture. In reality, crack growth caused the delayed failure for the SCRB210 tubes. These tubes contained large defects which altered the delayed failure mechanism from creep rupture (observed when testing small specimens) to crack growth.

ACKNOWLEDGMENTS

The author would like to thank professors R. E. Tressler and J. C. Conway, Jr. (penn State University), and J. J. Mecholsky, Jr. (University of Florida) for their helpful suggestions during the course of this research. Special appreciation is also extended to Dr. D. L. Shelleman (Penn State University) for his assistance in experimentation. This research was funded by the Gas Research Institute .

REFERENCES

1)   Grathwohl, G.,"Creep and Fracture of Hot-Pressed Silicon Nitride with Natural and Artificial Flaws," in Proceedings of the Second International Conference on Creep and Fracture of Engineering Materials and Structures, Part I, 1984.

2)   Dalgleish, B. J., Slamovich E. B., and Evans A. G.,"Duality in the Creep Rupture of a Ploycrystalline Alumina," Journal of American Ceramic Society, Vol. 6, No. 11, 1985, pp. 575-581.

3)   Bassani, J. L.," Creep and Fracture of Engineering Materials and Structures," in Creep and Fracture of Engineering Materials and Structures, 1981, pp. 329.

4)   Wiederhorn , S. M.,"Subcritical Crack Growth in Ceramics," in Fracture Mechanics of Ceramics, Plenum Press, Vol. 2, 1974, pp. 613-646.

5)   Quinn, G., "Review of Static Fatigue in Silicon Nitride and Silicon Carbide," Ceramic Proceedings, Vol. 3, Nos 1-2, 1982, pp77-98.

6)   Quinn, G., and Katz, R., "Time Dependence of the High Temperature Strength of Sintered Alpha Silicon Carbide," TN 79-5, U.S. Army Materials and Mechanics Research Center, Watertown, Mass., June 1979.

7)   Srinivisan, M., "Elevated Temperature Stress Rupture Response of Sintered Alpha Silicon Carbide ," American Ceramic Society Bulletin, Vol. 58, No. 3,1979, pp. 347.

8)   Govila, R., "High Temperature Strength Characterization of Sintered Alpha Silicon Carbide," TR 82-52, NTIS ADA 121437, U.S. Materials and Mechanics Research Center, Watertown, Mass., October 1982.

9)   Magida, M. B., Forrest, K. A., and Heslin, T. M., "Dynamic and Static Fatigue of
a Machinable Glass Ceramic," Methods of Assessing the Structural Reliability of
Brittle Materials, ASTM STP 844, S. W. Frieman and C. M. Hudson, Eds.,
ASTM, 1984, pp. 81-94.

10)   Chuang, T., "A Diffusive Crack-Growth Model for Creep Fracture," J. Amer.
Ceram. Soc., Vol. 65, No. 2, 93 - 103, 1982.

11)   Fett, Theo, and Munz, Dietrich, "Life Time Prediction of Silicon Nitride at High
Temperatures," ASTM STP 844, 154 - 176, 1984.

12)   Mecholsky, J. J., Jr.,"Quantitive Analysis of Fracture Origins in Glass," Lecture
notes, The Pennsylvania State University, 1988.

13)   Golemboski, J. E., "Flexural Strength of Low Cost Tubular SiC Materials After
Static and Cyclic Loading at Elevated Temperatures," M. S. Thesis, The
Pennsylvania State University, University Park, Pennsylvania, December 1987.

14)   Minford, E. J., "Flaw Behavior Near the Threshold Stress Intensity for Slow
Crack Growth in Silicon Carbide Ceramics at High Temperatures," Ph.D. Thesis,
The Pennsylvania State University, University Park, Pennsylvania, 1983.

15)   Yavuz, B. O., "Subcritical Crack Growth Behavior and Threshold Stress Intensity
for Crack Growth In Silicon Carbide Ceramics at Elevated Temperatures," Ph.D.
Thesis, The Pennsylvania State University, University Park, Pennsylvania, 1987.

16)   Evans, A. G., "Slow Crack Growth in Brittle Materials Under Dynamic Loading
Conditions," International Journal of Fracture, Vol. 10, No. 2, 251 - 259, 1974.

17)   Jadaan, O. M., and Tressler, R. E.,"Methodology to Predict Delayed Failure Due
to Slow Crack Growth in Ceramic Tubular Components Using Data from Simple
Specimens," to be published in the Journal of Engineering Materials and
Technology, ASME, July 1993.

18)   SAS User's guide: Basics, SAS Institute Inc., Gary, North Carolina.

19)   Chuang, T., Tressler, R. E., and Minford, E. J., "On the Static Fatigue Limit at
Elevated Temperatures," Materials Science and Engineering, Vol. 82, 187 - 195,
1986.

20)   Jadaan, O. M., "Fast Fracture and Lifetime Prediction of Ceramic Tubular
Components," Ph.D. Thesis, The Pennsylvania State University, University Park,
Pennsylvania, 1990.

21)   Jadaan, O. M., Shelleman, D. L., Conway, J. C., Jr., Mecholsky, J. J., Jr., and
Tressler, R. E.,"Prediction of the Strength of Ceramic Tubular Components: Part I-
Analysis," Journal of Testing and Evaluation, Vol. 19, No. 3, 181-191, May 1991.

22)   McHenry, K. D., "Elevated Temperature Slow Crack Growth in Hot Pressed and
Sintered Silicon Carbide," Ph.D. Thesis, The Pennsylvania State University,
University Park, Pennsylvania, 1978.

23)   Quinn, G., and Katz, R. N., "Time-Dependent High Temperature Strength of Sintered Alpha SiC," J. Amer. Ceram. Soc., Vol. 63, No. 1 - 2, 117 - 119, 1980.

24)   Govila, R. K., "Flexural Stress Rupture Strength of Sintered α-SiC," Time Dependent Failure Mechanisms and Assessment Methodologies, pp. 100 - 110, J. G. Early, R. Shives, and J. H. Smith (Eds.), Cambridge University Press, 1982.

25)   Shelleman, D. L., Jadaan, O. M., Butt, D. P., Tressler, R. E., Hellman, J. R., and Mecholsky, J. J., Jr.," High Temperature Tube Burst Test Apparatus," Journal of Testing and Evaluation, Vol. 20, No. 4, 275-284, July 1992.

26)   Shelleman, D. L.,"Test Methodology for Tubular Ceramic Components (Fast Fracture Strength Study)," Ph.D. Thesis, The Pennsylvania State University, University Park, Pennsylvania, May 1991.

27)   Monkman, F. C., and Grant, N. J., "An Empirical Relationship Between Rupture Life and Minimum Creep Rate in Creep-Rupture Tests," ASTM Proceedings, Vol. 56, 593 - 605, 1956.

28)   Deglano, G. N., and Swanson, J. A., ANSYS: User's Manual.Swanson Analysis System, Inc., Houston, PA, 1988.

29)   Wiederhorn, S. M., et al., "Test Methodology for Tubular Components," pp. 257 - 276 in Projects Within the Center for Advanced Materials, annual report to Gas Research Institute, Center for Advanced Materials, The Pennsylvania State University, May 1989.

30)   Shelleman, D. L., Jadaan, O. M., Conway, J. C., Jr., Mecholsky, J. J. , "Prediction of the Strength of Ceramic Tubular Components: Part II-Experimental Verification," Journal of Testing and Evaluation, Vol. 19, No. 3, 192-200, May 1991.

31)   Newman, J. C., Jr., and Raju, I. S., "Analysis of Surface Cracks in Finite Plates Under Tension or Bending Loads," NASA Technical Paper 1578, 1979.

32)   Quinn, G. D.,"Characterization of Turbine Ceramics After Long-Term Environmental Exposure," Report No. AMMRCTR 80-15. U.S. Army Materials Technology Laboratory, April 1980.

33)   Quinn, G. D.,"Static Fatigue of a Siliconized Silicon Carbide," Report No. TR 87-20. U.S. Army Materials Technology Laboratory, 1981.

34)   Govila, R. K., "Phenomenology of Fracture in Sintered Alpha Silicon Carbide," J. Materials Science, Vol. 19, 2111 - 2120, 1984.

35)   Quinn, G., "Stress Rupture of Sintered Alpha SiC," Technical Report AMMRC TN 81 - 4, 1981.

36)   Walton, M. A., "Dynamic Fatigue of SiC at Elevated Temperatures," M.S. Thesis, The Pennsylvania State University, University Park, Pennsylvania, 1980.

Winfried G.T. Kranendonk[1] and Sido Sinnema[1]

EVALUATION OF TESTS FOR MEASURING THE STRENGTH OF CERAMIC TUBES.

REFERENCE: Kranendonk, W. G. T. and Sinnema, S., "Evaluation of Tests for Measuring the Strength of Ceramic Tubes," Life Prediction Methodologies and Data for Ceramic Materials, ASTM STP 1201, C. R. Brinkman and S. F. Duffy, Eds., American Society for Testing and Materials, Philadelphia, 1994.

Abstract: To measure the strength of ceramic tubes the C- and O-ring tests have been proposed. In the analysis of these tests, assumptions have been made, which can not always be fulfilled in practice. With computer simulations we investigated the stress distribution in C- and O-rings. We found that for C- and O-rings with a length to wall-thickness ratio of 6 the plane stress assumption fails and that axial stresses are not negligible. Also we investigated the influence of the load angle on the stress distribution in an O-ring test.

Keywords: failure probability, C-ring test, O-ring test, plane stress approximation, finite-element method

---

INTRODUCTION

Ceramic tubes are utilized in many industrial processes, taking advantage of their resistance to high temperature and chemical corrosion. Examples are heat exchangers and filter tubes. In many of these applications both thermal and mechanical stresses determine the lifetime and, therefore, the reliability of these constructions. Furthermore, tubes are among the simplest geometries for calculations. These practical and theoretical arguments support the selection of

---

[1]   Research scientist and manager, respectively, Corporate Research Laboratory, Hoogovens Groep BV, P.O. box 10.000, 1970 CA IJmuiden, The Netherlands

tubes for further strength analysis.

Although ceramics have excellent properties as mentioned above, they have one disadvantage: ceramics are brittle. Furthermore, the flaws, which act as the initiators of the cracks, differ in size and orientation from part to part and, as a consequence, the strength can vary considerably from part to part. Due to this high variation of strength, a statistical approach of the estimation of the strength is necessary. For this purpose several models have been proposed [1,2] and computer programs are available based on these methods.

Usually, the values of the input data for these models are determined with mechanical tests, in which the probability distribution of the mechanical strength is estimated. The most common of these tests are the three- and four-point flexure tests for uniaxial loading experiments and ball-on-ring tests for biaxial loading. However, these tests are especially suitable for flat, plane-parallel samples and they are difficult or impractical to apply to tubes and other cylindrically shaped specimens.

In this paper we discuss some aspects of two tests for measuring the mechanical strength of tubes. These tests are: 1) the O-ring test, 2) the C-ring test. These tests are shown schematically in figure 1.

In practice, application of these tests gives rise to several problems. In this paper we discuss some of these problems and we present the first results of experiments and computer simulations.

The first problem arises from the fact that it is difficult to cut thin C-rings from macroporous and therefore very brittle ceramic tubes with a small outer diameter (15 mm or less). Furthermore, especially for thin C-rings it is difficult to realize a proper alignment of the C-rings during loading in the fracture experiment. Because of these practical restrictions, we are forced to use relatively long C-

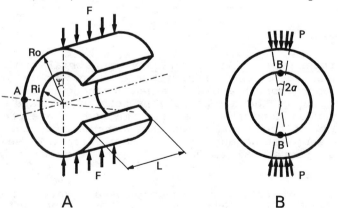

A   B

FIG. 1 Schematic outline of the C-ring test (A) and O-ring test (B).

rings, i.e., the length is relatively long as compared to the thickness of the tube. The plane stress assumption, which is normally used for the analysis of the test results, may not hold, and the question arises how large is the departure from the real stress state, and what is the influence of this departure on the failure predictions.

The second problem we discuss in more detail concerns the O-ring test. In this test the load is usually not concentrated on a small part of the ring, because this would cause an undesirable stress-concentration near the load. Instead the load is applied to a small area at the circumference of the ring. The size of this area can be defined by the load angle α, as shown in figure 1B. Especially for tubes with a small outer diameter this load angle can be in the order of 10° to 15°. The influence of the load angle on the probability of failure has already been investigated for the Brazilian disk test by Vardar and Finnie [3]. They found a considerable variation of the mean value of pressure at fracture as a function of the load angle. For tubes with an outer diameter of 13.5 mm, which we used in our investigation and with a load area defined by a width of 3 mm, the load angle is 12.5 degrees. As a second problem we investigate the influence of the load angle on the stress distribution and on the failure strengths in O-ring tests.

THEORY

Many models for the calculation of the probability of failure P_f or the probability of survival P_s are based on the empirical (cumulative) Weibull distribution (1):

$$P_f = 1 - P_s = 1 - \exp\left[-\int_V \left(\frac{\sigma_e - \sigma_u}{\sigma_0}\right)^m dV\right] \tag{1}$$

where σ_e is an effective, uniaxial stress and σ_u is a threshold stress. In many applications of the Weibull distribution the threshold stress is assumed to be zero and we adopt the same assumption. σ_0 is the Weibull scale parameter and m is the Weibull modulus. The relation between the effective stress and the biaxial or triaxial stress state is defined by a fracture criterion and a model for the shape of the defects.

To calculate the probability of failure one needs the stress distribution of the specimen under investigation. Jadaan and coworkers [4] recently published the stress distributions for the C-ring in compression and tension and for the O-ring in diametral compression. The stress distributions for the C- and O-rings are both calculated for the plane stress condition. In both tests stress concentrations occur, where the tensile tangential stress is the stress component with the highest value. In the C-ring test the locus of the highest tangential stress is in the horizontal symmetry plane at the outer radius, denoted by point A in figure 1A. For the O-ring test the locus of the highest tangential stress is in the vertical symmetry plane at the inner radius, at points B in figure 1B.

The distribution of tangential stresses in a C-ring under compressive loading was already discussed by Ferber and coworkers [5]. Using formulae for the bending of curved beams, it was shown:

$$\sigma_\theta = \frac{F}{A} \frac{R(r-r_a)}{r(R-r_a)} \cos\psi$$

$$r_a = \frac{(r_o+r_i)}{2}$$

$$R = \frac{(r_o-r_i)}{\ln(r_o/r_i)}$$

$$A = L(r_o-r_i)$$

(2)

where σ_θ is the tangential stress, F is the load, r is the radius, r_i the inner radius, r_o the outer radius, L the length of the C-ring and ψ the angle between r and the vertical plane as defined in figure 1A.

Using expression (2) Ferber and coworkers derived the probability of failure by assuming the maximum tensile stress criterion as the fracture criterion:

$$\ln[(1-P_f)^{-1}] = \left(\frac{\sigma_\theta^{max}}{\sigma_o}\right)^m L r_a^2 I_r I_\theta$$

(3)

where I_r and I_θ are the integrals:

$$I_\theta = 2\int_0^{\frac{\pi}{2}} \cos^m \psi \, d\psi$$

(4)

$$I_r = \int_1^{r_o/r_a} \left(\frac{1-1/x}{1-r_a/r_o}\right)^m x\,dx$$

and σ_θ^{max} is derived from (2) for $\psi=\pi/2$ and $r=r_o$.

Expression 3 gives the expression for the probability of failure, when volume flaws control the strength.

An analytical expression for the tangential stresses in an O-ring under compressive loading (in the limit $\alpha \rightarrow 0$) was obtained from elasticity theory by Jadaan and coworkers [4]:

$$\sigma_\theta = \frac{F}{L\pi r_o} Q(r,\psi)$$

$$Q(r,\psi) = 2A_0 - B_0 \left(\frac{r}{r_o}\right)^{-2}$$

$$+ \sum_{n=2,4,6,\ldots}^{\infty} \left[2(n+1)(n+2)A_n \left(\frac{r}{r_o}\right)^n + 2n(n-1)B_n \left(\frac{r}{r_o}\right)^{n-2} \right.$$

$$\left. + 2n(n+1)C_n\, x^n \left(\frac{r}{r_o}\right)^{-n-2} + 2(n-1)(n-2)D_n\, x^{n-2} \left(\frac{r}{r_o}\right)^{-n} \right] n \cos(n\psi)$$

$$A_0 = \frac{1}{2(1-x^2)} \qquad B_0 = -\frac{x^2}{1-x^2} \tag{5}$$

$$A_n = \frac{1}{2R_n}\left(\frac{1-x^{2n}}{n} - \frac{1-x^{2n-2}}{n+1}\right) \qquad B_n = \frac{1}{2R_n}\left(\frac{1-x^{2n}}{n} - \frac{1-x^{2n+2}}{n-1}\right)$$

$$C_n = \frac{x^n}{2R_n}\left(-\frac{1-x^{2n}}{n} + \frac{1-x^{2n-2}}{n+1}x^2\right) \qquad D_n = \frac{x^n}{2R_n}\left(-\frac{1-x^{2n}}{n}x^2 + \frac{1-x^{2n+2}}{n-1}\right)$$

$$R_n = (1-x^{2n})^2 - n^2 x^{2n-2}(1-x^2)^2$$

$$\text{where} \qquad x = \frac{r_i}{r_o} \qquad \text{and} \qquad F : \text{load}$$

Note the differences in the coefficients A_n and B_n in this paper and in reference [4].

EXPERIMENTAL PART

We investigated the tests both by computer simulation and by experiment.
The computer simulation analysis consisted of two steps.
The first step was a multiaxial stress analysis for samples under load, using the finite element method. Stress analysis was performed using the FEM code ANSYS (version 4.4A). In order to obtain a high accuracy for the prediction of the calculated stresses, the three-dimensional models were meshed with 3D 20-node isoparametric solid elements (STIF95) and the mesh was so refined that a result with an average energy error of less than two percent was obtained. For this measure of the accuracy the energy error estimation of Zhu and Zienkiewicz was used [6].
The second step was the calculation of the failure probability from the stresses, obtained in the first step. The failure probability was estimated with the program CARES (version 2), using the Weibull and Batdorf fracture theories. The program offers the possibility to model the defects with Griffith and penny-shaped cracks and to choose

from several fracture criteria [7]. In this study we applied only the penny-shaped crack as a model for the defects and we used the normal stress criterion as fracture criterion.

To test the predictions from the Batdorf theory we performed the C-ring compression and the O-ring compression tests on tube segments. All tubes were macroporous ceramic tubes, made of 30μm α-Al$_2$O$_3$ powder. The tubes were fabricated by extruding a paste, drying the green product and sintering it at about 1600°C. This process resulted in a mean pore size of 6 μm. The fracture experiments were performed with C- and O-rings, cut from the same tube.

The C-ring and O-ring tests were done on an INSTRON bank (type 8561) with a cross head speed of 0.5 mm/min. The force acting on the C- and O-ring was measured with a 1 kN load cell. As already mentioned, it is important to prevent asymmetric stress concentrations in the O-ring during loading. The same problem arises in the diametral compression test of discs and two solutions have been proposed. The first solution is grinding two small parallel planes on the circumference of the ring opposite to each other. This solution has the disadvantages that it introduces extra surface flaws and that it is difficult to apply to porous material. In the second method, which we applied, the force is transferred to the O-ring via a soft medium. For this purpose materials like wood, rubber, fiber frax for high temperature testing and even computer punch cards have been used in the past. We used cardboard with a thickness of 1 mm to transfer the force to the O-ring: small slips with a width of 3 mm and a length, which is at least the length of the O-ring, were glued on the O-ring opposite to each other.

To increase the accuracy of the positioning of these slips a simple mould was made. On these slips small square pieces of the same cardboard, the 'feet' were glued. These 'feet' had

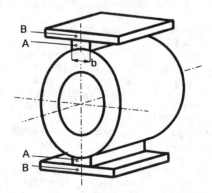

FIG. 2 Sketch of a complete preparation for an O-ring test. The load is transmitted to the ring via small slips (A) with a width of b=3 mm. The 'feet' (B) prevent the ring from rolling during the fracture experiment.

again a length of at least the length of the O-rings. Figure 2 shows a sketch of a complete O-ring preparation: the O-ring itself with the force-transmitting slips and 'feet' glued on it. These 'feet' had as a second goal, that they prevent the O-rings from rolling before and during the fracture experiment. Unfortunately, this method of preparing the O-ring also has the disadvantage that during the fracture experiment the O-ring can be pushed out of position when the small slips are not glued exactly opposite to each other.

To estimate the Weibull parameter m and the scale parameter σ_0 from the experimental results of the C-ring tests the maximum tangential stress σ_θ^{max} at the symmetry plane (point A in figure 1A) was calculated from equation 2. The maximum tangential stresses σ_θ^{max} of all the C-ring tests were arranged in ascending order. The failure probability P_f was defined with the expression (6):

$$P_f = \frac{j-0.5}{N} \qquad (6)$$

where j is the rank of a particular sample and N is the total number of fractured specimens.

Equation (1) can be rewritten as (7):

$$\ln(-\ln(1-p_f)) = m \ln\sigma_\theta^{max} - m \ln\sigma_0 + \ln\int_V \left(\frac{\sigma}{\sigma_\theta^{max}}\right)^m dV \qquad (7)$$

$$where \qquad KV = \int_V \left(\frac{\sigma}{\sigma_\theta^{max}}\right)^m dV \qquad (7a)$$

is the effective volume. From a linear least squares fit of $\ln(-\ln(1-P_f))$ against $\ln \sigma_\theta^{max}$ the Weibull modulus can be obtained from the slope. To calculate the scale parameter σ_0 is necessary to evaluate the effective volume. For the C-ring analysis the effective volume was calculated from equation (2) using the method described by Ferber et al [5]. For the O-ring analysis the effective volume was calculated using numerical integration after substituting equation (5) in equation (7a).

To determine, whether either volume defects or surface defects initiated the fracture, the fracture surface of some broken C-ring samples were inspected using SEM (Scanning Electron Microscopy). No conspicuous structures at the outer radius of the C-rings, where in the experiment the highest tensions occur, were noted. However, on the fracture surface several voids up to 100 μm could be seen. A typical example is shown in figure 3a. Also some samples showed macroscopic flute marks on the fracture surface (already visible with the naked eye), originating from the production process.

A   B

FIG 3 SEM examination of the fracture surfaces of the C-rings: figure 3A shows subsurface volume flaws with a size of 100 μm and figure 3B shows a detail of a flute mark.

Figure 3b shows a detail of such an imperfection. From these facts we infer that the fracture initiated at volume defects.

RESULTS AND DISCUSSION

The plane stress approximation is applicable to components, where the thickness is very small as compared to the other dimensions of the component. For C- and O-rings it means that the thickness, in this case the length of the rings should be small as compared to the difference between the outer and inner radius. With this approximation the axial stresses are assumed to be negligible as compared to the radial and tangential stresses. The plane strain approximation on the other hand can only be applied to objects, for which the thickness or length is very large as compared to the other sizes of the object. In the case of C- or O-rings it is difficult to estimate a priori for which length the approximations will break down.
Figures 4 and 5 show the ratio of the axial stress $\sigma_z(z/L)$ and tangential stress $\sigma_\theta^{max}(z/L)$ for the points, where the tangential stress has its maximum value, as a function of z/L. z is the coordinate along the axis of rotation. The analysis of two C-rings, one with a length of 20 mm and the other with a length of 40 mm were plotted in figure 4.

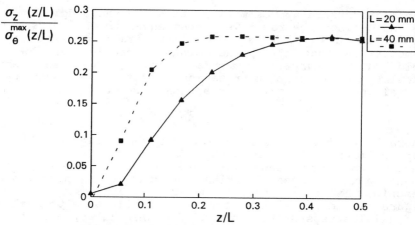

FIG 4. The ratio of the axial stress and the tangential stress for C-rings on the line, where the tangential stresses are maximum. The ratios are plotted for C-rings with two different lengths, L= 20 mm and L= 40 mm (see legend)   and an inner radius r_i =4.0 mm and outer radius r_o= 6.75 mm.

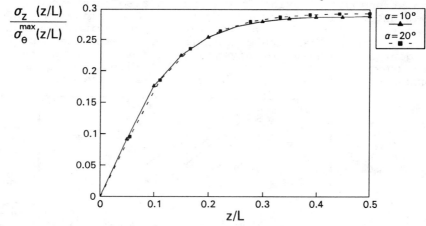

FIG. 5 The ratio of the axial stress and the tangential stress for O-rings on the line parallel to the axis of rotation, where the tangential stresses have their maximum at the inner radius. The results of two O-rings with different load angles α (see legend) are shown. L=20.0 mm, r_o= 6.75 mm, r_i= 4.0 mm.

Figure 5 shows the analysis of two O-rings, both with a length of 20 mm but different load angles. In both analyses the inner radius was 4.0 mm and the outer radius 6.75 mm. If the plane stress approximation should hold, the ratio should be (nearly) zero. Both figures, however, show that

the plane stress approximation fails. In the middle of the
C- and O-rings the ratio is almost equal to the Poisson
ratio, which is 0.25 for Al_2O_3. The simulation results also
demonstrate that the plane strain approximation does not
hold. At the symmetry plane at z/L=0.5, the plane perpendicular to the axis of rotation, relation (8) should hold
for all nodal points in this approximation.

$$\sigma_z = \nu(\sigma_r + \sigma_\theta) \tag{8}$$

where ν is the Poisson ratio. However, relation (8) only
(approximately) holds for the locus with the highest tangential stresses. Thus, neither the plane stress nor the
plane strain stress distribution can, even approximately,
describe the stress state in our C- and O-rings.
Figures 4 and 5 also show that we actually have to deal with
a biaxial stress state instead of a uniaxial stress state as
in the case of plane stress (the radial stresses are always
negligible in comparison to the tangential stresses). This
means that a mixed-mode fracture criterion is probably
better than the maximum tensile stress criterion, which is
used by Ferber and coworkers [5].
The stress distribution in O-rings is also determined by the
load angle. Figure 6 gives the ratio of the maximum tangential stress for a load angle of α to the maximum tangential
stress for a load angle α equals to zero. The ratio of
tangential stresses is given for different ratios between
the outer and inner radius of the O-ring. The tangential
stresses for a load angle larger than zero were calculated

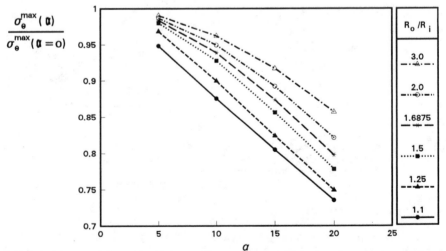

FIG. 6 The ratio of the maximum tangential stresses σ_θ^{max} at
the inner radius for a load angle α and for a load angle α=0
for different r_o/r_i ratio's (see legend). The maximum tangential stress for α=0 was obtained from expression (5).

with the finite element method in the plane stress approximation, while the tangential stresses for $\alpha=0$ were calculated from equation (5).
The influence of the load angle is largest the smaller the ratio of the outer and inner radius is. For our test specimens with $r_o/r_i=1.6875$ and an estimated load angle of 12.5° the maximum tangential stress decreases approximately 15% with respect to an O-ring test with a load angle of 0°. The results of series of fracture experiments on C-rings and O-rings (both tests are compression tests) are shown in figure 7. Since an adequate analysis for C- and O-rings with a biaxial stress distribution is not available yet, we used the analysis of Ferber and coworkers for the C-ring test. From the data of the C-ring test we obtained a Weibull-modulus m= 8.71 and a scale parameter σ_0=8.66 MPa(mm)$^{3/m}$. For the O-ring test the Weibull modulus was significantly higher: m=13.21 and a scale parameter of σ_0=8.04 MPa(mm)$^{3/m}$.

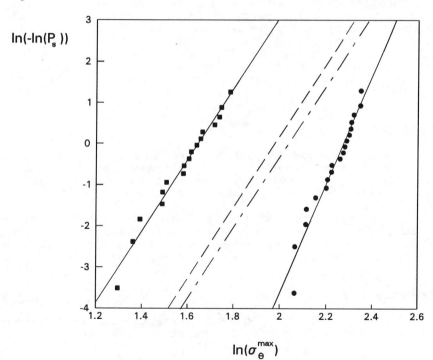

FIG. 7 Results of the C-ring test (squares) and O-ring test (circles). The solid lines are the Weibull plots. For the C-ring test the Weibull parameters are: m=8.7 and σ_0=8.7 MPa(mm)$^{3/m}$ and for the O-ring test: m=13.2 and σ_0=8.04 MPa(mm)$^{3/m}$. The predicted Weibull plots of the O-ring test, using the parameters of the C-ring test are given by the dashed line (α=10°) and the dashed-dotted line (α=15°).

These parameters were obtained with linear regression as described in the previous section. The difference in the Weibull moduli of both tests suggested that the flaw distribution near the inner radius may differ from the flaw distribution near the outer radius. The Weibull plot for the C-ring and O-ring test is also shown in figure 7.

With the parameters, obtained from the C-ring test, the strength of the O-ring was estimated. In figure 7 the prediction for the strength of the O-ring with a load angle of 10° is denoted by the dashed line and with a load angle of 15° by the dashed-dotted line. The predictions for the strength of the O-ring is a factor two lower than for the experimental strength. Following the above discussion several arguments may qualitatively explain this large difference: firstly, because the tensile axial stresses are not negligible, they will lower the strength of both the C-ring as well as the O-ring, but it is unknown how large the decrements for both tests will be. This emphasizes the need for an analysis of the probability of failure, taking into account the axial stresses. Secondly, the influence of the load angle was not taken into account in the analysis of the O-ring test and finally the difference in the Weibull moduli of the C- and O-ring test indicate a possible difference in the flaw distribution near the outer radius and near the inner radius.

CONCLUSIONS

For practical reasons C- and O-rings, cut from macroporous tubes have a large length to wall-thickness ratio. In these cases the plane stress approximation fails and the plane-strain approximation only holds for test-specimens with a large length.

The load angle in the O-ring test changes the maximum stresses significantly, which leads to small changes in the mean strength, especially when the width of the load area is large as compared to the thickness of the ring.

ACKNOWLEDGMENTS

The authors like to thank Mr. Tamis for drawing the figures and Messrs. Leering and Kooy for technical assistance.

REFERENCES

[1] Batdorf, S.B. and Crose, J.G., "A Statistical Theory for the Fracture of Brittle Structures Subjected to Nonuniform Polyaxial Stresses", Journal of Applied Mechanics, Vol. 41, No. 2, 1974, pp 459-464.

[2] Stanley, P., Fessler, H. and Sivill, A.D., "An Engineer's Approach to the Prediction of Failure Probabili-

ty of Brittle Components", *Proceedings of the British Ceramic Society*, Vol. 22, 1973, pp 453-487.

[3] Vardar,Ö. and Finnie, I., "An analysis of the Brazilian disk fracture test using the Weibull probabilistic treatment of brittle strength", *International Journal of Fracture*, Vol. 11, No. 3, 1975, pp 495-508.

[4] Jadaan, O.M., Shelleman, D.L., Conway, J.C., Mecholsky, J.J. and Tressler, R.E., "Prediction of the Strength of Ceramic Tubular Components: Part I-Analysis", *Journal of Testing and Evaluation*, Vol. 19, No. 3, 1991, pp 181-191.

[5] Ferber, M.K., Tennery, V.J., Waters, S.B. and Ogle, J., "Fracture strength characterization of tubular ceramic materials using a simple c-ring geometry", *Journal of Materials Science*, Vol. 21, 1986, pp 2628-2632.

[6] Kohnke, P.C., "ANSYS Theoretical Manual", Swanson Analysis Systems, Inc., 1989.

[7] Nemeth, N.N., Manderscheid, J.M. and Gyekenyesi, J.P., "Ceramics Analysis and Reliability Evaluation of Structures (CARES) Users and Programmers Manual", NASA Technical Parper 2916, 1990.

Nemeth, N.N., Manderscheid, J.M. and Gyekenyesi, J.P., "Designing Ceramic Components with the CARES Computer Program", *Ceramic Bulletin*, Vol. 68, No. 12, 1989, pp 2064-2072.

Angelika Brückner-Foit , Armin Heger[1] , Dietrich Munz[2]

EFFECT OF PROOF TESTING ON THE FAILURE PROBABILITY OF MULTIAXIALLY LOADED CERAMIC COMPONENTS

REFERENCE: Brückner-Foit, A., Heger, A., and Munz, D., "Effect of Proof Testing on the Failure Probability of Multiaxially Loaded Ceramic Components," *Life Prediction Methodologies and Data for Ceramic Materials, ASTM STP 1201*, C. R. Brinkman and S. F. Duffy, Eds., American Society for Testing and Materials, Philadelphia, 1994.

ABSTRACT: The effect of proof testing on the probability of fast fracture of ceramic components subject to multiaxial loading is analyzed using multiaxial Weibull theory. A local risk of fracture is defined which corresponds to the probability that failure of the component considered is caused by a flaw located at a given point in the structure. The failure behaviour after proof testing is discussed using two simple examples. It is shown in which way the efficiency of the proof test is related to the local risk of fracture during the proof test and under operating conditions.

KEYWORDS: Reliability, Weibull theory, multiaxial loading, proof testing, local risk of fracture, failure probability

INTRODUCTION

Ceramic components can be used in engineering structures only if a minimum level of reliability can be guaranteed. This can be achieved either by suitable design procedures by which the stresses in the components are minimized or by eliminating those components with the most dangerous flaws using an appropriate procedure for non-destructive testing.

Highly sophisticated methods for identifying natural flaws in ceramic materials have been developed which are mainly based on visualizing the flaws using X-rays or ultrasonics. The resolution power of these methods has increased to a level where even the small flaws contained in high performance ceramics are detectable. However, non-destructive inspection has several drawbacks which - at the time being - prevent the method from being used in routine tests. The first problem is that the amount of time needed for a complete inspection of a component can be quite large for complicated structures even if suitable inspection levels are defined such as proposed by Matsuo [1] et al.. The second problem is that flaws are detected with a certain probability which depends on the size, the shape, the physical nature and the location of

---

[1]   Scientist, University of Karlsruhe, Institute for Reliability and Failure Analysis

[2]   Professor, University of Karlsruhe, Institute for Reliability and Failure Analysis, P.O. Box 3640, W 7500 Karlsruhe 1, Germany

a flaw. Under unfavourable conditions, the detection probability may be quite small even for very large and dangerous flaws. The third problem concerns sizing of the flaws, i.e. the relation between the measured flaw size and the physical flaw size.

An alternative method is to perform proof tests in order to eliminate those components which contain potentially dangerous flaws. This can be performed as a routine test during the manufacturing process. Two major problems may be encountered during proof testing. The first one is that the natural flaws may grow stably under the sustained load of the proof test and that the strength of the components is reduced due to the test [2]. The second problem is that the stress distribution during testing may be different from the stress distribution during operating conditions and hence the most dangerous flaw in the test is not necessarily the most dangerous flaw under service loading.

The failure behaviour of components subject to proof testing is addressed in this paper. In the first section, the Weibull theory for determining the failure probability of ceramic components subject to a multiaxial stress state is summarized, and the effect of a proof test on the failure probability is analysed. The last part of this section deals with the local risk of fracture, i.e. the probability the worst flaw triggering fracture is located at a given point of the component. The numerical method used is briefly described in the second section. Finally, two examples are discussed which are supposed to illustrate the typical problems encountered during proof testing.

FAILURE OF CERAMIC COMPONENTS

Weibull theory for multiaxial loading

The failure behaviour of ceramic components subject to a multiaxial stress state can be assessed using the extended Weibull theory as it was developed by Batdorf et al. [3], [4], Evans [5], and Matsuo [6]. It is assumed that failure is caused by unstable extension of natural flaws with random size, location and orientation with respect to the principal stress axes. The worst flaw, i.e. the flaw for which the most unfavourable combination of size, location and orientation is obtained, will propagate unstably and will cause catastrophic failure.

The probability that the size a of a given flaw will exceed the critical crack size a_c is given by:

$$Q_1 = \int_V f_{\vec{x}}(\vec{x}) \int_\Omega f_\Omega(\Omega) \int_{a_c(\vec{x},\,\Omega)}^\infty f_a(a)\, da\, d\Omega\, d\vec{x} \ , \tag{1}$$

where V denotes the volume of the component considered, Ω the surface of a unit radius sphere, $\vec{x}$ the coordinate vector, and $f_{\vec{x}}(\vec{x})$, $f_\Omega(\Omega)$, $f_a(a)$ are the probability density functions of the corresponding random variables.

The critical crack size can be determined using fracture mechanics, if the natural flaws can be approximated by planar cracks. Within the framework of this model, a multiaxial stress state gives rise to a mixed mode loading of a crack, and the critical crack size is a function of the mode I - mode III stress intensity factors K_I, K_{II}, K_{III}:

$$a_c = a_c(K_I, K_{II}, K_{III}) \tag{2}$$

with

$$K_I = \sigma_n\sqrt{a} \cdot Y_I$$
$$K_{II} = \tau\sqrt{a} \cdot Y_{II} \tag{3}$$
$$K_{III} = \tau\sqrt{a} \cdot Y_{III}$$

with the correction factors Y_I, Y_{II} and Y_{III}. The stress σ_n normal to the crack plane is given by the following relation

$$\sigma_n = (\sigma_1 \cos^2\phi + \sigma_2 \sin^2\phi) \cdot \sin^2\theta + \sigma_3 \cos^2\theta \;, \tag{4}$$

where σ_1, σ_2, σ_3 are the principal stresses and ϕ, θ are the polar angles determining the orientation of the crack plane relative to the principal axes. The shear stress τ in the crack plane is given by

$$\tau = \sqrt{\tau_{r\phi}^2 + \tau_{r\theta}^2} \tag{5}$$

with

$$\tau_{r\phi} = (\sigma_2 - \sigma_1) \cdot \sin\phi \cos\phi \sin\theta \tag{6}$$

and

$$\tau_{r\theta} = (\sigma_1 \cos^2\phi + \sigma_2 \sin^2\phi - \sigma_3) \cdot \sin\theta \cos\theta \;. \tag{7}$$

An equivalent mode I stress intensity factor K_{Ieq} can be introduced with

$$K_{Ieq} = \sigma_{eq} \cdot \sqrt{a} \cdot Y_I \;, \tag{8}$$

where the equivalent stress σ_{eq} depends on σ_n, τ and on Y_I, Y_{II}, and Y_{III}. The critical crack size is then given by:

$$a_c = \left(\frac{K_{Ic}}{\sigma_{eq} \cdot Y_I} \right)^2 \tag{9}$$

where K_{Ic} denotes the fracture toughness. A variety of multiaxiality criteria are given in the literature leading to different expressions for σ_{eq}. A summary can be found in [7]. An example of one of these criteria is [8]:

$$\sigma_{eq} = \sqrt{\sigma_n^2 + \left(\frac{Y_{II}}{Y_I} \right)^2 \cdot \tau^2 + \frac{1}{1-v} \cdot \left(\frac{Y_{III}}{Y_I} \right)^2 \cdot \tau^2} \tag{10}$$

which is derived using the assumption that value of the energy release rate in the crack plane determines the onset of unstable crack propagation.

In Weibull theory the following expression is used for the integral over the crack size in Eq.(1) [7]:

$$\int_{a_c(\vec{x}, \Omega)}^{\infty} f_a(a)\, da = \left(\frac{\sigma_{eq}(\vec{x}, \Omega)}{\sigma_0} \right)^m \tag{11}$$

where Eq.(9) was used for the critical crack size. The parameters m, σ_0 in Eq.(11) depend on the toughness of the matrix material and on the statistical properties of the flaw size distribution.

If all locations of flaws and all orientations occur with equal probability, the material is homogeneous and isotropic, and $f_{\vec{x}}(\vec{x})$ and $f_\Omega(\Omega)$ can be replaced by uniform distributions. Hence Q_1 in Eq.(1) is equal to:

$$Q_1 = \frac{1}{V} \int_V \frac{1}{4\pi} \int_\Omega \left(\frac{\sigma_{eq}}{\sigma_0} \right)^m d\Omega \, d\vec{x} \ . \tag{12}$$

The number n of cracks in the volume V is also a random variable and can be described by Poisson's distribution. The probability of having exactly n cracks in V is given by:

$$p_n = \frac{M^n \cdot e^{-M}}{n!} \ , \tag{13}$$

where M is the average number of cracks in V. The following relation is obtained for the failure probability P_f from eqs.(12), (13) [7]:

$$P_f = 1 - \exp(-M \cdot Q_1) \ . \tag{14}$$

Effect of proof testing

Proof tests are performed in order to reduce the failure probability by eliminating the components with the most dangerous flaws prior to putting a component in service. Frequently it is not possible to impose the same stress distribution during the proof test as the one which is present under operating conditions. This is especially true for thermal loading. Hence the most dangerous flaw in the proof test and the most dangerous flaw under operating conditions may differ from each other, and the failure probability for components having survived the proof test may be almost as high as without proof testing. For example, surface cracks normal to the loading direction have the highest failure risk in a uniaxial stress state which is a typical test load, whereas the critical crack size is independent of the orientation of the crack plane for an equibiaxial stress state caused by thermal loading.

In the Weibull theory summarized in the previous section the orientation of the crack planes is given in terms of the polar angles ϕ, θ which are defined relative to the principal stress axes. In general, the principal stress axes are oriented in a different way for operating conditions and for proof testing, respectively. The equivalent stress σ_{eq}, Eq.(8), defines the loading of a flaw and depends on the orientation angles ϕ, θ via the normal stress σ_n, Eq.(4), and the shear stress τ, Eqs.(5)-(7). If the orientation of the crack plane is defined with respect to a global coordinate system where the stress tensor is non-diagonal, the formulae for σ_n and $\tau_{r\phi}$, $\tau_{r\theta}$ read

$$\sigma_n = (\sigma_{11} \cos^2\phi + \sigma_{22} \sin^2\phi) \cdot \sin^2\theta + \sigma_{33} \cos^2\theta + 2\sigma_{12} \cos\phi \sin\phi \sin^2\theta \\ + 2\sigma_{13} \cos\phi \sin\theta \cos\theta + 2\sigma_{23} \sin\phi \sin\theta \cos\theta \ , \tag{15}$$

$$\tau_{r\phi} = (\sigma_{22} - \sigma_{11}) \cdot \sin\phi \cos\phi \sin\theta \\ + \sigma_{12} \cos 2\phi \sin\theta - \sigma_{13} \sin\phi \cos\theta - \sigma_{23} \cos\phi \cos\theta \ , \tag{16}$$

and

$$\tau_{r\theta} = (\sigma_{11} \cos^2\phi + \sigma_{22} \sin^2\phi - \sigma_{33}) \cdot \sin\theta \cos\theta \\ + \sigma_{12} \sin\phi \cos\phi \sin\theta \cos\theta + \sigma_{13} \cos\phi \cos 2\theta + \sigma_{23} \sin\phi \cos 2\theta \ , \tag{17}$$

where σ_{ij} denotes the components of the stress tensor. Inserting Eqs.(16),(17) into Eq.(5) yields the shear stress τ. The equivalent stress is again given by a suitably selected multiaxiality criterion such as Eq.(10).

The probability that a given flaw will survive the proof test and fail during service is equal to the probability that its size is less than the critical crack size of the proof test $a_{c,pr}$, but exceeds the critical crack size of corresponding to operational loading $a_{c,op}$. Obviously this makes only sense in those cases where $a_{c,op} \leq a_{c,pr}$:

$$Q_1 = \frac{1}{V} \int_V \frac{1}{4\pi} \int_\Omega \int_{a_{c,op}(\vec{x},\Omega)}^{a_{c,pr}(\vec{x},\Omega)} f_a(a) \, da \cdot H(a_{c,pr}(\vec{x},\Omega) - a_{c,op}(\vec{x},\Omega)) \, d\Omega \, d\vec{x} \ , \qquad (18)$$

where H is the Heavyside step function with

$$H = \begin{cases} 1 & \text{for } x \geq 0 \\ 0 & \text{for } x < 0 \end{cases} . \qquad (19)$$

Using Eq.(9) to introduce the corresponding values of the equivalent stress, $\sigma_{eq,op}$ and $\sigma_{eq,pr}$, for the critical crack sizes $a_{c,op}$ and $a_{c,pr}$, respectively, and integrating over the crack size as in Eq.(11), Q_1 in Eq.(18) can be rewritten as:

$$Q_1 = \frac{1}{V} \int_V \frac{1}{4\pi} \int_\Omega \left(\left(\frac{\sigma_{eq,op}}{\sigma_0} \right)^m - \left(\frac{\sigma_{eq,pr}}{\sigma_0} \right)^m \right) \cdot H(\sigma_{eq,op} - \sigma_{eq,pr}) \, d\Omega \, d\vec{x} \ , \qquad (20)$$

where $\sigma_{eq,op}$ and $\sigma_{eq,pr}$ are defined with respect to the same coordinate system.

The failure probability for a component having survived the proof test can be derived using the same line of arguments as in Eqs.(12)-(14), and is given by inserting Eq.(20) into Eq.(14). This relation was first given by Hamanaka et al. [9].

Local risk of fracture

The probability that a flaw is located in a subvolume V_s and propagates unstably is equal to

$$Q_1(V_s) = \frac{1}{V} \int_{V_s} \frac{1}{4\pi} \int_\Omega \left(\frac{\sigma_{eq}}{\sigma_0} \right)^m \, d\Omega \, d\vec{x} \ . \qquad (21)$$

The survival probability of a flaw located in V_s or $V-V_s$ is given by:

$$R_1(V) = 1 - Q_1(V) \ . \qquad (22)$$

If n flaws are present in V, none of them propagates unstably with probability

$$R_n(V) = (1 - Q_1(V))^n \ . \qquad (23)$$

Hence fracture is triggered by one flaw in V_s and all other n-1 flaws remain stable with probability

$$S_n(V_s) = n \cdot Q_1(V_s) \cdot (1 - Q_1(V))^{n-1} \ . \qquad (24)$$

Combining Eq.(24) with the probability p_n, Eq.(13), of having n flaws in V and summing over n yields the probability that there is exactly one unstable flaw contained in the subvolume V_s and an arbitrary number of non-propagating flaws in V:

$$S(V_s) = \sum_{n=0}^{\infty} p_n \cdot S_n(V_s) \ , \qquad (25)$$

i.e.

$$S(V_s) = M \cdot Q_1(V_s) \cdot \exp(-M \cdot Q_1(V)) \ . \qquad (26)$$

The normalized quantity

$$P_0(V_s) = \frac{S(V_s)}{S(V)} \qquad (27)$$

is the quotient of the fracture risks associated with volumes V_s and V, respectively, and can be written as:

$$P_o(V_S) = \frac{Q_1(V_S)}{Q_1(V)} \quad ; \tag{28}$$

$P_o(V_S)$ is independent of the applied load level.

For a very small subvolume, $V_s = d\vec{x}$, the integral over V_s can be replaced by the integrand times $d\vec{x}$, and the probability is obtained that failure is caused by a flaw at point $\vec{x}$ in the interval $[\vec{x}, \vec{x} + d\vec{x}]$:

$$p_o(\vec{x}) \cdot d\vec{x} = \frac{\int_{\Omega} (\sigma_{eq})^m \, d\Omega}{\int_V \int_{\Omega} (\sigma_{eq})^m \, d\Omega \, d\vec{x}} \cdot d\vec{x} \tag{29}$$

The quantity p_o corresponds to the marginal probability density function of the location of the fracture origins obtained by Matsuo et al. [10] using competing risk theory, and P_o in Eq.(28) is the corresponding probability distribution function, if V_s is defined as an interval in three-dimensional space. p_o also correponds to the local risk of fracture.

Eq.(21) can be modified for components undergoing proof tests using the same line of arguments as in the previous section. The corresponding probability density funcion of the location of fracture origins is:

$$p_o(\vec{x}) \cdot d\vec{x} = \frac{\int_{\Omega} ((\sigma_{eq,op})^m - (\sigma_{eq,pr})^m) \cdot H(\sigma_{eq,op} - \sigma_{eq,pr}) \, d\Omega}{\int_V \int_{\Omega} ((\sigma_{eq,op})^m - (\sigma_{eq,pr})^m) \cdot H(\sigma_{eq,op} - \sigma_{eq,pr}) \, d\Omega d\vec{x}} \cdot d\vec{x} \tag{30}$$

NUMERICAL EVALUATION OF THE FAILURE PROBABILITY

The five-dimensional integral Eq.(12) or, in the case of proof-testing, Eq.(18) has to be evaluated in order to determine the failure probability. Conventional numerical integration procedures such as Gauss integration can be used for the integration over the orientation angles, whereas the integration over the volume is more difficult.

The stress tensor for real structures is generally determined by a finite element analysis which yields the values of σ_{ij} at a limited number of points within each element, the so-called Gauss points. Because of the high value of m which is obtained for advanced ceramic materials, the gradients of the integrand within each element are very steep, and a high number of evaluation points is needed in order to achieve convergence of the numerical integration. A very fine finite element mesh could be used in order to solve this problem, but then the computational costs of the stress analysis may become prohibitive.

An arbitrary number of evaluation points can be generated, if the shape functions of the finite elements are used to interpolate the stress field between the Gauss points of the finite element mesh. Using these additional points conventional numerical integration methods can again be applied for calculating the volume integral Eq.(12) or Eq.(18). This procedure is implemented in the computer code STAU and its extensions (STAULE for lifetime evaluation and STAUP for proof testing) [11].

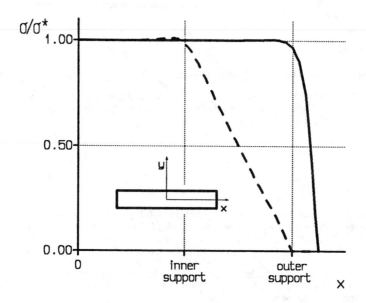

Figure 1. : Stress distribution on the tensile surface of the straight bar; —— $\sigma_{xx} = \sigma_{zz}$ service loading; - - - σ_{xx} proof test

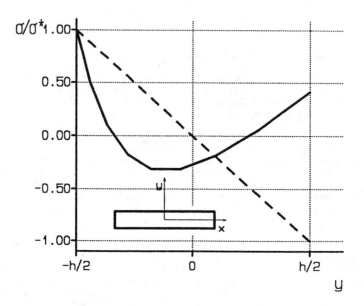

Figure 2. : Stress distribution in thickness direction of the straight bar; —— $\sigma_{xx} = \sigma_{zz}$ service loading; - - - σ_{xx} proof test

With this numerical procedure it is also possible to evaluate the corresponding integrals for surface flaws with high numerical accuracy. There is no need to introduce an additional layer of surface elements as it was proposed elsewhere [12].

EXAMPLES

Eqs.(14), (18) determine the failure probability of a component having survived a proof test, whereas Eq.(30) gives the corresponding local risk of fracture. In the following the effect of proof testing on the fracture behaviour is illustrated by two examples.

The first example is a straight bar. Four-point bend loading, i.e. a uniaxial stress state, represents the proof test, whereas the service loading is caused by thermal loading. The bar is assumed to be heated up to a given temperature level and is suddenly cooled down on one side which leads to a biaxial stress state. The stress distributions on the tensile surface and across the thickness of the bar at the beginning of the cooling are shown in Figures 1,2.

A circular disk is considered in the second example. The proof test is assumed to be the concentric ring test [13], whereas the service load is again caused by a thermal shock. This implies that a biaxial stress state is present in both cases. Figure 3 shows the stress distributions on the tensile surface. The decrease of the tensile stresses in thickness direction is very similar to that given in Figure 2.

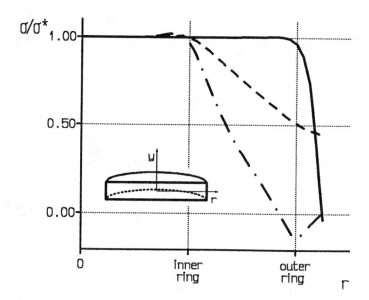

Figure 3. : Stress distribution on the tensile surface of the circular disc; —— $\sigma_{rr} = \sigma_{\theta\theta}$ service loading; - - - $\sigma_{\theta\theta}$ proof test; - · - · σ_{rr} proof test.

The efficiency of the proof test can be illustrated by comparing the number of specimens lost during the proof test to the number of specimens which fail after having survived the proof test. Since the failure probability for the proof test increases

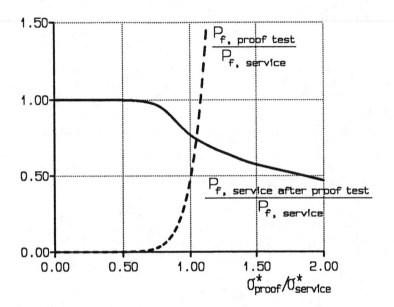

Figure 4. : Efficiency of proof test in dependence of the proof load, straight bar

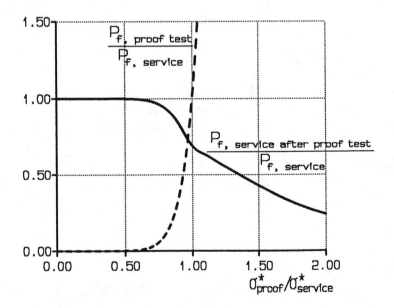

Figure 5. : Efficiency of proof test in dependence of the proof load, circular disk

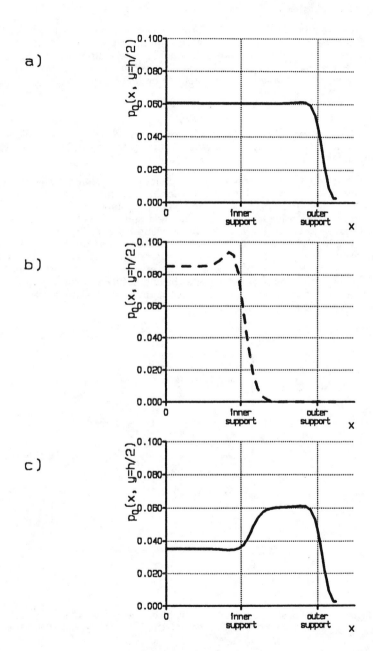

Figure 6. : Local risk of fracture at the tensile surface for the straight bar, $m = 10$, $\sigma^*_{proof}/\sigma^*_{service} = 1$. a) service loading; b) proof test; c) service loading after proof test.

with the proof load and the failure probability of those components which survived the proof test decreases with the proof load, an optimum value should be chosen for the proof load. In general, this value depends on the load level during operating conditions. For a given service load level, the weighted sum of the failure probabilities should be minimized in order to obtain an optimum proof load:

$$w_1 \cdot P_{f,proof\ test} + w_2 \cdot P_{f,service\ after\ proof\ test} \rightarrow \min \quad . \tag{31}$$

The weighting factors w_1, w_2 depend on the example considered, e.g. it is desirable to have few failures during proof testing for expensive components (high value of w_1) or to have a very small number of failures during service for safety relevant structures (high value of w_2).

Figures 4 and 5 show the dependence of the failure probabilities for the proof test and during service after proof test for both examples. In both cases, the plots are non-dimensionalized by dividing the failure probabilities by the failure probability obtained for service loading without proof test and by dividing the outer fiber stress of the proof test σ^*_{proof} by the corresponding value $\sigma^*_{service}$ of the service load. The Weibull exponent m was 10, and the Weibull parameter σ_0 was chosen in such a way that the failure probability for the service load was equal to 10%. In both cases the failure probability for the proof test increases dramatically with the proof load, whereas the failure probability during service after having survived the proof test is only reduced by a factor of two (straight bar) or three (circular disc). The curve $P_{f,service\ after\ proof\ test}$ has a characteristic kink at $\sigma^*_{proof}/\sigma^*_{service} = 1$.

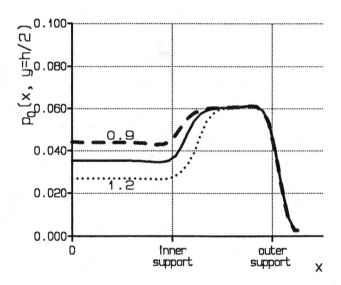

Figure 7. : Local risk of fracture during service loading after proof test at the tensile surface for the straight bar, $m = 10$, - - - - $\sigma^*_{proof}/\sigma^*_{service} = 0.9$, —— $\sigma^*_{proof}/\sigma^*_{service} = 1$, - · - · $\sigma^*_{proof}/\sigma^*_{service} = 1.2$

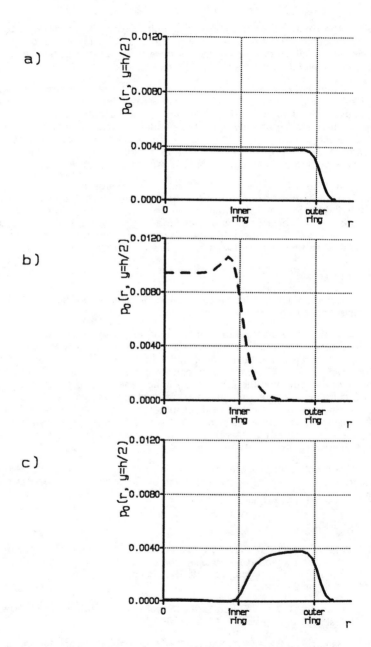

Figure 8. : Local risk of fracture at the tensile surface for the circular disk, $m=10$, $\sigma^*_{proof}/\sigma^*_{service}=1$. a) service loading; b) proof test; c) service loading after proof test.

The relatively small gain in reliability due to the proof test can be understood from the behaviour of the local risk of fracture $p_0(\bar{x})$. In all cases the highest contribution to the failure probability comes from flaws located at the tensile surface of the specimens which is considered in the following.

The local risk of fracture p_0 evaluated at the tensile surface for $\sigma^*_{proof}/\sigma^*_{service} = 1$ is shown in Figure 6. Due to the uniaxial stress field all flaws which are parallel or slightly inclined to the stress axis will not be affected by the proof loading, even though they are potentially dangerous under operating conditions. Outside the inner span of the bending fixture the risk of fracture falls off rapidly, whereas the thermal stresses are still present. An increase of the load level in the proof test mainly affects the contribution from inside the inner span, as shown in Figure 7.

In Figure 8 the local risk of fracture is considered for the circular disk. A vanishing contribution to the failure probability comes from those flaws which are located within the inner ring of the double-ring fixture, if the outer fibre stress of the proof test is equal to the thermal stress on the surface. However, a significant contribution to the failure probability can be attributed to the flaws which are located outside the inner ring because they experience a rather low proof load.

CONCLUSIONS

The local risk of fracture can be used to analyse the failure behaviour of ceramic components subject to multiaxial loading.

In the case of proof testing two different stress fields with different degrees of multiaxiality have to be compared with each other, namely the stress field caused by the proof load and the stress field under operating conditions. The local efficiency of the proof test can be analysed by calculating the local risk of fracture for the components having survived the proof test. An increase of the proof load affects the local risk of fracture significantly only in those areas where both stress states are similar. Above a certain level of the proof load, all potentially dangerous flaws in those areas are eliminated by the proof test, i.e. any flaw which would propagate unstably under service loading extends unstably under proof loading, but the flaws located elsewhere will still contribute to the failure probability.

An optimum proof load can be defined by minimizing the weighted sum of the failure probabilities of the proof test and of the service loading after surving the proof test.

REFERENCES

[1]   Matsuo, Y., Kitakami, K., and Kimura, S., "A new theory of non-destructive inspection based on fracture mechanics and fracture statistics", in: Fracture Mechanics of Ceramics, Vol. 10, R.C. Bradt et al., Eds., Plenum Press, New York, 1992, pp 317-327.

[2]   Evans, A. G., and Fuller, E. R., "Proof testing - the effect of slow crack growth", Mat. Sci. Engng Vol.19, 1975, pp 69-77.

[3]   Batdorf, S. B., and Crose, J. G., "A statistical theory for the fracture of brittle structures subjected to nonuniform stress", J. Appl. Mechanics, Vol.41, 1974, pp 459-461.

[4]   Batdorf, S. B., and Heinisch, H. L., "Weakest link theory reformulated for arbitrary fracture criterion", J. Amer. Ceram. Soc., Vol.61, 1978, pp 355-358.

[5]   Evans, A. G., "A general approach for the statistical analysis of multiaxial fracture", J. Am. Ceram. Soc., Vol.61, 1978, pp 302-308.

[6]   Matsuo, Y., "A probabilistic analysis of fracture loci under bi-axial stress state", Bull. JSME, Vol.24, 1981, pp 290-294.

[7]   Thiemeier, T., Brückner-Foit, A., and Kölker, H., "Influence of the fracture criterion on the failure probability of ceramic components", J. Am. Ceram Soc., Vol.74, 1991, pp 48-52.

[8]   Paris, P. C., and Sih, G. C., "Stress Analysis of Cracks", in: Fracture Toughness Testing and Its Applications, ASTM STP 381, American Society of Testing of Materials, Philadephia, PA, 1965, pp 30-83.

[9]   Hamanaka, J., Suzuki, A., and Sakai, K., "Structural reliability evaluation of ceramic components", J. Europ. Ceram. Soc., Vol.6, 1990, pp 375-381.

[10] Matsuo, Y., and Kitakami, K., "On the statistical theory of fracture location combined with competing risk theory", in: Fracture Mechanics of Ceramics, Vol.7, R.C. Bradt et al., Eds., Plenum Press, New York, 1986, pp 223-235.

[11] Heger, A., Brückner-Foit, A., and Munz, D., "STAU - ein Programm zur Berechnung der Ausfallwahrscheinlichkeit mehrachsig beanspruchter keramischer Komponenten als Post-Prozessor für Finite-Elemente-Programme", Internal Report, Institute for Reliability and Failure Analysis, University of Karlsruhe, Germany, 1991.

[12] Nemeth, N. N., Manderscheid, J. M., and Gyekenyesi, J. P., "Ceramic Analysis and Reliability Evaluation of Structures (CARES) User's and Programmer's Manual, NASA TP-2916, Washington, DC, 1989.

[13] Giovan, M. N., and Sines, G., "Biaxial and uniaxial data for statistical comparison of a ceramic's strength", J. Am. Ceram. Soc., Vol.62, 1979, pp 510-515.

Jonathan A. Wade[1], Charles S. White[1], and Francisco J. Wu[2]

PREDICTING CREEP BEHAVIOR OF SILICON NITRIDE COMPONENTS USING FINITE ELEMENT TECHNIQUES[3]

REFERENCE: Wade, J. A., White, C. S., and Wu, F. J., **"Predicting Creep Behavior of Silicon Nitride Components Using Finite Element Techniques,"** Life Prediction Methodologies and Data for Ceramic Materials, ASTM STP 1201, C. R. Brinkman and S. F. Duffy, Eds., American Society for Testing and Materials, Philadelphia, 1994.

ABSTRACT: The creep of silicon nitride tensile specimens has been modeled and incorporated into finite element software to predict the behavior of structural components. The experimental results are for the creep deformation of HIP'ed, yttria-doped silicon nitride at temperatures up to 1400°C. Results are for both homogeneous and joined specimens. This experimental data base was modeled using two approaches: Arrhenius law representation of the steady state phase, and theta projection method representation of both primary and secondary stages. The Arrhenius law has been incorporated into commercial finite element software and used to predict the creep deformation behavior and time to failure for a simulated component represented by a notched tensile specimen.

KEYWORDS: silicon nitride, finite element method, creep experiments, creep modeling, high temperature deformation

---

INTRODUCTION

The strengths of structural ceramics typically diminish at elevated temperatures (T > 0.6 T_{melt}). At very high stress levels failure can occur by fast fracture and failure predictions must be assessed using statistical methods and, perhaps, fracture mechanics. On the other hand, at typical design stress levels failure can more commonly be attributed to creep deformation leading to rupture after a finite service life. In contrast to its behavior in metals and alloys, this creep rupture can occur when very small amounts of deformation are

---

[1]Research Engineer and Senior Research Engineer, respectively, Saint-Gobain/Norton Industrial Ceramics Corporation, Goddard Rd., Northboro, MA 01532-1545.
[2]Formerly Senior Research Engineer, Saint-Gobain/Norton Industrial Ceramics Corporation, currently PDA Engineering, 67 South Bedford St.,Burlington, MA 01803.
[3]Research sponsored by the Ceramic Technology Project, DOE Office of Transportation Technologies, under contract DE-AC05-84OR213400 with Martin Marietta Energy Systems, Inc.

observed. This inherent brittleness of ceramics, even at elevated temperature, makes the need for robust design and prediction paramount.

Creep and creep rupture of structural ceramics are generally controlled by the physical and chemical properties of the grain and grain boundaries. The nucleation, growth and coalescence or fracture of cavities provides a mechanism for creep deformation since it constitutes the means for grain sliding and accommodation. Much work continues in the determination of exact mechanisms for the observed creep behavior of ceramics. The microphysical and chemical aspects of creep of ceramics are beyond the scope of this paper.

Typical creep behavior can be characterized by the material strain response over time to a constant applied stress. Generally the material responds with an initial, time-independent elastic strain. The material will also start deforming in creep with a constantly decreasing strain rate during an initial transient which is termed primary creep. Once the creep strain rate has reached its minimum, strain will accumulate at a constant rate. This region of nearly constant strain rate is called the secondary or steady state creep regime. Finally the material may experience an acceleration of strain rate leading to final fracture. This final phase is called tertiary creep and may or may not be present dependent upon the particular material. The generic steady state creep behavior has been studied widely and extensive work has been done to model its behavior [1,2]. Usually the steady state creep rate is described by a power function of stress, such as Norton's law, and an Arrhenius form function of temperature. In this paper, creep experiments are presented for NCX-5100 Si_3N_4. These experiments are modeled using Norton's law and incorporated in a finite element study of a notched tensile member. This presents a methodology for the design of high temperature components.

MATERIAL

The silicon nitride that was tested consisted of IPA milled Si_3N_4 containing 4 wt% Y_2O_3 (NCX-5100). This composition was chosen for its high temperature strength and creep resistance. The specimens consisted of two types: homogeneous NCX-5100 material, and joined billets of NCX-5100 having a slip interlayer at the center of the gauge section. Both joined and homogeneous specimens were tested to investigate whether any difference in creep behavior could be discerned due to the presence of the join. The join was made using a suspension of the parent material to join the green billets which had been obtained by cold isostatic pressing of the powder. The joined billets were hot isostatically pressed with the resulting joins being visually indistinguishable from the parent material.

EXPERIMENTAL PROCEDURE

A total of 23 NCX-5100 samples were tested. The preferred method for characterizing high temperature properties is the tension test (Figure 1). The uniaxial test geometry consists of a flat dog-bone with tapered holes to account for the relief of out-of-plane alignment. This type of specimen was used by Wiederhorn et al. [3] to characterize high temperature structural ceramics and has been modified for our NCX-5100 silicon nitride. A detailed finite element study was conducted to optimize the specimen geometry of [3]. The concern was that the highest creep strain rates should be confined to the gauge section rather than the area around the pin-loaded holes. As shown in Figure 1, the newly designed specimen has a uniform maximum principal stress in the gauge section with no more than a 3% increase above that value near the tapered transition region.

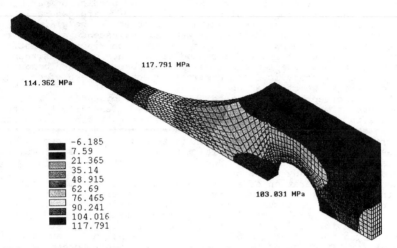

FIG. 1--Contours of maximum principal stress for tensile specimen.

Specimen Preparation

Densified billets of NCX-5100 silicon nitride were sliced through the thickness of the billets using a 150 grit diamond wheel on an automatic surface grinder. Prior to testing, the edges within the gauge section of each specimen were hand ground to remove stress intensifying machine damage. After a few passes on each of the four edges, the specimen was wiped clean and visually inspected for chipping at 10X to 20X using a stereoscope.

Mechanical Testing

Prepped tensile creep specimens were measured for average gauge section width and thickness using a dial caliper and recorded to the nearest 0.0254 mm (0.001 inch). The initial tensile stress was determined at 25% of the predicted tensile failure stress at 1370°C. The stress was used to calculate the applied load and the load was then programmed into the ramp profile of the creep testing supervisory computer/controller.

Load Profile--The typical profile includes a pre-load of 111 N. The preload was maintained throughout the temperature ramp and presoak. After 20 hours the load was ramped to achieve the test stress at a rate of 17 MPa/min. The stress was maintained within 1.4 MPa for the duration of the test.

Temperature Profile--The furnaces were heated to 1200°C at 30°C per minute. The ramp continued at 10-15°C per minute to the test temperature where it was held within 2°C to the conclusion of the test.

Extensometry--The strain data was obtained through laser dimensional sensors with standard modifications for measuring hot objects and a passline extension for ten to twelve inch transmitter to object separation to accommodate the furnace. The system measurement resolution was 10 microstrain and the measurement precision was +/- 30

FIG. 2--Laser targets used for extensometry.

microstrain over the full measurement range. The laser transmitter and
receiver were mounted on precision linear motion tables with one inch
barrel micrometer manual drives. Laser targets made of the same
material as the specimen were cantilevered off the gauge section (Figure
2). The measurement data was interpreted, conditioned and formatted in
the Zygo-1100 processor [4].

    Data Acquisition--The testing supervisory computer was linked to
the Zygo-1100 processor via RS-232 synchronous communication line. The
test system control program prompts the 1100 processor for the average
of the 150 most recent measurements which were in a moving queue. The
data were parsed and logged into fields with a time stamp and the
current specimen temperature. The load was recorded on a strip chart.
Following a test, the data were run through an RPL procedure in RS/1 to
check for proper column entry and deletion of errors and empty fields.

    The list of experiments and environmental conditions is given in
Table 1. Notice that the test temperatures range from 1250°C to 1400°C
and the applied stress levels range from 100 MPa to 250 MPa. The tests
ranged in time from a few hours to almost 1700 hours. Experiments
conducted on both continuous and on joined specimens are included. The
current data set is incomplete. Further tests are ongoing to expand the
range of testing conditions, especially of continuous specimens. Not
all of the tests were run to failure. Some of them were suspended for
various reasons including power failures. No conclusions should be
drawn about the tests that were suspended versus those tested to
failure. The results are summarized in the table by listing the quasi
steady state creep rate and the time to failure. The quasi steady state
creep rate was determined by fitting a straight line through the quasi-
linear, minimum creep portion of the creep-time response curve. Most of
the tests showed a pronounced primary creep region followed by a region
of almost constant steady state creep rate (Figure 3). The tests did
not show a tertiary region. The specimens failed while still showing
almost linear strain accumulation with time. This lack of a tertiary
component has also been seen in certain sintered silicon nitrides [5]
but not in others [6].

Figure 3 shows the creep curves for four experiments that were all conducted at the same conditions (σ=120 MPa, T=1370°C). Notice that there is considerable scatter in the creep curves. The accumulated strain can differ by almost a factor of 2 for identical tests. This variation in the experimental results makes modeling quite difficult. Three of the tests in Fig. 3 contain joins and one of the tests is for a monolithic specimen. Not enough monolithic (control) specimens have been tested to make statistically significant statements about the differences with the joined specimens but such comparisons will be made in the future as the test matrix is expanded.

TABLE 1--Experimental Tests.

Temperature (C)	Stress (MPa)	Specimen Type	Minimum Strain Rate (1/hours)	Test Time (hours)	Status at Test Completion
1400	100	join	1.775E-05	475	suspended
1400	100	join	1.829E-05	444	suspended
1400	100	continuous	2.555E-05	486	failed
1400	120	join	2.430E-05	202	failed
1400	120	join	2.190E-05	449	suspended
1400	120	join	2.010E-05	384	failed
1400	120	continuous	2.615E-05	407	failed
1370	120	join	8.900E-06	475	suspended
1370	120	join	1.130E-05	601	failed
1370	120	continuous	1.097E-05	504	suspended
1370	120	join	8.260E-06	669	suspended
1370	150	join	4.030E-05	93	failed
1370	150	continuous	4.364E-05	141	failed
1300	175	join	2.970E-06	819	suspended
1300	175	join	1.270E-06	1100	suspended
1300	200	join	1.200E-05	233	failed
1300	225	join	4.930E-05	54	failed
1300	225	join	4.110E-05	43	suspended
1300	225	continuous	3.205E-05	70	failed
1250	225	join	1.090E-06	1692	suspended
1250	225	join	1.650E-06	507	failed
1250	250	join	2.690E-06	890	failed
1250	250	join	2.250E-06	936	suspended

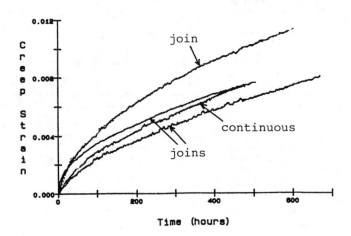

FIG. 3--Creep curves for 120 MPa at 1370°C

MATERIAL MODELS

The creep deformation behavior was modeled in two ways. The first model involved just the steady state creep rate for those tests which showed a quasi-secondary region. This modeling followed the classically accepted procedures that have become common in creep modeling of metals. This method will be discussed in detail below and incorporated into the ANSYS finite element code. The second model used the theta projection method [7,8]. This procedure involves a representation of all of the data and is presented in Appendix A as a method of reproducing the entire creep curves with a minimum of parameters.

Arrhenius Model of Steady State Creep Rate

In order to compare the experimental results seen here with other results in the literature we apply the simple Arrhenius form of Norton's equation.

$$\dot{\varepsilon}_s = A\sigma^n e^{-Q/RT} \tag{1}$$

Here $\dot{\varepsilon}_s$ is the steady state strain rate, A and n are material constants, σ is the applied stress, Q is the apparent activation energy for creep, R is the universal gas constant, and T is the absolute temperature. This equation has received wide acceptance for high temperature creep behavior in the literature for both ceramics and metals [3]. Consider the natural logarithm of Equation 5.

$$\ln\dot{\varepsilon}_s = \ln A + n\ln\sigma - \frac{Q}{RT} \tag{2}$$

The values of Q, n and A can be determined from plotting the logarithm of the steady state creep strain rate (Table 1) against various parameters. The value of Q was determined from plotting $\ln(\dot{\varepsilon}_s)$ versus 1/T at constant stress (Figure 4). The value of Q is just the negative of the slope of that curve. From Figure 4 we find a value of Q as 650 KJ/mole when evaluated at an applied stress of 120 MPa. This value is consistent with other silicon nitride results in the literature [4,5]. Our preliminary results show that Q has a stress dependence since limited data at higher stresses gives a higher value of Q. This also is consistent with what was observed by Wiederhorn et al. [3]. Further experiments are underway to quantify this stress dependence. In the meantime the value of n can be found by plotting $\ln(\dot{\varepsilon}_s e^{Q/RT})$ against $\ln(\sigma)$ using the value of Q determined in Figure 4. This use of a temperature dependent creep strain rate can be used to plot all of the experimental creep data with the slope of the best fit line giving n (Figure 5). The value of 3.6 for n that is found here falls in the range of 2.0 to 6.9 reported in the literature [3,5,6]. Note the considerable scatter from the best fit line. This gives an indication of the level of inconsistency in the experimental data base.

Failure Modeling

The time to failure for each steady state strain rate is plotted in Figure 6. This correlation follows the classically accepted Monkman-Grant [10] relationship given below.

$$\dot{\varepsilon}_s^\beta t_f = C \tag{3}$$

Here t_f is the time to failure and the Monkman-Grant parameters β and C are 1.012 and 2.670e-3 $hr^{-.012}$. Since β is very nearly equal to unity Equation 3 shows us that the creep strain to failure is nearly a

constant. This can be used as the criterion for failure in the finite
element code.

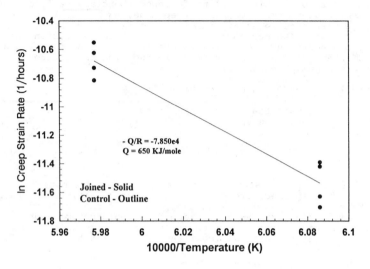

FIG. 4--Determination of Q at 120 MPa.

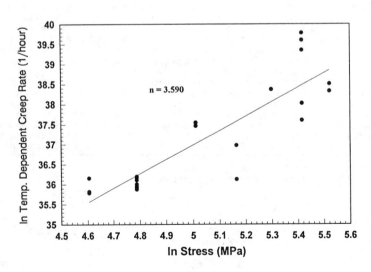

FIG. 5--Determination of the stress exponent n.

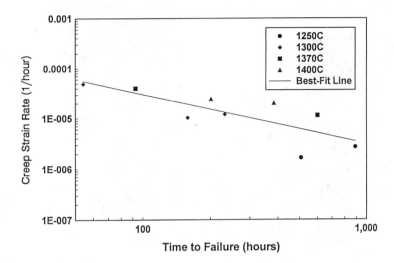

FIG. 6--Monkman-Grant relationship for failure times.

FINITE ELEMENT IMPLEMENTATION

The development of material models only has usefulness insofar as they can be applied to problems of technological interest. The finite element method has certainly become the most popular manner of applying advanced constitutive modeling to "real" situations. In this section the Arrhenius equation for steady state creep is used in conjunction with a commercial finite element package (ANSYS) to predict the life of a notched tensile component.

In choosing the Arrhenius equation (Eq. 1) for implementation, we are ignoring, for now, the primary creep region. In this material it is recognized that primary creep is a very important contributor to the total creep strain but further understanding is needed of the differences observed in the primary creep for identical test conditions before a suitable model can be put in place. By first modeling just the steady state creep, the methodology for evaluation of service life can be validated.

ANSYS[11] is a commercial finite element package having many of the features common to modern analysis programs including nonlinear materials and thermal stresses. The one dimensional material law in Equation 1 can be generalized for stress analysis using the concepts of equivalent stress and equivalent strain.

$$\dot{\varepsilon}_{ij}^{cr} = A\sigma_e^n e^{-Q/RT} \frac{3}{2} \frac{\sigma'_{ij}}{\sigma_e} \qquad (4)$$

Here, σ'_{ij} are the components of the deviator of the stress tensor, and σ_e is the equivalent stress ($\sigma_e = (3/2 \; \sigma'_{ij} \bullet \sigma'_{ij})^{1/2}$). The total strain rate is given as the sum of the creep strain rate of Equation 8 and the elastic strain rate.

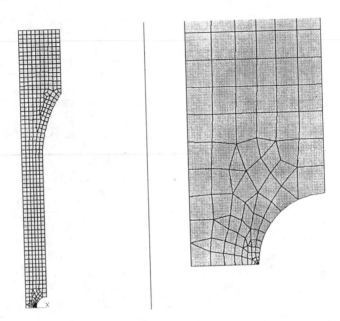

FIG. 7--Geometry and mesh for analysis of notched tensile specimen.

The component that was chosen to be simulated is very simple. A geometry was chosen that could be tested in the laboratory. A notched tensile specimen was selected since it is loaded primarily in tension and has a strong stress and strain gradient. The notch provides a universal stress and strain concentrator. Figure 7 shows the geometry and mesh of one quarter of the specimen. This was a two dimensional analysis since it has axisymmetry and also possesses symmetry about its cross section through the center of the notch. The mesh, which contained 529 eight node, axisymmetric elements was verified by a simple convergence study. The sample was modeled to be at 1368°C and to have a uniform far field stress of 10 MPa applied to the top face. Its life was predicted by monitoring the creep strain in the gauge section and determining when the Monkman-Grant equation was satisfied. Since the Monkman-Grant exponent was very close to unity the failure criterion reduces to a critical creep strain criterion. The simulation took about 100 time steps, each with 3 to 4 iterations needed to reach convergence.

Figure 8 shows the stress distribution immediately upon loading and just prior to failure. At time equal to 10 hours we see the stress concentration effect due to the notch with the maximum vertical stress occuring at the notch root and quickly falling off. After 500 hours we see that the stress has redistributed due to the creep deformation and the maximum vertical stress occurs at the center of the section and not at the notch root. Notice that the creep strain has distributed the stress to be more uniform across the notch ligament. Figure 9 shows the distribution of creep strain at 500 hours. At early times the creep strain is quite small and located right at the notch root. After 500 hours the creep strain magnitude has increased and propagated away from the notch root but the maximum accumulated strain of 0.267% is right at the notch root. Notice the unusual behavior that after 500 hours, the maximum vertical normal stress and the maximum creep strain are not at the same place in the component. As Equation 4 describes, the maximum equivalent strain rate should be co-located with the maximum effective

stress. After 500 hours the maximum creep strain that has accumulated
is 0.267%. This is the magnitude of the creep strain from the Monkman-
Grant equation so this corresponds to the time that we would predict
failure. To date, no experiments have been conducted to verify these
predictions but they are planned for the future.

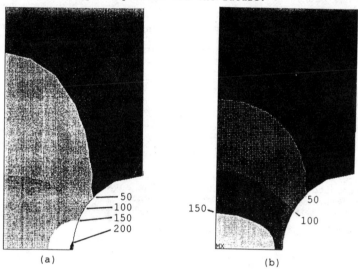

FIG. 8--Vertical normal stress at times equal to (a) ten and
(b) five hundred hours (in MPa).

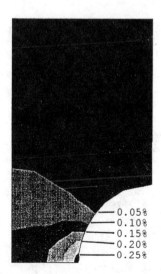

FIG. 9--Accumulated creep strains after five hundred hours.

CONCLUSIONS

This study illustrates a procedure for determining the life of ceramic components undergoing high temperature creep. An experimental data base has been generated for the creep deformation and fracture of HIP'ed silicon nitride with 4% yttria doping. Two methods have been shown for representing the data: theta projection analysis and Arrhenius modeling of the steady state creep rate. The steady state model has been used in conjunction with a finite element code to predict the deformation and life of a simulated ceramic tensile member. Future experiments are planned to verify this modeling.

ACKNOWLEDGEMENTS

Research sponsored by the U.S. Department of Energy, Assistant Secretary for Conservation and Renewable Energy, Office of Transportation Tehnologies, as part of the Ceramic Technology Project of the Materials Development Program, under contract DE-AC05-84OR213400 with Martin Marietta Energy Systems, Inc.

The authors would like to acknowledge the support of the project's technical monitor Dr. Michael Santella. The authors would also like to thank Ara Vartabedian for assistance in preparing the figures and tables, David Collette for sample preparation and Thomas Trostel for technical support. Also fruitful discussions and encouragement are acknowledged with Dennis Tracey and Glenn Sundberg.

APPENDIX A

Theta Projection Method

The Norton's equation approach to modeling the quasi steady state creep that was taken in the body of this paper. This method has received the most attention in the literature but is by no means the only creep model available.

An attempt has been made by Evans and Wilshire [7,8] to develop equations which adequately describe the shape of typical creep curves and also quantify the variation of such curves with stress and temperature. An expression for the time dependent creep strain can be described as a function of various shape terms, θ_i which reproduce the creep curve at a specific stress and temperature.

$$\varepsilon^{cr} = \varepsilon^{cr}(t, \theta_1, \theta_2, \ldots, \theta_n) \qquad (A.1)$$

Different forms of Equation A.1 have been used in the literature. The form which is used here can be understood as the sum of two terms. One term represents the decaying primary component and the other the accelerating tertiary component of creep strain. The chosen equation takes the form below.

$$\varepsilon^{cr} = \theta_1(1 - e^{-\theta_2 t}) + \theta_3(e^{\theta_4 t} - 1) \qquad (A.2)$$

Here θ_1 and θ_3 are strain like components representing the maximum magnitude of primary and tertiary creep. The θ_2 and θ_4 are components describing the rate of the controlling processes. The form of Equation A.2 describes the shape of a creep curve for ductile metals quite well but might not be expected to work as well for ceramics lacking a 'onounced tertiary region. One goal here is to investigate that 'respondance. Maximum likelihood fits are used to calculate the θ_i's

for each experimental creep curve. Interpolation between testing
conditions is provided by writing the coefficients as below.

$$\theta_i = \theta_i(T, \sigma) \tag{A.3}$$

The numerical model is capable of calculating the creep strain for
any stress/temperature combination once the variations in these
coefficients are determined. The curve fit that was used is a simple
exponential factor expansion in stress and temperature.

$$\ln\theta_i = A_i + B_i\sigma + C_iT + D_i\sigma T \tag{A.4}$$

Here i ranges from 1 to 4. This curve fit reduces the experimental data
base to a total of 16 constants. Evans et al. [9] applied the concept
to different pressureless sintered silicon nitride ceramics produced
using MgO, CeO$_2$, and Y$_2$O$_3$ additives. The theta projection method showed
reasonable temperature interpolation capabilities for design
calculations involving continuously varying stress and temperature
conditions.
The θ_i's have been fit to Equation A.4 and the coefficients are
listed (Table A.1). A comparison of the fit with experiment for creep
at 1370°C and 120 MPa is shown in Figure A.1. Since the experiments did
not show a well defined tertiary region there is some flexibility in
determining the θ_3's and θ_4's. In comparing this method to Norton's

TABLE A.1--Theta projection coefficients.

	A	B	C	D
$\theta 1$	-13.927	8.680E-02	5.912E-03	-6.054E-05
$\theta 2$	30.275	-4.172E-01	-2.308E-02	2.755E-04
$\theta 3$	-13.307	1.339E-01	9.189E-03	-8.559E-05
$\theta 4$	-36.268	-2.319E-01	1.105E-02	1.710E-04

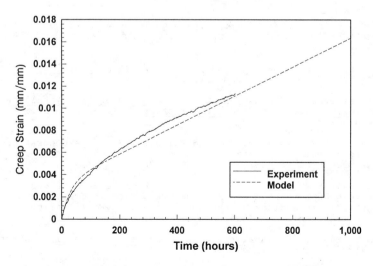

FIG. A.1--Theta projection versus experiment at 1370°C and 120 MPa

equation it is useful to consider the minimum creep rate predicted by
the theta projection method. There is no true "steady state" creep rate
but for a specimen that fails prior to a well defined tertiary region
then this minimum rate is the assymptotic final creep rate before
failure. The minimum creep rate is given by

$$\frac{de}{dt} = \theta_1\theta_2 e^{-\theta_2 t_m} + \theta_3\theta_4 e^{\theta_4 t_m}$$
(A.5)

and t_m the time where the minimum creep rate occurs is

$$t_m = \frac{1}{\theta_2+\theta_4}\ln\frac{\theta_1\theta_2^2}{\theta_3\theta_4^2}.$$
(A.6)

Overall, the theta projection method provides a useful way not only to
represent the experiments but to interpolate to other testing conditions
as well.

REFERENCES

[1]   Riedel, H., "Fracture at High Temperatures," Materials Research
      and Engineering, Springer-Verlag Berlin, Heidelberg, 1987.

[2]   Evans, H. E., Mechanisms of Creep Failure, Elsevier Applied
      Science Publishers LTD., 1984.

[3]   Wiederhorn, S. M., Krause, R. and Cranmer, D. C., "Tensile Creep
      Testing of Structural Ceramics," Proceedings of the Annual
      Technology Development Contractor's Coordination Meeting,
      held at Dearborn, MI, 1991.

[4]   Zygo Corporation, Laurel Brook Road, Middlefield, CT 06455.

[5]   Chakraborty, D. and Mukhopadhyay, A. K., "Creep of Sintered
      Silicon Nitride," Ceramics International, Vol. 15, 1989, pp 237-
      245.

[6]   Todd, J. A. and Xu, Z.-Y., "The High Temperature Creep Deformation
      of Si_3N_4-$6Y_2O_3$-$2Al_2O_3$," Journal of Materials Science, Vol. 24,
      1989, pp. 4443-4452.

[7]   Evans, R. W. and Wilshire B., Creep of Metals and Alloys, The
      Institute of Metals, London, 1985.

[8]   Wilshire, B. and Evans, R. W., Creep Behavior of Crystalline
      Solids, Progress in Creep and Fracture, Vol. 3, Pineridge Press,
      Swansea, U.K., 1985.

[9]   Evans, R. W., Murakami, T. and Wilshire, B., "The Generation
      of Long-Term Creep Data for Silicon Nitride Ceramics," British
      Ceramics Proceedings, No. 39, Dec. 1987.

[10] Monkman, F. C. and Grant, N. J., "An Empirical Relationship
      Between Rupture Life and Minimum Creep Rate in Creep-Rupture
      Tests," Proceedings of the American Society for Testing and
      Materials, Vol. 56, 1956, pp. 593-620.

[11] ANSYS Engineering Analysis System Theoretical Manual, Swanson
      Analysis Systems, Inc., 1986.

Jerry B. Sandifer[1], Michael J. Edwards[1], Thomas S. Brown III[1],
Stephen F. Duffy[2]

HIGH TEMPERATURE LIFE PREDICTION OF
MONOLITHIC SILICON CARBIDE HEAT EXCHANGER TUBES

REFERENCE: Sandifer, J. B., Edwards, M. J., Brown, T. S., and Duffy,
S. F., "High Temperature Life Prediction of Monolithic Silicon Carbide
Heat Exchanger Tubes," Life Prediction Methodologies and Data for
Ceramic Materials, ASTM STP 1201, C. R. Brinkman and S. F. Duffy, Eds.,
American Society for Testing and Materials, Philadelphia, 1994.

ABSTRACT: The need for improved performance in high temperature
environments is prompting industry to consider the use of structural
ceramic materials in heat exchanger tubes and other high temperature
components. In recognition of this need, the U. S. Department of
Energy has supported work for the development of nondestructive
methods for evaluating flaws in monolithic ceramic components, and the
associated establishment of criteria for the acceptance of flawed
components. Under this development of flaw assessment criteria, DOE
supported the work being presented in this paper.

The approach to developing the life prediction model combines
finite element predictions, considering creep behavior, with continuum
damage mechanics and Weibull reliability statistics. ABAQUS is used
to predict time dependent creep response of the component based on
experimental creep data. A continuity parameter is then calculated at
each time step following continuum damage mechanics methods. Finally,
Weibull statistics are used with the resulting continuity parameter to
predict the reliability at each time step, through the use of the
NASA-Lewis computer program CARES, interfaced to ABAQUS with ABACARES.

There is very limited data available to characterize the creep,
continuum damage and reliability behavior of the material. For the
life prediction model reported, it is assumed that the material
damages isotropically. Directional effects of the damage can be added
as material databases improve.

KEYWORDS: silicon carbide, continuum damage, Weibull statistics,
reliability analysis, creep, life prediction.

---

1 Babcock & Wilcox, Research and Development Division, Alliance, OH.
2 Cleveland State University, Dept. of Civil Engrg., Cleveland, OH.
  (NASA Resident Research Associate at Lewis Research Center)

INTRODUCTION

The use of structural ceramics in heat exchanger tubes and other high temperature components is being considered in the power generation and other industries to extend the temperature limitations imposed by traditional materials.   To encourage this use, the U. S. Department of Energy has supported work on the development of nondestructive methods for evaluating flaws in monolithic ceramic components [1], and the establishment of a criteria for the acceptance of flawed components. As a part of the development of a flaw assessment criteria, the DOE supported this work on the "High Temperature Life Prediction of Monolithic Silicon Carbide Heat Exchanger Tubes".

At low temperature, fast fracture reliability is an issue, but at higher temperature, time dependent deformation and damage of structural ceramics may become the dominant concern. Wiederhorn[3] has shown that at high temperature, two phase structural ceramics exhibit a time dependent "creep" damage behavior that eventually leads to structural failure. It is the high temperature time dependent behavior that is the subject of this paper.

The approach to developing the life prediction model combines finite element analysis (FEA) stress predictions considering creep behavior, with continuum damage mechanics, and Weibull reliability statistics. The FEA program ABAQUS [2] is used to predict the time dependent stress states throughout the component considering the redistribution of stress resulting from creep behavior. A continuity parameter is then calculated at each time step following continuum damage mechanics methods [3]. Finally, Weibull statistics are applied to an "effective" stress that is based on the continuity of the material as damage progresses. These statistics are used to predict the reliability at each time step, through the use of the NASA-Lewis computer program CARES [4], interfaced to ABAQUS with the B&W developed ABACARES [5].

The continuum damage modeling is basically the same as reported by NASA Lewis [3] which is a one dimensional analysis. The purpose of this paper is to extend this work to two-dimensional cases so it can be applied to tubular components that are commonly required in the power generation and process industries. This extension to two-dimensions will be accomplished by representing the two-dimensional state of stress in the tubular component by the Tresca stress [6].

In the report that follows, a methodology for high temperature life prediction based on creep rupture will be described, and an application will be presented. As with most developing materials in structural ceramics, the experimental database is limited. There has been some work in characterizing creep and creep rupture for two phase structural ceramics[3], and other work in the open literature presents Weibull parameters at ambient temperature [7]. However, the method proposed in this paper requires further application in order to assess the appropriateness of the approach and the validity of the continuum damage and statistical failure assumptions.

ANALYTICAL APPROACH

The analytical approach is summarized in Figure 1. As shown, a creep analysis is performed using the finite element code ABAQUS.

---

3   Wiederhorn, S.M. to Sandifer, J.B., Personal communication, October, 1992.

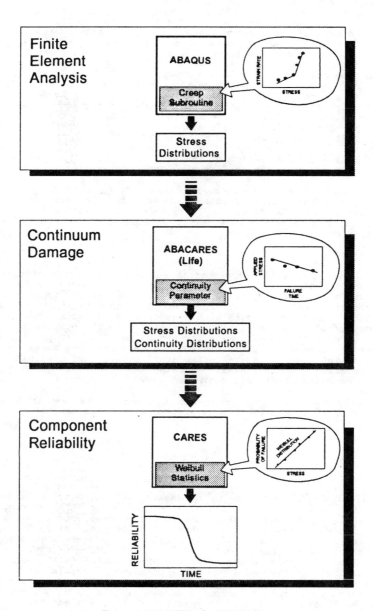

Figure 1 ANALYTICAL APPROACH

The steady state thermo-mechanical load is applied and the analysis is performed for a prescribed number of time steps. The stress is redistributed based on the resulting creep strain at the end of each step in the creep analysis. A state variable (continuity) is then defined to quantify the damage state [3] of the ceramic material throughout the component volume. This continuity parameter is calculated through the application of a power law that identifies the rate of damage. The power law is extended to two dimensional problems by using the stress state from the creep analysis, and an effective stress is then calculated based on the continuity of the material as the damage progresses. Finally, a Weibull analysis is performed on the effective stress using the CARES [4] code to predict the fast fracture reliability of each element in the model at the end of each time step. This Weibull procedure assumes that the developing crack growth or flaw enlargement does not significantly change the Weibull parameters. For this analysis, CARES is interfaced with ABAQUS through the use of a modified version of ABACARES [5] which computes the effective stress and accounts for time dependent reliability [6].

The heat exchanger tubes being evaluated in this DoE project were made of a slip cast siliconized silicon carbide material. This is a two phase ceramic similar to the Coors RBSC 210 evaluated in the NIST studies[3], so the data from that report will be used in an analysis demonstration.

Creep Analysis

Industrial ceramic components in service at high temperature can experience creep[3]. The stress field in these components may re-distribute as a result of creep deformations, and this complicates the life prediction calculations. To accommodate this behavior, the life prediction methodology begins with a creep analysis which uses the creep deformations to redistribute the stress in the component before the damage calculations are made for the current time step. Explicit integration methods are used in ABAQUS to determine the stress redistribution over the current time step in the creep analysis. Due to the relatively small inelastic strains involved, this technique was found to be quite stable and convergent.

To represent the creep behavior, a power law is used in an ABAQUS user subroutine to express steady-state creep strain rate as a function of stress, i.e.,

$$\dot{\varepsilon} = c_o \sigma^m \tag{1}$$

Wiederhorn's test results[3] demonstrate that the creep behavior (strain rate vs. stress) is piecewise continuous as shown in Figure 2. From the Figure, it can be seen that the creep constants c_o and m for the siliconized silicon carbide material can have two different values in compression and another value in tension. These variations in creep behavior are handled in the ABAQUS user subroutine by identifying the stress regime in each element and selecting the appropriate values from an input table of creep constants. To determine whether the two-dimensional state of stress is in tension or compression, the normal principle stresses are broken down into two sets — the "hydrostatic" (or dilatoric) component [8] and the deviatoric. The sign of the hydrostatic stress then determines whether the element is in tension or compression.

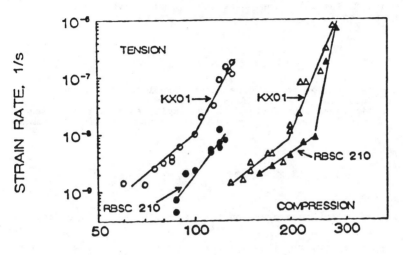

SILICONIZED SILICON CARBIDE
CREEP AT 1300C

Figure 2 STRAIN RATE VS STRESS

The material constants c_0 and m are determined as described above, the thermo-mechanical loads are applied and creep behavior is calculated by ABAQUS for a specified number of time steps. ABAQUS redistributes the stress at the end of each time step based on the calculated creep strain. The results of the creep analysis are then temperature, volume, creep strain and the redistributed stress for each element at each time step.

<u>Damage Calculation</u>

It is assumed that the evolution of the micro defects represents an irreversible thermodynamic process [3]. On the continuum level this requires the introduction of an internal state variable that serves as a measure of accumulated damage. In a one dimensional (uniaxial) case, let A_0 represent the cross sectional area in an undamaged state. Then denote A as the current cross sectional area in a damaged state where material defects exist in the cross section. A state variable, continuity, can then be defined to quantify the damage state as follows:

$$\psi = A/A_0 \tag{2}$$

In our two dimensional case, we consider A_0 to represent the elemental area carrying the load in the undamaged state, and A the elemental area carrying the load in the damaged state. With the definition in Equation 2, the continuity is 1 in the undamaged state, and 0 at failure.

For this time dependent analysis, the rate of change of continuity must be specified. The following power law for one dimensional problems was chosen [3]:

$$\dot{\psi} = -B(\sigma/\psi)^n \tag{3}$$

The power law is extended to two dimensional problems by representing the stress in Equation (3) as the Tresca stress from the creep analysis. The incremental reduction in continuity as time progresses from t_{i-1} to t_i for one time step, is determined by integrating Equation (3):

$$\int_{\psi_{i-1}}^{\psi_i} \psi^n \, d\psi = -B \, \sigma_i^n \int_{t_{i-1}}^{t_i} dt \tag{4}$$

resulting in the following expression for continuity after time step i:

$$\psi_i = [\psi_{i-1}^{n+1} - B\sigma_i^n(n+1)(t_i - t_{i-1})]^{1/(n+1)} \tag{5}$$

From Equation (5), the continuity can now be calculated for each element for each time step, assuming element stresses to be constant over the time step.

Fast Fracture Reliability

Monolithic ceramic materials exhibit an inherent scatter in strength that can be suitably characterized by the weakest link theory using Weibull statistics [3]. Using Weibull methods, the reliability of a component is [3]:

$$R_i = \exp\left\{-V_i\left(\frac{\sigma_{i,eff}}{\beta}\right)^\alpha\right\} \tag{6}$$

Where $\sigma_{i,eff} = \sigma_i / \psi_i$,

σ_i is the stress for element i,
ψ_i is the continuity of element i,
$\sigma_{i,eff}$ is the effective stress, increased by the damage,
V_i is the volume of element i,
α is the Weibull modulus, and
β is the Weibull scale parameter.

Finally, using Equation (6), the fast fracture reliability of a component can be calculated for each element at each time step as the product of all the element reliabilities. A modified Weibull based treatment of the failure probability for these cases with non-monolithic increasing stress has been accounted for based on Reference 9.

DEMONSTRATION OF THE METHODOLOGY

There is limited data available on creep behavior of two phase ceramic materials. Some of the most appropriate information available is the creep and creep rupture data from tensile coupon specimens developed by NIST for the Coors RBSC 210 material[3]. This data will be used to characterize the creep and damage behavior of the material, and the Weibull parameters will be estimated from the literature [7]. This database is limited, and the database of component applications where creep has been observed or measured is even more limited. Due to this lack of component structural performance data, a preliminary demonstration of the methodology will be presented using a simple C-ring test performed at high temperature during the DoE program [1].

Fracture toughness tests were performed on C-rings, but an assessment of the time dependent behavior of the material was beyond the scope of the project. As a result, the elaborate test set up and measurements required to accurately record creep behavior were not performed. However, for our demonstration case, we will make predictions of the C-ring behavior in the DoE tests [1] using the creep/continuum damage methodology described in this paper, and demonstrate that the predictions are consistent with the observed C-ring Behavior. This is not an ideal test case since the measurements in the C-ring tests were not refined enough to observe time dependent behavior, but it will serve to demonstrate the method and give some confidence in the predictions. The criteria used for comparison with test data was the computed reliability during the loading history. Rapid reductions in computed reliability would be used to predict failure conditions.

C-ring Test Conditions

The C-ring was a 3" (76.2 mm) OD, 2.5" (63.5 mm) ID ring that was 0.5" (12.7 m) wide with a .010" (0.254 mm) deep flaw machined on the O.D. as shown in Figure 3. It was tested at a uniform temperature of 1350°C. The C-ring was heated to temperature gradually so the thermal stresses could be neglected in the analysis and only the steady mechanical diametral load would need to be considered. The load was applied in increments according to the following schedule (Table 1).

TABLE 1 -- Load Schedule

Load	Time at load (hrs)
46 lb (205 N)	18
51 lb (227 N)	6
56 lb (249 N)	17
61 lb (271 N)	5
66 lb (294 N)	14
70 lb (311 N)	3.5
77 lb (343 N)	2
84 lb (374 N)	1 (slow steps until break at 84 lb)

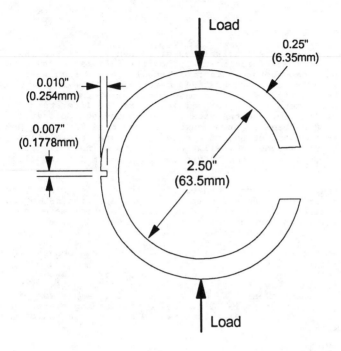

Figure 3 Tested C-Ring with .010" (0.254mm) seeded flaw

Material Creep Properties

The creep behavior of the material is represented by the data in Figure 2. As shown in the Figure, the behavior is different in tension than in compression. This behavior was modeled as a piecewise continuous function in ABAQUS's user Creep subroutine. To curve fit the tension data for the Coors RBSC 210 material we assume a power law for the creep behavior as follows:

$$\dot{\varepsilon} = c_o \sigma^m \tag{1}$$

Now, we can determine m as the slope of the plot of ln $\dot{\varepsilon}$ vs. ln σ for RBSC 210 in Figure 2. This gives a value of:

$$m = 7.43 \quad \text{(dimensionless)}$$

From equation 1 and the curve in Figure 2, we can determine the corresponding value of c_o to be:

$$c_o = 9.8E\text{-}21 \quad \left(\frac{1}{hr \cdot MPa^m}\right)$$

A similar procedure was then used to fit the two sections of the compressive data.

Material Damage Properties

The material damage parameters were defined by the power law in Equation 3 that relates the rate of change of continuity to the effective stress. The constants can be evaluated by integrating Equation 3 and recognizing that there is complete continuity at the beginning of the analysis before any damage has taken place.

The constants B and n in Equation 3 are determined as follows. Integrate Equation (3) from 0 to t and observe that at t = 0, ψ = 1.

$$\int_{1}^{\psi} \psi^n \, d\psi = -B\sigma_i^n \int_{0}^{t} dt \tag{7}$$

we obtain the following expression for continuity:

$$\psi = [1-B\sigma_i^n (n+1) \ t \ ]^{1/(n+1)} \tag{8}$$

Now, find the time to failure by setting ψ = 0 in Equation (8).

$$t_f = [B\sigma_i^n (n+1)]^{-1} \tag{9}$$

Taking the natural log of Equation (9), we can obtain the values of B and n for Coors RBSC 210 from Figure 4, which is a plot of ln(applied stress) vs. ln (failure time)[3].

$$\ln \sigma_i = - \ 1/n \ \{\ln t_f - \ln[B(n+1)] \} \tag{10}$$

The slope of the line in Figure 4 is -1/n. This results in the following value of n:

$$n = 10.08 \quad \text{(dimensionless)}$$

Now, from Equation (9) and the curve in Figure 4:

$$B = 1.54E-24 \ \left(\frac{1}{hr \cdot MPa^n}\right)$$

Weibull Parameters

Statistical fracture properties of two grades of siliconized silicon carbide material were determined [7] at ambient temperature at EG&G Idaho as part of a program of evaluating structural ceramics with tubular geometries for heat exchanger applications. These results will be used to estimate the Weibull parameters for this calculation. The two materials are Norton's NC-430, and Coors SC-AXI. For our analysis we will assume our material (Coors RBSC 210) is closer to the Coors SC-AXI and use the appropriate Weibull parameters α and β (Equation 6). Reference 7 reports ambient temperature C-ring test data from which siliconized SiC Weibull parameters can be estimated. The general expression for reliability of a component having a varying stress field like a C-ring is as follows.

$$R = \exp \{-\int_v (\sigma/\beta)^\alpha dV\} \tag{11}$$

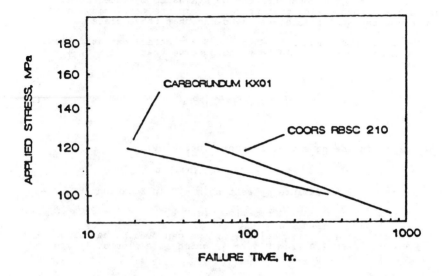

Figure 4 Applied Stress vs Failure Time

Using the Reference 7 data and Equation 11, the following Weibull parameters were determined:

$$\alpha = 15.25 \text{ (dimensionless)}$$
$$\beta = 236.8 \text{ (MPa} \bullet \text{(mm)}^{3/\alpha})$$

During the review of this paper, the authors have learned that Jadaan [10] has published high-temperature Weibull parameters for Coors RBSC 210. The use of these material properties provides better results as shown in Reference 6.

Summary of Material Properties

Table 2 is a summary of the material properties used in the C-ring test analysis.

TABLE 2 -- Material Properties

	PROPERTY		UNITS
Elastic	E = 255E+03		MPa
	μ = .3		dimensionless
Creep	m = 27.67 C_o = 5.29E-71	σ < -240MPa	m = dimensionless
	m = 3.74 C_o = 4.28E-14	-240MPa $\leq \sigma \leq$ 0MPa	$C_o = \dfrac{1}{hr \cdot MPa^m}$
	m = 7.43 C_o = 9.80E-21	σ > 0MPa	
Continuity	n = 10.08		dimensionless
	B = 1.54E-24		$\dfrac{1}{hr \cdot MPa^n}$
Weibull	α = 15.25		dimensionless
	β = 236.8		MPa $\bullet$ $(mm)^{3/\alpha}$

ANALYSIS RESULTS

The finite element model of the C-ring with the .010" (0.254 mm) flaw is shown in Figure 5. Taking advantage of symmetry, one quarter of the C-ring was modeled and the load was applied as shown. The Figure shows an isometric view of the model and a close-up of the area around the flaw. An analysis was performed using the material properties and load schedule defined above. Three analyses were run to characterize the failure behavior of the C-ring test.

Fast Fracture Reliability Without Creep

Fast fracture reliability was assessed with ABAQUS and CARES without considering creep or continuum damage. Reliability was plotted vs. applied load in Figure 6 for the C-ring with the .010" (0.254 mm) flaw compared with the same C-ring with no flaw. The test C-ring failed at an 84 lb (374 N) load after following the load schedule described above. The results in Figure 6 show that the fast fracture reliability is zero for 84 lb (374 N) load, so a failure is expected when the 84 lb (374 N) load is applied. Two assumptions should be noted with regards to Figure 6. The Weibull parameters are based on ambient temperature data for a similar SiC material while the actual C-ring test was performed at a temperature of 1300°C. In addition, this fast fracture analysis prediction does not account for any stress redistribution which would be expected to increase the fast fracture reliability in the direction of the un-flawed C-ring analysis results. The actual C-ring loading was applied in steps over more than 65 hours, thus permitting stress redistributions not accounted for in this analysis.

Continuum Damage Reliability With Creep at Constant Load

This analysis examines the behavior when both creep and continuum damage are considered in the reliability calculation.

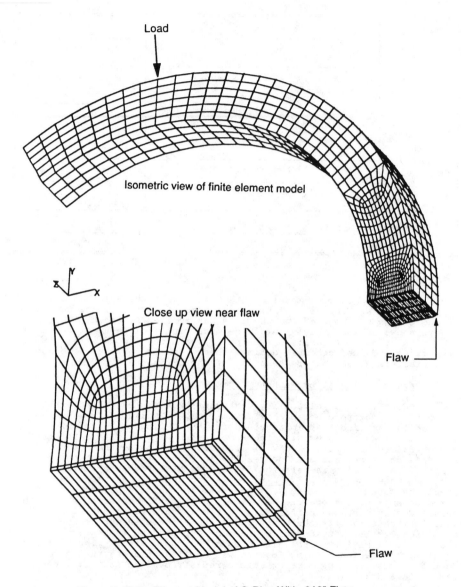

Load

Isometric view of finite element model

Close up view near flaw

Flaw

Flaw

**Figure 5 Finite Element Model of C-Ring With .010" Flaw
(Quarter Symmetry)**

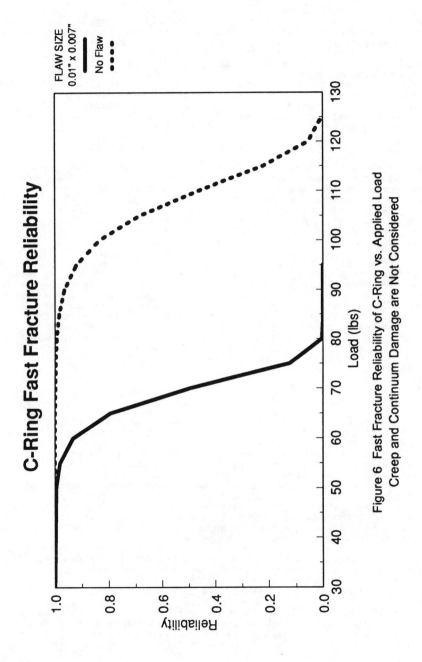

Figure 6 Fast Fracture Reliability of C-Ring vs. Applied Load
Creep and Continuum Damage are Not Considered

Figure 7 is a plot of reliability vs. time for four different loads considering both stress redistribution and damage accumulation. The four loads that were chosen to cause failure in a reasonable time period were 55 lb (245 N), 60 lb (267 N), 65 lb (289 N), and 70 lb (311 N). As shown in the figure, an analysis with a 60 lb (267 N) load predicted an initial reliability of 0.93. This value slowly drops to 0.91 after 90 hours before quickly dropping to zero. Each of these curves display this trend of an initial reliability which drops sharply at failure. Additionally, Figure 7 shows that the time to failure is highly dependent upon the load (stress state). This observation suggests that stress redistribution, due to creep effects, are very important for accurate predictions of component life.

Reliability Prediction With the Actual Load History

For the third analysis the experimental loading conditions were applied and creep and continuum damage effects were included. This represents the complete analysis sequence being proposed in this paper. The results are shown in Figure 8 as a plot of reliability vs. time. A plot of fast fracture reliability with creep effects, but no continuum damage is shown for comparison, and the load vs. time sequence is also shown on the plot for reference. This Figure presents several interesting results.

First, it shows that the continuum damage reliability drops from 0.92 after 63.5 hours, to zero after 66 hours, when the C-ring failed. The corresponding drop in fast fracture reliability only reaches a low of 0.76. Next, in both the continuum damage and fast fracture cases, creep effects, which redistribute stresses, are evident in the reliability predictions. This effect is best illustrated when the C-ring was loaded to 70 lb (311 N) during the time from 60 hours to 63.5 hours. The continuum damage model predicts a reliability of 0.92, while fast fracture reliability is 0.98. However, if creep was not considered then the fast fracture reliability for a 70 lb (311 N) load would be 0.5 (see Figure 6).

CONCLUSIONS

A methodology is presented that predicts the reliability of monolithic structural ceramics that are subject to time dependent (creep) damage. The results of the demonstration analysis with a C-ring cut from a heat exchanger tube show that the predicted behavior is somewhat physically appealing. Accurate predictions were not anticipated in this case since a good material property database was not available for the specific material at the temperature tested, however, the results were sufficiently close to encourage the continued assessment of the method. Additional appropriate test cases need to be developed, and appropriate material properties need to be thoroughly determined to provide an adequate evaluation of the creep/continuum damage approach presented.

With refinements to better address the effects of bi-axial stress fields, and actual material properties, this methodology appears to be quite general and readily applicable to many structural ceramic components and materials. An inherent assumption which may require clarification is the notion that an effective stress can be used with undamaged material Weibull parameters to predict long term fracture reliability. The evolution and coalescence of voids and/or flaw growth in preferential directions may alter the Weibull parameters since the nature of the underlying flaw population as well as applied stress influence these parameters.

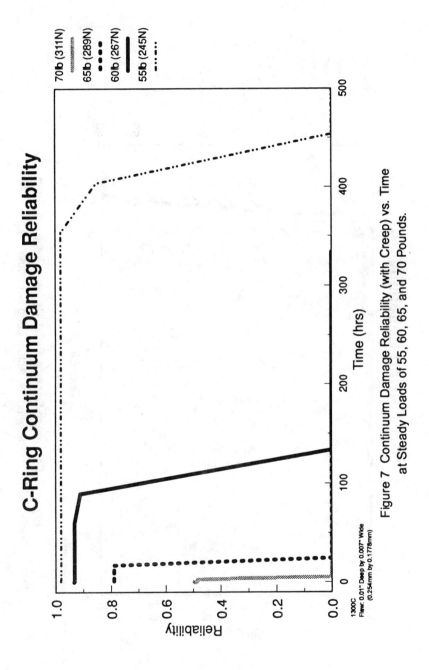

Figure 7 Continuum Damage Reliability (with Creep) vs. Time at Steady Loads of 55, 60, 65, and 70 Pounds.

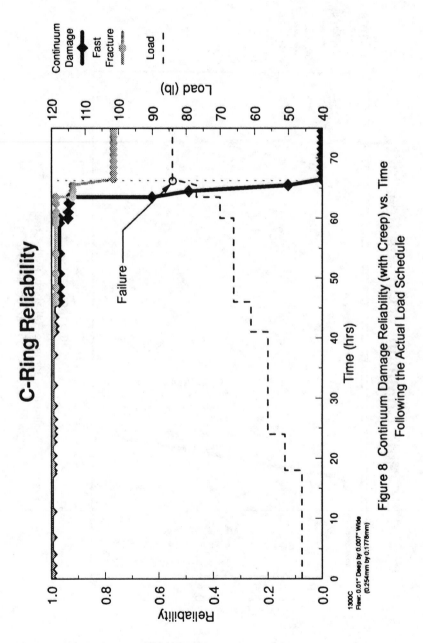

Figure 8 Continuum Damage Reliability (with Creep) vs. Time
Following the Actual Load Schedule

REFERENCES

[1]   U.S. Department of Energy Cooperative Agreement DE-FC07-89ID12887. "Assessment of Strength Limiting Flaws in Ceramic Heat Exchanger Components", 1993.

[2]   Hibbitt, H.D., Karlsson, G.I., Sorensen, E.P., 1984b; ABAQUS User's Manual. Providence: Hibbitt, Karlsson & Sorensen, Inc.

[3]   Duffy, S.F. and Gyekenyesi, J.P., "Time Dependent Reliability Model Incorporating Continuum Damage Mechanics for High-Temperature Ceramics" NASA Technical Memorandum 102046, May, 1989.

[4]   Nemeth, N.N., Manderscheid, J.M. and Gyekenyesi, J.P., 1990, "Ceramic Analysis and Reliability Evaluation of Structures (CARES). User's and Programmer's Manual", NASA TP-2916.

[5]   Edwards, M.J., Powers, L.M., and Stevenson, I., "ABACARES, A Program which Provides CARES (Ceramic Analysis and Reliability Evaluation of Structures) Analytical Techniques to the ABAQUS User", Proceedings of the ABAQUS Users' Conference, May 27 - 29, 1992, Newport, Rhode Island.

[6]   Edwards, M.J., Sandifer, J.B., Brown, T.S., Duffy, S.F., "The Implications of a Continuum Damage Model for Life Prediction of Two Phase Structural Ceramics", Current Capabilities for Non-destructive Testing and Lifetime Prediction Symposium, The American Ceramic Society, October, 1993.

[7]   Landini, D.J., Flinn, J.E., and Kelsey, P.V., Jr., "The Slit-Ring Test for Evaluating Fracture in Tubular Cross Sections", Advances in Ceramics - Volume 14, "Ceramics in Heat Exchangers", Ed., Bryan D. Foster and John B. Patton, The American Ceramic Society, 1985.

[8]   Shigley, J.E. and Mischke, C.R., Mechanical Engineering Design, Fifth Edition, McGraw Hill Book Company, New York, 1989.

[9]   Stanley, P., Chau, F.S., "A Probabilistic Treatment of Brittle Fracture Under Non-Monotonically Increasing Stresses", International Journal of Fracture, Vol. 22, (1983), p. 187-202.

[10]   Jadaan, O., "Fast Fracture and Lifetime Prediction for Ceramic Tubular Components", Ph,D. thesis, 1990, Pennsylvania State University.

Noel N. Nemeth[1], Lynn M. Powers[2], Lesley A. Janosik[1], and
John P. Gyekenyesi[1]

TIME-DEPENDENT RELIABILITY ANALYSIS OF MONOLITHIC CERAMIC COMPONENTS
USING THE CARES/LIFE INTEGRATED DESIGN PROGRAM

REFERENCE: Nemeth, N. N., Powers, L. M., Janosik, L. A., and Gyekenyesi,
J. P., "Time-Dependent Reliability Analysis of Monolithic Ceramic Compo-
nents Using the Cares/Life Integrated Design Program," Life
Prediction Methodologies and Data for Ceramic Materials, ASTM STP 1201,
C. R. Brinkman and S. F. Duffy, Eds., American Society for Testing and
Materials, Philadelphia, 1994.

ABSTRACT: The computer program CARES/LIFE calculates the time-dependent
reliability of monolithic ceramic components subjected to thermomechanical
and/or proof test loading. This program is an extension of the CARES
(Ceramics Analysis and Reliability Evaluation of Structures) computer
program. CARES/LIFE accounts for the phenomenon of subcritical crack
growth (SCG) by utilizing either the power or Paris law relations. The
two-parameter Weibull cumulative distribution function is used to characterize the variation in component strength. The effects of multiaxial
stresses are modeled using either the principle of independent action
(PIA), Weibull's normal stress averaging method (NSA), or Batdorf's
theory. Inert strength and fatigue parameters are estimated from rupture strength data of naturally flawed specimens loaded in static,
dynamic, or cyclic fatigue. Two example problems demonstrating component reliability analysis and fatigue parameter estimation are
included.

KEY WORDS: CARES, CARES/LIFE, ceramic design, fatigue, reliability,
Batdorf, Weibull, subcritical crack growth

Advanced ceramic components designed for gasoline, diesel, and
turbine heat engines are leading to lower engine emissions, higher fuel
efficiency, and more compact designs. Ceramic materials are also used
for wear parts (nozzles, valves, seals, etc.), cutting tools, grinding
wheels, bearings, and coatings. Among the many requirements for the
successful application of advanced ceramics are the proper characterization of material properties and the use of a mature and validated
brittle material design methodology.
        Ceramics are brittle and the lack of ductility leads to low strain
tolerance, low fracture toughness, and large variations in observed
fracture strength. The material as processed has numerous inherent randomly distributed flaws. The observed scatter in fracture strength is
caused by the variable severity of these flaws. The ability of a
ceramic component to sustain a load also degrades over time. This is
due to a variety of effects such as oxidation, creep, stress corrosion,
and cyclic fatigue. Stress corrosion and cyclic fatigue result in a

[1]Research engineer and manager, respectively, Structural Integrity
Branch, NASA Lewis Research Center, Cleveland, OH 44135.

[2]Research associate, Department of Civil Engineering, Cleveland
State University, Cleveland, OH 44115.

phenomenon called subcritical crack growth (SCG). SCG initiates at a pre-existing flaw and continues until a critical length is reached causing catastrophic propagation. SCG failure is a load-induced phenomenon over time. It can also be a function of chemical reaction, environment, debris wedging near the crack tip, and deterioration of bridging ligaments. Because of this complexity, models that have been developed tend to be semi-empirical and approximate the behavior of subcritical crack growth phenomenologically.

The objective of this paper is to present a description of the integrated design computer program, CARES/LIFE (Ceramics Analysis and Reliability Evaluation of Structures LIFE prediction program), which predicts fast-fracture and/or time-dependent reliability of monolithic ceramic components. This program is an extension of the CARES [1,2] program, which predicts fast-fracture reliability of monolithic ceramic components.

Two examples are provided to illustrate some of the capabilities of the CARES/LIFE program: (a) time-dependent reliability analysis of a rotating annular disk under constant and sinusoidal cyclic angular speeds, and (b) evaluation of fatigue parameters for a ring-on-ring square plate specimen under dynamic loading.

PROGRAM CAPABILITY AND DESCRIPTION

The CARES/LIFE integrated design computer program predicts the probability of failure of a monolithic ceramic component as a function of its service life. CARES/LIFE couples commercially available finite element programs, such as MSC/NASTRAN, ANSYS, and ABAQUS, with probabilistic design methodologies to account for material failure from subcritical crack growth of preexisting flaws. The code is written in FORTRAN 77 and is divided into three separately executable modules which perform: (1) statistical analysis and characterization of experimental data obtained from the fracture of laboratory specimens; (2) neutral data base generation from results of MSC/NASTRAN, ANSYS, and ABAQUS finite element analysis programs; and (3) time-dependent reliability evaluation of thermomechanically loaded ceramic components (including proof test effects on a survived component and the consequences of off-axis and multiaxial loading). Finite element heat transfer and linear-elastic stress analyses are used to determine temperature and stress distributions in the component. Component reliability for volume (intrinsic) flaws is determined from finite element stress, temperature, and volume output from either two-dimensional, three-dimensional, or axisymmetric elements. Reliability for surface (extrinsic) flaws is calculated from shell element stress, temperature, and area data. CARES/LIFE produces an optional PATRAN data file containing risk-of-rupture intensities (a local measure of reliability) for graphical rendering of a structure's critical regions.

The phenomenon of subcritical crack growth is modeled with the power law and the Paris law relations. The most commonly used method is the power law [3,4], which describes crack velocity as a function of stress intensity factor. For cyclic fatigue, the Paris law [5,6,7] is used to model subcritical crack growth. The Paris law relates crack growth per load cycle to the range in the crack tip stress intensity factor. The power and Paris laws require two experimentally determined parameters which are material/environmental constants. Steady-state cyclic loading is accounted for by using the Paris law or by employing g-factors [8] in conjunction with the power law. The g-factor approach equates variable cyclic loadings to equivalent static loadings. CARES/LIFE includes sinusoidal, square, and sawtooth loading waveforms. Typically, the use of g-factors is appropriate for flat R-curve materials.

The probabilistic nature of material strength and the effects of multiaxial stresses are modeled by using either the principle of independent action (PIA) [9,10], Weibull's normal stress averaging (NSA)

method [11], or Batdorf's theory [12,13]. Batdorf's theory combines
linear elastic fracture mechanics with the weakest-link mechanism. It
requires a user-selected flaw geometry and a mixed-mode fracture criterion to describe volume or surface strength limiting defects. The
combination of a particular flaw shape and fracture criterion results in
an effective stress, which is a function of the far-field stresses, and
acts on the crack plane. CARES/LIFE includes the total strain energy
release rate theory, which assumes coplanar crack extension [13]. Out-
of-plane crack extension criteria are approximated by a simple semi-
empirical equation [14,15]. Available flaw geometries include the
Griffith crack, penny-shaped crack, semi-circular crack, and notch
crack. If the normal stress acting on the flaw plane is compressive,
then no crack growth is assumed to occur.

Weibull parameters, the Batdorf crack density coefficient, and
fatigue parameters are estimated from rupture strength data of naturally
flawed specimens. CARES/LIFE includes closed form solutions for pure
tensile, three- or four-point loaded bars (beams) under isothermal
conditions. For other specimen geometries, material parameters can be
estimated via effective volume and area calculations (a finite element
model of the specimen geometry and loading is required).

CARES/LIFE retains all the capabilities of the original CARES code
for fast-fracture [1,2,16]. These include least-squares and maximum
likelihood Weibull parameter estimation, Kolmogorov-Smirnov and
Anderson-Darling goodness-of-fit testing, Kanofsky-Srinivasan 90-percent
confidence bands, 90-percent confidence bounds on Weibull parameters,
and the Stefansky outlier test.

CARES/LIFE estimates the fatigue parameters (N and B) from
naturally flawed specimens ruptured under static, cyclic, or dynamic
loading. These parameters can be calculated using either the median
value technique, a least squares regression technique, or a median
deviation regression method which is somewhat similar to trivariant
regression [17]. The inert strength distribution Weibull modulus, m,
and characteristic strength, σ_θ, are optionally estimated from the
fatigue data for a failure time of one second with constant stress rate
loading (or a lifetime of $1/(N+1)$ cycles). The Weibull modulus is a
dimensionless quantity which, for a 2-parameter distribution, measures
the degree of strength dispersion of the flaw population. The characteristic strength corresponds to the stress level at which 63.21 percent
of specimens would fracture. The fatigue data is manipulated to find
the underlying fast-fracture strengths. This enables goodness-of-fit
testing and using the outlier test. The resulting goodness-of-fit statistics are applied to the original fatigue data.

To ensure compatibility of failure probabilities for surface and
volume flaw specimens, relationships between the fatigue parameters and
various failure criteria have been established. From test specimen data
(uniaxial tension, 3- and 4-point bend), compatibility is derived by
equating the risk of rupture of the uniaxial Weibull model to the risk
of rupture of the PIA, NSA, or the Batdorf shear-sensitive models. This
satisfies the requirement that for a uniaxial stress state, all models
produce the same probability of failure as the uniaxial Weibull model.

Finite element analysis is an ideal mechanism for obtaining stress
distributions necessary to calculate a structure's survival probability.
Each element can be made arbitrarily small, such that the stresses can
be taken as constant throughout each element (or subelement). In
CARES/LIFE the reliability calculations are performed at the Gaussian
integration points of the element or, optionally, at the element
centroid [2]. Using element integration points enables the element to
be divided into sub-elements, where integration point sub-volumes, sub-
areas, and sub-temperatures are calculated. The location of the
Gaussian integration point in the natural space of the finite element,
as well as corresponding weight functions, are considered when the
subelement volume/area is calculated. The number of subelements in each
element depends on the integration order chosen, and the element type.
Probability of survival for each element is assumed to be a mutually

exclusive event.   Overall component reliability is the product of all the calculated element (or subelement) survival probabilities.

Proof test methodology is incorporated into the PIA, Weibull normal stress averaging, and Batdorf theories, accounting for the effect of multiaxial stresses. With Weibull normal stress averaging and the Batdorf theory, a proof test load need not closely simulate the actual service conditions on a component. This is important because it allows reliability analysis to be performed when proof test stresses have not been applied in the same direction and/or location as the service load stresses. CARES/LIFE simultaneously processes two finite element analysis neutral files containing the stress analysis results for the proof test and the service load conditions. The duration of the proof test and service load are also considered and each load situation can have different material-environmental constants (Weibull and fatigue parameters).

For steady-state cyclic load, component reliability analysis can be performed if the ratio of the minimum cycle stress to the maximum cycle stress is constant throughout the component. When this ratio is not constant, two finite element result files are required for the reliability analysis. These two files represent the extremes of the cyclic loading range. The capability of performing transient analysis is also planned for a future update of CARES/LIFE. If temperature cycling is present resulting in a variation of material properties as a function of time, the most severe set of parameters will be used in the analysis with the Paris law.

THEORY

Time-dependent reliability is based on mode I equivalent stress distribution transformed to its equivalent stress distribution at time $t=0$.   Investigations of mode I crack extension [18] have resulted in the following relationship for the equivalent mode I stress intensity factor

$$K_{Ieq} = \sigma_{Ieq}(\Psi, t) \ Y \ \sqrt{a(t)} \tag{1}$$

where σ_{Ieq} is the equivalent mode I stress on the crack, Y is a function of crack geometry, $a(t)$ is the appropriate crack length, and Ψ represents a location x,y,z (within the body) and the orientation α, β of the crack. In some models such as the Weibull and PIA, Ψ represents a location only. Y is a function of crack geometry; however, herein it is assumed constant with subcritical crack growth. Crack growth as a function of equivalent mode I stress intensity factor is assumed to follow a power law relationship

$$\frac{da(t)}{dt} = A \ K_{Ieq}^{N} \tag{2}$$

where A and N are material/environmental constants.   The transformation of the equivalent stress distribution at the time of failure, t_f, to its critical effective stress distribution at time equals zero is expressed as [19,20]

$$\sigma_{Ieq,0}(\Psi, t_f) = \left[\frac{\int_0^{t_f} \sigma_{Ieq}^{N}(\Psi, t) \ dt}{B} + \sigma_{Ieq}^{N-2}(\Psi, t_f) \right]^{\frac{1}{(N-2)}} \tag{3}$$

where $\sigma_{I_{eq}}(\Psi, t_f)$ is the equivalent stress distribution in the component at time $t=t_f$, and B is the material/environmental fatigue parameter

$$B = \frac{2}{A \ Y^2 \ K_{IC}^{N-2} \ (N-2)} \tag{4}$$

The dimensionless fatigue parameter N is independent of fracture criteron. B is adjusted to satisfy the requirement that for a uniaxial stress state, all models produce the same probability of failure. Parameter B has units of stress^2 × time.

Volume Flaw Analysis

The probability of failure for a ceramic component using the Batdorf model [12,13,21] for volume flaws is expressed as

$$P_{fV} = 1 - \exp\left\{-\int_V \left[\int_0^{\sigma_{e_{max}}} \frac{\Omega}{4\pi} \frac{dn_V(\sigma_{cr})}{d\sigma_{cr}} d\sigma_{cr}\right] dV\right\} \tag{5}$$

where $\sigma_{e_{max}}$ is the maximum value of $\sigma_{I_{eq},0}$ for all values of Ψ, V is the volume, n_V is the crack density function, and Ω is the area of a solid angle projected onto a unit radius sphere in principal stress space containing all crack orientations for which the effective stress is greater than or equal to the critical mode I strength, σ_{cr}. The crack density distribution is a function of the critical effective stress distribution. For volume flaw analysis, the crack density function is expressed as

$$n_V(\sigma_{cr}(\Psi)) = k_{BV} \ \sigma_{cr}^{m_V} \tag{6}$$

where k_{BV} and m_V are material constants. The solid angle is expressed as

$$\Omega = \int_0^{2\pi} \int_0^\pi H(\sigma_{I_{eq},0}, \sigma_{cr}) \ \sin\alpha \ d\alpha \ d\beta \tag{7}$$

where

$$H(\sigma_{I_{eq},0}, \sigma_{cr}) = \begin{cases} 1 & \sigma_{I_{eq},0} \geq \sigma_{cr} \\ 0 & \sigma_{I_{eq},0} < \sigma_{cr} \end{cases} \tag{8}$$

and α and β are the radial and azimuthal angles on the unit sphere, respectively. The transformed equivalent stress $\sigma_{I_{eq},0}$ is dependent on an appropriate fracture criterion, crack shape, and time t_f. Equation (5) can be simplified by performing the integration of σ_{cr} [21] yielding the time-dependent probability of failure

$$P_{fV}(t_f) = 1 - \exp\left[-\frac{k_{BV}}{2\pi} \int_V \int_0^{2\pi} \int_0^{\frac{\pi}{2}} \left[\sigma_{I_{eq},0}(\Psi, t_f)\right]^{m_V} \sin\alpha \ d\alpha \ d\beta \ dV\right] \tag{9}$$

Fracture criteria and crack shapes available for time-dependent analysis are identical to those used for fast-fracture in CARES [1,2]. These fracture criteria include Weibull normal stress averaging (a shear-insensitive case of the Batdorf theory), total coplanar strain energy release rate, and the noncoplanar (Shetty) criterion.

For a stressed component, the probability of failure is calculated from equation (9). The finite element method enables discretization of the component into incremental volume elements. CARES/LIFE evaluates the failure probability at the Gaussian integration points of the element or optionally at the element centroid. Subelement volume is defined as the contribution of the integration point to the element volume in the course of the numerical integration procedure. The volume of each subelement (corresponding to a Gauss integration point) is calculated using shape functions inherent to the element type [2].

Surface Flaw Analysis

The probability of failure for a ceramic component using the Batdorf model for surface flaws is

$$P_{fs} = 1 - \exp\left\{-\int_A \left[\int_0^{\sigma_{e_{max}}} \frac{\omega}{\pi} \frac{dn_s(\sigma_{cr})}{d\sigma_{cr}} d\sigma_{cr}\right] dA\right\} \tag{10}$$

where A is surface area, n_s is the crack density function, and ω is the length of an angle α projected onto a unit radius semi-circle in principal stress space containing all of the crack orientations for which the effective stress is greater than or equal to the critical stress. Analogous to the argument for volume flaws, equation (10) can be reformulated, yielding [21]

$$P_{fs}(t_f) = 1 - \exp\left[-\frac{k_{BS}}{\pi}\int_A \int_0^\pi [\sigma_{Ieq,0}(\Psi, t_f)]^{m_s} d\alpha \, dA\right] \tag{11}$$

The transformed equivalent stress $\sigma_{Ieq,0}$ is dependent on the appropriate fracture criterion, crack shape, and time t_f. The criteria and crack shapes available for time-dependent analysis are identical to those used for fast fracture. These fracture criteria include Weibull normal stress averaging (a shear-insensitive case of the Batdorf theory), total coplanar strain energy release rate, and the noncoplanar (Shetty) criterion.

The finite element method enables discretization of the surface of the component into incremental area elements. CARES/LIFE evaluates failure probability at the Gaussian integration points of shell elements or optionally at the element centroids. The area of each subelement (corresponding to a Gaussian integration point) is calculated using shape functions inherent to the element type [2].

Static Fatigue

Static fatigue is defined as the application of a nonvarying load over a period of time. For this case $\sigma_{Ieq}(\Psi, t)$ is independent of time (denoted by $\sigma_{Ieq}(\Psi)$) and integration of equation (3) with respect to time leads to

$$\sigma_{Ieq,0}(\Psi, t_f) = \sigma_{Ieq}(\Psi)\left[\frac{t_f \sigma_{Ieq}^2(\Psi)}{B} + 1\right]^{\frac{1}{N-2}} \tag{12}$$

where N, B, and t_f are as previously defined.

Dynamic Fatigue

Dynamic fatigue is defined as the application of a constant stress rate $\dot{\sigma}(\Psi)$ over a period of time t. Assuming the applied stress is zero at time equal to zero then

$$\sigma_{Ieq}(\Psi, t) = \dot{\sigma}(\Psi) \, t \tag{13}$$

Substituting equation (13) into equation (3) results in an expression for effective stress at the time of failure

$$\sigma_{Ieq,0}(\Psi, t_f) = \left[\frac{\sigma_{Ieq}(\Psi, t_f)^N \, t_f}{(N+1) \, B} + \sigma_{Ieq}(\Psi, t_f)^{N-2} \right]^{\frac{1}{N-2}} \tag{14}$$

where $\sigma_{Ieq}(\Psi)$, N, B, and t_f are as previously defined.

Cyclic Fatigue

Cyclic fatigue is the repeated application of a loading sequence. Analysis of the time-dependent probability of failure for a component subjected to various cyclic boundary load conditions is simplified by transforming that type of loading to an equivalent static load. The conversion satisfies the requirement that both systems will cause the same crack growth [8]. Implicit in this conversion is the validity of the crack growth equation (2). The probability of failure is obtained with respect to the equivalent static state.

Evans [22] and Mencik [8] defined g-factors, for various types of cyclic loading, that are used to convert the cyclic load pattern to an equivalent static load. For periodic loading, T is the time interval of one cycle. $\sigma_{Ieq}(\Psi)$ is the equivalent static stress acting over the same time interval, t_1, as the applied cyclic stress, $\sigma_{Ieqc}(\Psi, t)$, at some location Ψ. The equivalent static stress is related to the cyclic stress by

$$\sigma_{Ieq}^N(\Psi) \, t_1 = \int_0^{t_1} \sigma_{Ieqc}^N(\Psi, t) \, dt = t_1 \left[\frac{\int_0^T \sigma_{Ieqc}^N(\Psi, t) \, dt}{T} \right] \tag{15}$$

so that

$$\sigma_{Ieq}^N(\Psi) \, t_1 = g(\Psi) \, \sigma_{Ieqc_{max}}^N(\Psi) \, t_1 \tag{16}$$

and the g-factor is

$$g(\Psi) = \left[\frac{\int_0^T \left(\frac{\sigma_{Ieqc}(\Psi, t)}{\sigma_{Ieqc_{max}}(\Psi)} \right)^N \, dt}{T} \right] \tag{17}$$

CARES/LIFE uses the maximum cyclic stress, $\sigma_{Ieqc_{max}}(\Psi)$, of the periodic load as a characteristic value to define the g-factor. For a periodic load over a time t_1 the mode I static equivalent stress distribution is

$$\sigma_{Ieq,0}(\Psi, t_f) = \sigma_{Ieqc_{max}}(\Psi) \left[\frac{g(\Psi) \, t_f \, \sigma_{Ieqc_{max}}^2(\Psi)}{B} + 1 \right]^{\frac{1}{N-2}} \tag{18}$$

The use of g-factors for determining component life is an unconservative practice for materials prone to cyclic damage. The Paris law [5], which has traditionally been used in metals design, has been suggested as a model of fatigue damage for some ceramic materials [6,7]. The Paris law describes the crack growth increment per cycle (da/dn) as

$$\frac{da(n)}{dn} = A \ \Delta K_{Ie}^N \tag{19}$$

where

$$\Delta K_{Ie} = \Delta \sigma_{Ieqc}(\Psi) \ Y \ \sqrt{a(n)} \ = [\sigma_{Ieqc_{max}}(\Psi) - \sigma_{Ieqc_{min}}(\Psi)] \ Y \ \sqrt{a(n)} \tag{20}$$

The subscripts max and min indicate the maximum and minimum cycle stress, respectively. Integration of equation (19) parallels that of equation (2), yielding the cyclic fatigue equivalent stress distribution

$$\sigma_{Ieq,0}(\Psi, n_f) = \sigma_{Ieqc_{max}}(\Psi) \left[\frac{\sigma_{Ieqc_{max}}^2(\Psi) \ (1 - R(\Psi))^N \ n_f}{B} + 1 \right]^{\frac{1}{N-2}} \tag{21}$$

where

$$R(\Psi) = \frac{\sigma_{Ieqc_{min}}(\Psi)}{\sigma_{Ieqc_{max}}(\Psi)} \tag{22}$$

The number of cycles to failure is n_f and B is now expressed in units of stress^2 × cycle (B and N are determined from cyclic data).

Evaluation of Fatigue Parameters from Inherently Flawed Specimens

Lifetime reliability of structural ceramic components depends on the history of the loading, component geometry, distribution of pre-existing flaws, and the parameters (N and B) that characterize subcritical crack growth. These crack growth parameters must be measured under conditions representative of the service environment. When determining fatigue parameters from rupture data of naturally flawed specimens, the statistical effects of the flaw distribution must be considered along with the strength degradation effects of subcritical crack growth. In the following discussion fatigue parameter estimation methods are described for surface flaw analysis using the power law formulation for constant stress rate loading (dynamic fatigue). Analogous formulations for volume flaws, static fatigue, and cyclic fatigue have also been developed [23].

For the uniaxial Weibull distribution the probability of failure is expressed as

$$P_{fs}(t_f) = 1 - \exp\left[-k_{ws} \int_A [\sigma_{1,0}(\Psi)]^{m_s} \ dA \right] \tag{23}$$

where $\sigma_{1,0}$ denotes the transformed uniaxial stress and is analogous to $\sigma_{Ieq,0}$ as defined in equation (3). The Weibull crack density coefficient is given by

$$k_{wS} = \frac{1}{\sigma_{oS}^{m_S}} \tag{24}$$

In equation (23) Ψ represents a location (x,y). The Weibull scale parameter, σ_{oS}, corresponds to the stress level where 63.21 percent of specimens with unit area would fail and has units of stress × area$^{1/m_S}$. CARES/LIFE normalizes the various fracture criteria to yield an identical probability of failure for the uniaxial stress state. This is achieved by adjusting the fatigue constant B. For the uniaxial Weibull model this adjusted value is denoted by B_w and for the Batdorf model it is denoted by B_B. From the dynamic fatigue equation (14), substituting B_{wS} for B, N_S for N, the uniaxial stress σ_1 for σ_{Ieq}, and rearranging equation (23) while assuming that

$$\frac{\sigma_1^2(\Psi, t_f) \, t_f}{(N_S+1) \, B_{wS}} >> 1 \tag{25}$$

the median behavior of the dynamic fatigue experimental data can be described by

$$\sigma_{f_{0.5}} = A_d \, \dot{\sigma}^{\,1/(N_S+1)} \tag{26}$$

where $\sigma_{f_{0.5}}$ is the median rupture stress of the specimen and $\dot{\sigma}$ represents the stress rate at the location of maximum stress. The constant A_d is

$$A_d = \left\{ \frac{(N_S + 1) \, B_{wS} \, \sigma_{oS}^{N_S-2}}{\left[\dfrac{A_{ef}}{\ell n\left(\dfrac{1}{1-0.50}\right)} \right]^{1/\tilde{m}_S}} \right\}^{1/(N_S+1)} \tag{27}$$

where

$$\tilde{m}_S = \frac{m_S}{N_S - 2} \tag{28}$$

and

$$A_{ef} = \int_A \left(\frac{\sigma_1(\Psi, t_f)}{\sigma_f} \right)^{\tilde{m}_S N_S} dA \tag{29}$$

The constants A_d and A_{ef} are obtained by equating risks of rupture. A_{ef} is a modified effective area required for the time-dependent formulation. Equation (29) is formulated for the uniaxial Weibull distribution, where $\sigma_1(\Psi, t_f)$ denotes the maximum principal stress distribution. For multiaxially stressed components, the Batdorf technique is used to evaluate fatigue parameters. The analogous formulation for A_{ef} is then

$$A_{ef} = \frac{2\bar{k}_{BS}}{\pi} \int_A \left[\int_0^{\frac{\pi}{2}} \left(\frac{\sigma_{Ieq}(\Psi, t_f)}{\sigma_f} \right)^{\tilde{m}_s N_s} d\alpha \right] dA \qquad (30)$$

where the normalized Batdorf crack density coefficient is $\bar{k}_{BS} = k_{BS}/k_{wS}$. Equation (27) is applicable except that B_{BS} replaces B_{wS}. The relationship between B_{BS} and B_{wS} for a uniaxial load is established by equating the risk of rupture of the Batdorf model with that of the uniaxial Weibull model [23]

$$\frac{B_{wS}}{B_{BS}} = \left[\frac{\pi \int_A \sigma_1^{\tilde{m}_s N_s} (\Psi, t_f) \; dA}{2 \, \bar{k}_{BS} \int_A \int_0^{\frac{\pi}{2}} \sigma_{Ieq}^{\tilde{m}_s N_s} (\Psi, t_f) \; d\alpha \; dA} \right]^{1/\tilde{m}_s} \qquad (31)$$

As N_s becomes large, equation (31) approaches unity.

The terms A_d and N_s in equation (26) are determined from experimental data. Taking the logarithm of equation (26) yields

$$\ln \sigma_{f_{0.5}} = \ln A_d + \frac{1}{N_s + 1} \ln \dot{\sigma} \qquad (32)$$

Linear regression analysis of the experimental data is used to solve equation (32). The median value method is based on least squares linear regression of median data points for various stress rates. Another technique uses least squares linear regression on all the data points. The third option for estimating fatigue parameters is a modification to a method used by Jakus [17] called trivariant regression analysis. In this procedure fatigue parameters are determined by minimizing the median deviation of the logarithm of the failure stress. The median deviation is the mean of the residuals, where the residual is defined as the absolute value of the difference between the logarithm of the failure stress and the logarithm of the median value. In CARES/LIFE this minimization is accomplished by maximizing m_s (N_s + 1) estimated from the data versus the fatigue exponent.

To obtain A_d based on the median line of the distribution the following steps are taken. Experimental data at a sufficient number of discrete levels of applied stressing rate are transformed to equivalent failure times t_{Ti} at a fixed stress rate $\dot{\sigma}_T$ (equating failure probabilities using equation (23))

$$t_{Ti} = t_{fi} \left(\frac{\dot{\sigma}_i}{\dot{\sigma}_T} \right)^{N_s/(N_s+1)} \qquad (33)$$

where the subscript T indicates a transformed value and $t_{fi} = \sigma_{fi} / \dot{\sigma}_i$. In CARES/LIFE the value of $\dot{\sigma}_T$ is taken as the lowest level of stressing rate in the data set. With the data defined by a single distribution, Weibull parameter estimation is performed on the transformed data using

$$P_{fi} = 1 - \exp\left[-\left(\frac{t_{Ti}}{t_{T\theta s}} \right)^{\tilde{m}_s (N_s+1)} \right] \qquad (34)$$

where the characteristic time is

$$
t_{T\theta s} = \left[\frac{(N_S + 1)\ B\ \sigma_{oS}^{N_S-2}}{A_{ef}^{1/\tilde{m}_S}\ \dot{\sigma}_T^{N_S}} \right]^{1/(N_S+1)}
\tag{35}
$$

CARES/LIFE performs least squares or maximum likelihood Weibull parameter estimation as described in reference [16] to solve equation (34) for $m_S\ (N_S + 1)$ and $t_{T\theta s}$. Substituting $\dot{\sigma}_T$ for $\dot{\sigma}$ in equation (26) and solving for the rupture stress in equation (34) corresponding to a 50% probability of failure, $\sigma_{T_{0.50}}$, yields

$$
A_d = t_{T\theta s} \left\{ \dot{\sigma}_T^{N_S} \left[\ell n \left(\frac{1}{1-0.50} \right) \right]^{1/\tilde{m}_S} \right\}^{1/(N_S+1)}
\tag{36}
$$

This value of A_d is used with the fatigue exponent N_S estimated either with the least squares (using all experimental rupture stresses) or median deviation method. The fatigue constant B is obtained from equations (27) or (35).

Information on the underlying inert strength distribution can also be obtained from dynamic fatigue data. Using equations (33) and (34), the estimated value of $\tilde{m}_S\ (N_S + 1)$ and the characteristic time $t_{T\theta s}$ are obtained. The inert strength (fast-fracture) Weibull modulus is

$$
m'_S = \tilde{m}_S\ (N_S - 2)
\tag{37}
$$

where superscript ' denotes a fast-fracture parameter estimated from fatigue data. CARES/LIFE calculates a characteristic strength, $\sigma'_{\theta s}$, based on an extrapolation of the fatigue data to a specific failure time. This time is arbitrarily fixed at a value of one second for a constant stress rate load. Using equation (33) yields

$$
\sigma'_{\theta s} = \dot{\sigma}_T\ t_{T\theta s}^{(N_S+1)/N_S}
\tag{38}
$$

With the calculated quantities $\sigma'_{\theta s}$ and m'_S, the inert strength distribution is defined using

$$
P_{fi} = 1 - \exp\left[-\left(\frac{\sigma'_{fi}}{\sigma'_{\theta s}} \right)^{m'_S} \right]
\tag{39}
$$

Equating the risk of rupture of equation (34) to that of equation (39), the fatigue data is transformed to its inert equivalent stress by

$$
\sigma'_{fI_i} = \sigma'_{\theta s} \left(\frac{t_{T_i}}{t_{T\theta s}} \right)^{(N_S+1)/(N_S-2)}
\tag{40}
$$

where σ'_{fI_i} represents the i'th transformed specimen inert fracture strength. The Weibull scale parameter σ_{oS} is obtained from the characteristic strength, $\sigma_{\theta s}$, using

$$
\sigma_{oS} = \sigma_{\theta s}\ A_e^{1/m_S}
\tag{41}
$$

The characteristic strength is defined as the level of uniaxial stress where 63.21 percent of specimens would fail. The effective area, A_e, is

$$A_e = \int_A \left(\frac{\sigma_1(\Psi)}{\sigma_f} \right)^{m_g} dA \tag{42}$$

or, equivalently, for the Batdorf technique

$$A_e = \frac{2\bar{k}_{BS}}{\pi} \int_A \left[\int_0^{\frac{\pi}{2}} \left(\frac{\sigma_{Ieq}(\Psi)}{\sigma_f} \right)^{m_g} d\alpha \right] dA \tag{43}$$

and $\sigma_{Ieq}(\Psi)$ is the effective stress distribution on the unit circle for the fast-fracture loading condition.

EXAMPLE 1 - PARAMETER ESTIMATION

This example demonstrates the determination of Weibull and fatigue parameters from multiaxially loaded specimen rupture data. In addition, Weibull parameters evaluated using fast-fracture data will be contrasted to Weibull parameters estimated from fatigue data. Ring-on-ring loaded square plate specimens made from soda-lime glass were prepared and fractured under dynamic fatigue loading in a distilled water environment at the NASA Lewis Research Center. Ring-on-ring loading induces an equibiaxial stress state within the inner loading ring. The plate specimens measured 50 mm x 50 mm and had an average thickness, h, of 1.50 mm.

The maximum fracture stress at the specimen center was computed using the fracture load, P, and the equation given by Shetty et al. [24]

$$\sigma_f = \frac{3\,P}{4\,\pi\,h^2} \left[2\,(1+\nu)\,\ell n \left(\frac{r_o}{r_i} \right) + \frac{(1-\nu)\,(r_o^2 - r_i^2)}{R_s^2} \right] \tag{44}$$

where the diagonal half length, R_s, was 35.92 mm, the inner radius, r_i, was 5.02 mm, the outer radius, r_o, was 16.09 mm, and Poisson's ratio, ν, was 0.25. Figure 1 shows the fracture strengths of the 121 specimens loaded at the stressing rates of 0.02, 0.20, 2.00, and 20.00 MPa/s. In addition, 30 specimens were tested in fast fracture in an inert environment of silicon oil. All flaw origins resided on the specimen surface (surface flaws).

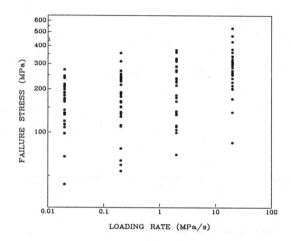

FAILURE STRESS (MPa)

LOADING RATE (MPa/s)

FIG. 1--Dynamic fatigue fracture strengths for the ring-on-ring loaded soda-lime glass square plate specimens.

Using the CARES/LIFE parameter estimation module, the fast-fracture Weibull parameters and the fatigue parameters were determined. For the fast-fracture strengths measured in silicon oil the Weibull modulus, m_S, and characteristic strength, $\sigma_{\theta S}$, were estimated to be 2.871 and 394.2 MPa, respectively, using the maximum likelihood technique. The least-squares method produced estimates of 2.675 and 395.3 MPa, respectively. For the fatigue data, the crack growth exponent, N_S, was determined with the median value, least-squares, and the median deviation techniques. The inert strength Weibull modulus, m'_S, and a characteristic strength, $\sigma'_{\theta S}$, were estimated from the fatigue data using equations (37) and (38), respectively. Using equation (40) the fatigue data was transformed to an equivalent inert strength distribution. The ranked data (median ranking [16]) is plotted in figure 2, along with the ranked inert strengths of the fast-fracture specimens. The transformed inert strengths in the figure were determined using the parameters obtained from median deviation analysis utilizing equation (32) and the maximum likelihood method utilizing equation (34). The solid line represents the transformed fatigue distribution with $m'_S = 2.344$ and $\sigma'_{\theta S} = 387.4$ MPa. The dashed line represents the fast-fracture strengths with $m_S = 2.871$ and $\sigma_{\theta S} = 394.2$ MPa. In figure 2 both distributions yield similar Weibull slopes, which supports the assumption that the fatigue data and the inert strength data were generated from the same flaw population.

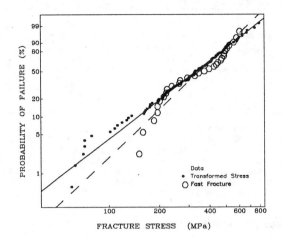

FIG. 2--Soda-lime glass fatigue data transformed to an equivalent inert strength distribution versus inert strengths measured in silicon oil. The solid line and the dashed are maximum likelihood estimations of the respective distributions.

The specimen effective areas, A_{ef} and A_e, were evaluated numerically using equations (30) and (43), respectively, with the CARES/LIFE component reliability analysis module and a finite element model of the specimen. One quarter of the total specimen was reproduced due to symmetry and the model consisted of brick and wedge elements. Quadrilateral and triangular shell elements were attached to the faces of the solid elements corresponding to the specimen external surfaces. The shell elements contributed negligible stiffness to the model. For this example, the Batdorf model with the G_T (total strain energy release rate) fracture criterion and a Griffith crack was used in the numerical evaluation. The scale parameter σ_{os} and the fatigue constant B_{BS} were calculated from equations (41) and (35), respectively. Table 1 lists the results of this analysis for the option using Weibull parameters estimated from the fatigue data. Table 2 lists the results for the option using the Weibull parameters from the fast-fracture data.

TABLE 1--Parameters estimated from fatigue data.

Method	Fatigue exponent, N_s	Fatigue constant, B_{BS} $MPa^2 \cdot s$	Weibull modulus, m'_s	Scale parameter, σ'_{os} $MPa \cdot mm^{2/m_s}$
Median Value	12.60	4 445	2.279	5 904
Least Squares	11.24	6 337	2.208	6 707
Med Deviation	11.88	5 982	2.344	5 443

TABLE 2--Parameters estimated from fatigue and fast-fracture data.

Method	Fatigue exponent, N_s	Fatigue constant, B_{Bs} $MPa^2 \cdot s$	Weibull modulus, m_s	Scale parameter, σ_{os} $MPa \cdot mm^{2/m_s}$
Median Value	12.60	3 225	2.675	3 915
Least Squares	11.24	8 527	2.675	3 915
Med Deviation	11.88	6 051	2.871	3 304

The scale parameter values in tables 1 and 2 show a significant variation. Because of the low value of the Weibull modulus, the small differences in the estimated values translate to large differences in the scale parameter. The characteristic strengths estimated from the fatigue data and the fast-fracture data were all within ±7.0 MPa of the value of 394.0 MPa (the fast-fracture value of $\sigma_{\theta s}$ from maximum likelihood estimation) for all methods of estimation.

EXAMPLE 2 - ROTATING ANNULAR DISK

This example illustrates the use of the CARES/LIFE program modules for the time-dependent reliability analysis of a silicon nitride (NC-132) rotating annular disk. The following static and cyclic loading conditions were analyzed, respectively: (1) constant rotational speeds of 60 000 r/min, 80 000 r/min, and 100 000 r/min and (2) sinusoidal cyclic angular speed varying from 60 000 r/min to 80 000 r/min over a period of 0.01 h. Volume flaw analysis was performed assuming a penny-shaped crack geometry, and surface flaw analysis assumed a semi-circular crack geometry. Probability of failure results as a function of time are presented for the Batdorf model using the noncoplanar strain energy release rate fracture criterion.

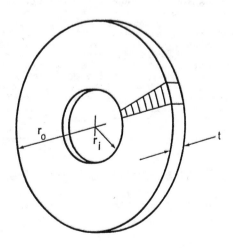

FIG. 3--Rotating annular disk with 15° sector finite element mesh. Inner disk radius, r_i, 6.35 mm; outer disk radius, r_o, 41.28 mm; disk thickness, t, 3.80 mm.

The disk dimensions are shown in figure 3. Other required
material properties include a Young's modulus of 289 GPa, a Poisson's
ratio of 0.219, and a material mass density of 3.25×10^3 kg/m^3. Analysis
was performed on the annular rotating disk at a uniform 1000° C
temperature. The material/environmental fatigue parameters, N=30 and
B=320 MPa^2 hr, were estimated using NC-132 data at 1000° C [25]. The
same fatigue parameters were used for both volume and surface flaw
analyses. For this example, the Weibull material parameters measured at
room temperature were assumed to be valid for the 1000° C analysis. For
volume flaw analysis, the Weibull modulus was m_v=7.65 and the scale
parameter was σ_{ov}=74.79 MPa m$^{0.3922}$; for surface flaw analysis, the
Weibull parameters were m_s=7.65 and σ_{os}=232.0 MPa m$^{0.2614}$ [1].

For volume flaw analysis the MSC/NASTRAN finite element model of
the disk consisted of brick elements and used the structured solution
sequence 114, static cyclic symmetry analysis. Eight HEXA elements were
used in one 15° sector of the model (Fig. 3). Each element spanned both
the thickness and circumferential directions. The stresses and element
volumes calculated using the finite element model were within
approximately 1 percent of the closed-form solution. For surface flaw
analysis the same MSC/NASTRAN finite element model was used, with the
addition of QUAD8 shell elements attached to the faces of the solid
elements to identify external surfaces. The shell elements shared
common nodes with the solid elements. The contribution of these
elements to the overall stiffness of the model was negligible.

A neutral file containing information obtained from the MSC/NASTRAN
finite element analysis was created for subsequent reliability analysis.
Weibull plots of the time-dependent probability of failure results are
shown in figures 4 and 5. As expected, for volume flaws the probability
of failure for the sinusoidal loading, between 80 000 and 60 000 RPM
(Fig. 5) falls between the probabilities of failure for the constant
80 000 and 60 000 RPM (Fig. 4) case. The same is true when comparing
surface flaw failure probabilities. For any given rotational speed and
time, the surface flaw failure probability was considerably less than
the volume flaw failure probability, indicating that failure would most
likely result from volume flaws for this example case.

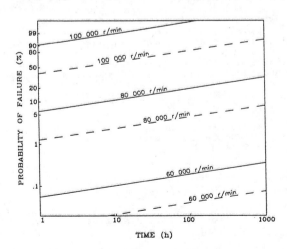

FIG. 4--Time-dependent probability of failure for annular disk
rotating at constant angular speeds. Solid lines denote volume flaw
analysis; dashed lines denote surface flaw analysis.

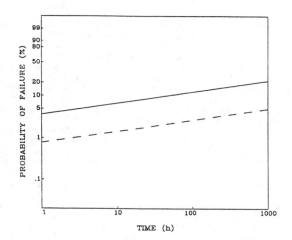

FIG. 5--Time-dependent probability of failure for annular disk rotating at sinusoidal cyclic speed varying from 60 000 r/min to 80 000 r/min (period=0.01 h). Solid line denotes volume flaw analysis; dashed line denotes surface flaw analysis.

CONCLUSION

The use of structural ceramics for high-temperature applications depends on the strength, toughness, and reliability of these materials. Ceramic components can be designed for service if the factors that cause material failure are accounted for. This design methodology must combine the statistical nature of strength controlling flaws with fracture mechanics to allow for multiaxial stress states, concurrent flaw populations, and subcritical crack growth. This has been accomplished with the CARES/LIFE public domain computer program for predicting the time-dependent reliability of monolithic structural ceramic components. Potential enhancements to the code include the capability for transient analysis, three-parameter Weibull statistics, creep and oxidation modeling, flaw anisotropy, threshold stress behavior, and parameter regression for multiple specimen sizes.

REFERENCES

[1]   Nemeth, N. N., Manderscheid, J. M., Gyekenyesi, J. P., "Ceramic Analysis and Reliability Evaluation of Structures (CARES)", NASA TP-2916, Aug. 1990.

[2]   Powers, L. M., Starlinger, A., Gyekenyesi, J. P., "Ceramic Component Reliability With the Restructured NASA/CARES Computer Program", NASA TM-105856, Sept. 1992.

[3]   Evans, A. G., and Wiederhorn, S. M., "Crack Propagation and Failure Prediction in Silicon Nitride at Elevated Temperatures", Journal of Material Science, Vol. 9, 1974, pp 270-278.

[4]   Wiederhorn, S. M.: Fracture Mechanics of Ceramics. Bradt, R. C., Hasselman, D. P., and Lange, F. F., eds., Plenum, New York, 1974, pp. 613-646.

[5] Paris, P. and Erdogan, F., "A Critical Analysis of Crack Propagation Laws", Journal of Basic Engineering, Vol. 85, 1963, pp. 528-534.

[6] Dauskardt, R. H., Marshall, D. B. and Ritchie, R. O., "Cyclic Fatigue Crack Propagation in Mg-PSZ Ceramic", Journal of the American Ceramic Society, Vol. 73, No. 4, 1990, pp 893-903.

[7] Dauskardt, R. H., James, M. R., Porter, J. R. and Ritchie, R. O., "Cyclic Fatigue Crack Growth in SiC-Whisker-Reinforced Alumina Ceramic Composite: Long and Small Crack Behavior", Journal of the American Ceramic Society, Vol. 75, No. 4, 1992, pp 759-771.

[8] Mencik, J., "Rationalized Load and Lifetime of Brittle Materials", Communications of the American Ceramic Society, Mar. 1984, pp C37-C40.

[9] Barnett, R. L.; Connors, C. L.; Hermann, P. C.; and Wingfield, J. R.: "Fracture of Brittle Materials Under Transient Mechanical and Thermal Loading," U. S. Air Force Flight Dynamics Laboratory, AFFDL-TR-66-220, 1967. (NTIS AD-649978)

[10] Freudenthal, A. M., "Statistical Approach to Brittle Fracture," Fracture, Vol. 2: An Advanced Treatise, Mathematical Fundamentals, H. Liebowitz, ed., Academic Press, 1968, pp. 591-619.

[11] Weibull, W. A., "The Phenomenon of Rupture in Solids," Ingenoirs Vetenskaps Akadanien Handlinger, 1939, No. 153.

[12] Batdorf, S. B. and Crose, J. G., "A Statistical Theory for the Fracture of Brittle Structures Subjected to Nonuniform Polyaxial Stresses", Journal of Applied Mechanics, Vol. 41, No. 2, June 1974, pp 459-464.

[13] Batdorf, S. B.; and Heinisch, H. L., Jr.: Weakest Link Theory Reformulated for Arbitrary Fracture Criterion. Journal of the American Ceramic Society, Vol. 61, No. 7-8, July-Aug. 1978, pp. 355-358.

[14] Palaniswamy, K., and Knauss, W. G., On the Problem of Crack Extension in Brittle Solids Under General Loading, Mechanics Today, Vol. 4, 1978, pp. 87-148.

[15] Shetty, D. K.: Mixed-Mode Fracture Criteria for Reliability Analysis and Design with Structural Ceramics. Journal of Engineering for Gas Turbines and Power, Vol. 109, No. 3, July 1987, pp. 282-289.

[16] Pai, S. S.; and Gyekenyesi, J. P.: Calculation of the Weibull Strength Parameters and Batdorf Flaw Density Constants for Volume and Surface-Flaw-Induced Fracture in Ceramics, NASA TM-100890, 1988.

[17] Jakus, K.; Coyne, D. C.; and Ritter, J. E.: Analysis of Fatigue Data for Lifetime Predictions for Ceramic Materials. Journal of Material Science, Vol. 13, 1978, pp. 2071-2080.

[18] Paris, P. C.; and Sih, G. C.: Stress Analysis of Cracks. ASTM STP 381, 1965, pp. 30-83.

[19] Thiemeier, T., "Lebensdauervorhersage fun Keramische Bauteile Unter Mehrachsiger Beanspruchung", Ph.D. dissertation, University of Karlesruhe, Germany, 1989.

[20] Sturmer, G., Schulz, A. and Wittig, S., "Lifetime Prediction for Ceramic Gas Turbine Components", ASME Preprint 91-GT-96, June 3-6, 1991.

[21]   Batdorf, S. B., "Fundamentals of the Statistical Theory of
       Fracture", Fracture Mechanics of Ceramics, Vol. 3, eds., Bradt, R.
       C., Hasselman, D. P. H. and Lange, F. F., Plenum Press, New York
       (1978), pp 1-30.

[22]   Evans, A. G., "Fatigue in Ceramics", International Journal of
       Fracture, Dec. 1980, pp 485-498.

[23]   Nemeth, N. N.; Powers, L. M.; Janosik, L. A.; and Gyekenyesi, J.
       P.: "Ceramics Analysis and Reliability Evaluation of Structures
       LIFE prediction program (CARES/LIFE)", to be published.

[24]   Shetty, D. K.; Rosenfield, A. R.; Bansal, G. K.; and Duckworth, W.
       H.: Biaxial Flexure Test for Ceramics, Journal of the American
       Ceramic Society, Vol. 59, No. 12, pp. 1193-1197, 1980.

[25]   Quinn, J. B., "Slow Crack Growth in Hot-Pressed Silicon Nitride",
       Fracture Mechanics of Ceramics. Bradt, R. C.; Evans, A. G.; and
       Hasselman, D. P. H.; Lange, F. F.; eds., Plenum Publishing Corp.,
       Vol. 6, 1983, pp. 603-636.

Author Index

B

Brehm, P., 291
Brinkman, C. R., 62
Brown, T. S., III, 373
Brückner-Foit, A., 346

C

Chao, L.-Y., 228
Choi, S. R., 84, 98
Chuang, T.-J., 207
Cuccio, J., 291

D

De With, G., 192
Ding, J.-L., 62
Dortmans, L. J., 192
Duffy, S. F., 207, 373

E

Edwards, M. J., 373

F

Fang, H., 291
Fett, T., 161
Fok, S. L., 143
Foley, M. R., 3

G

Gyekenyesi, J. P., 390

H

Heger, A., 346
Hild, F., 112

J

Jadaan, O. M., 309
Janosik, L. A., 390
Johnson, C. A., 250, 265, 291

K

Kelkar, A. D., 19
Khandelwal, P. K., 127
Kranendonk, W. G. T., 333
Kraus, R., 36
Krishnaraj, S., 19

L

Lamon, J. L., 175
Liu, K. C., 62

M

Margetsen, J., 280
Marquis, D., 112
Munz, D., 161, 346

N

Nemeth, N. N., 390

P

Palko, J. L., 98
Peralta, A., 291
Powers, L. M., 390
Pujari, V. K., 3

Q

Quinn, G. D., 36

S

Salem, J. A., 84, 98
Sales, L. C., 3
Sandifer, J. B., 373
Sankar, J., 19
Scholten, H. F., 192
Shetty, D. K., 228
Sinnema, S., 333
Smart, J., 143
Song, J., 291

Subject Index

411